STATA QUICK REFERENCE AND INDEX

RELEASE 11

A Stata Press Publication
StataCorp LP
College Station, Texas

Published by Stata Press, 4905 Lakeway Drive, College Station, Texas 77845
Typeset in TEX
Printed in the United States of America

10 9 8 7 6 5 4 3 2 1

ISBN-10: 1-59718-055-6
ISBN-13: 978-1-59718-055-9

The suggested citation for this software is

StataCorp. 2009. *Stata: Release 11*. Statistical Software. College Station, TX: StataCorp LP.

Table of contents

Combined subject table of contents 1

data types .... Quick reference for data types 27

estimation commands .... Quick reference for estimation commands 28

file extensions .... Quick reference for default file extensions 35

format .... Quick reference for numeric and string display formats 36

immediate commands .... Quick reference for immediate commands 37

missing values .... Quick reference for missing values 38

postestimation commands .... Quick reference for postestimation commands 39

prefix commands .... Quick reference for prefix commands 48

reading data .... Quick reference for reading non-Stata data into memory 50

Acronym glossary 53

Vignettes index 59

Author index 61

Subject index 87

Combined subject table of contents

This is the complete contents for all the Reference manuals except the *Mata Reference Manual*.

Every estimation command has a postestimation entry; however, the postestimation entries are not listed in the subject table of contents.

Getting Started

[GSM]   *Getting Started with Stata for Mac* ....
[GSU]   *Getting Started with Stata for Unix* ....
[GSW]   *Getting Started with Stata for Windows* ....
[U]   Chapter 3 .... Resources for learning and using Stata
[R]   help .... Display online help

Data manipulation and management

Basic data commands

[D]   clear .... Clear memory
[D]   clonevar .... Clone existing variable
[D]   codebook .... Describe data contents
[D]   compress .... Compress data in memory
[D]   data management .... Introduction to data-management commands
[D]   describe .... Describe data in memory or in file
[R]   display .... Substitute for a hand calculator
[D]   drop .... Eliminate variables or observations
[D]   edit .... Browse or edit data with Data Editor
[D]   egen .... Extensions to generate
[D]   generate .... Create or change contents of variable
[D]   list .... List values of variables
[D]   lookfor .... Search for string in variable names and labels
[D]   memory .... Memory size considerations
[D]   obs .... Increase the number of observations in a dataset
[D]   sort .... Sort data
[D]   varmanage .... Manage variable labels, formats, and other properties

Functions and expressions

[U]   Chapter 13 .... Functions and expressions
[D]   dates and times .... Date and time (%t) values and variables
[D]   egen .... Extensions to generate
[D]   functions .... Functions

Dates and times

[U]   Section 12.5.3 .... Date and time formats
[U]   Chapter 24 .... Working with dates and times
[D]   dates and times .... Date and time (%t) values and variables
[D]   functions .... Functions

Inputting and saving data

[GS]	Chapter 6 (GSM, GSU, GSW)	Using the Data Editor
[U]	Chapter 21	Inputting data
[D]	edit	Browse or edit data with Data Editor
[D]	fdasave	Save and use datasets in FDA (SAS XPORT) format
[TS]	haver	Load data from Haver Analytics database
[D]	infile	Overview of reading data into Stata
[D]	infile (fixed format)	Read ASCII (text) data in fixed format with a dictionary
[D]	infile (free format)	Read unformatted ASCII (text) data
[D]	infix (fixed format)	Read ASCII (text) data in fixed format
[D]	input	Enter data from keyboard
[D]	insheet	Read ASCII (text) data created by a spreadsheet
[D]	odbc	Load, write, or view data from ODBC sources
[D]	outfile	Write ASCII-format dataset
[D]	outsheet	Write spreadsheet-style dataset
[D]	save	Save datasets
[D]	sysuse	Use shipped dataset
[D]	use	Use Stata dataset
[D]	webuse	Use dataset from Stata web site
[D]	xmlsave	Save and use datasets in XML format

Combining data

[U]	Chapter 22	Combining datasets
[D]	append	Append datasets
[MI]	mi append	Append mi data
[D]	cross	Form every pairwise combination of two datasets
[D]	joinby	Form all pairwise combinations within groups
[D]	merge	Merge datasets
[MI]	mi merge	Merge mi data

Reshaping datasets

[D]	collapse	Make dataset of summary statistics
[D]	contract	Make dataset of frequencies and percentages
[D]	expand	Duplicate observations
[D]	expandcl	Duplicate clustered observations
[D]	fillin	Rectangularize dataset
[D]	obs	Increase the number of observations in a dataset
[D]	reshape	Convert data from wide to long form and vice versa
[MI]	mi reshape	Reshape mi data
[TS]	rolling	Rolling-window and recursive estimation
[D]	separate	Create separate variables
[D]	stack	Stack data
[D]	statsby	Collect statistics for a command across a by list
[D]	xpose	Interchange observations and variables

Labeling, display formats, and notes

[GS]	Chapter 7 (GSM, GSU, GSW)	Using the Variables Manager
[U]	Section 12.5	Formats: Controlling how data are displayed
[U]	Section 12.6	Dataset, variable, and value labels
[D]	format	Set variables' output format

[D]   label . . . . Manipulate labels
[D]   label language . . . . Labels for variables and values in multiple languages
[D]   labelbook . . . . Label utilities
[D]   notes . . . . Place notes in data
[D]   varmanage . . . . Manage variable labels, formats, and other properties

Changing and renaming variables

[GS]   Chapter 7 (GSM, GSU, GSW) . . . . Using the Variables Manager
[U]   Chapter 25 . . . . Working with categorical data and factor variables
[D]   clonevar . . . . Clone existing variable
[D]   destring . . . . Convert string variables to numeric variables and vice versa
[D]   encode . . . . Encode string into numeric and vice versa
[D]   generate . . . . Create or change contents of variable
[D]   mvencode . . . . Change missing values to numeric values and vice versa
[D]   order . . . . Reorder variables in dataset
[D]   recode . . . . Recode categorical variables
[D]   rename . . . . Rename variable
[D]   split . . . . Split string variables into parts
[D]   varmanage . . . . Manage variable labels, formats, and other properties

Examining data

[GS]   Chapter 6 (GSM, GSU, GSW) . . . . Using the Data Editor
[D]   cf . . . . Compare two datasets
[D]   codebook . . . . Describe data contents
[D]   compare . . . . Compare two variables
[D]   count . . . . Count observations satisfying specified conditions
[D]   describe . . . . Describe data in memory or in file
[D]   duplicates . . . . Report, tag, or drop duplicate observations
[D]   edit . . . . Browse or edit data with Data Editor
[D]   gsort . . . . Ascending and descending sort
[D]   inspect . . . . Display simple summary of data's attributes
[D]   isid . . . . Check for unique identifiers
[R]   misstable . . . . Tabulate missing values
[MI]   mi describe . . . . Describe mi data
[MI]   mi misstable . . . . Tabulate pattern of missing values
[D]   pctile . . . . Create variable containing percentiles
[ST]   stdescribe . . . . Describe survival-time data
[R]   summarize . . . . Summary statistics
[SVY]   svy: tabulate oneway . . . . One-way tables for survey data
[SVY]   svy: tabulate twoway . . . . Two-way tables for survey data
[P]   tabdisp . . . . Display tables
[R]   table . . . . Tables of summary statistics
[R]   tabstat . . . . Display table of summary statistics
[R]   tabulate oneway . . . . One-way tables of frequencies
[R]   tabulate twoway . . . . Two-way tables of frequencies
[R]   tabulate, summarize() . . . . One- and two-way tables of summary statistics
[XT]   xtdescribe . . . . Describe pattern of xt data

File manipulation

[D]   cd . . . . Change directory
[D]   cf . . . . Compare two datasets

[D] changeeol . . . . Convert end-of-line characters of text file
[D] checksum . . . . Calculate checksum of file
[D] copy . . . . Copy file from disk or URL
[D] dir . . . . Display filenames
[D] erase . . . . Erase a disk file
[D] filefilter . . . . Convert ASCII text or binary patterns in a file
[D] mkdir . . . . Create directory
[D] rmdir . . . . Remove directory
[D] type . . . . Display contents of a file

Miscellaneous data commands

[D] corr2data . . . . Create dataset with specified correlation structure
[D] drawnorm . . . . Draw sample from multivariate normal distribution
[R] dydx . . . . Calculate numeric derivatives and integrals
[D] icd9 . . . . ICD-9-CM diagnostic and procedure codes
[D] ipolate . . . . Linearly interpolate (extrapolate) values
[D] range . . . . Generate numerical range
[D] sample . . . . Draw random sample

Multiple imputation

[MI] mi add . . . . Add imputations from another mi dataset
[MI] mi append . . . . Append mi data
[MI] mi convert . . . . Change style of mi data
[MI] mi copy . . . . Copy mi flongsep data
[MI] mi describe . . . . Describe mi data
[MI] mi erase . . . . Erase mi datasets
[MI] mi expand . . . . Expand mi data
[MI] mi export . . . . Export mi data
[MI] mi export ice . . . . Export mi data to ice format
[MI] mi export nhanes1 . . . . Export mi data to NHANES format
[MI] mi extract . . . . Extract original or imputed data from mi data
[MI] mi import . . . . Import data into mi
[MI] mi import flong . . . . Import flong-like data into mi
[MI] mi import flongsep . . . . Import flongsep-like data into mi
[MI] mi import ice . . . . Import ice-format data into mi
[MI] mi import nhanes1 . . . . Import NHANES-format data into mi
[MI] mi import wide . . . . Import wide-like data into mi
[MI] mi merge . . . . Merge mi data
[MI] mi misstable . . . . Tabulate pattern of missing values
[MI] mi passive . . . . Generate/replace and register passive variables
[MI] mi ptrace . . . . Load parameter-trace file into Stata
[MI] mi rename . . . . Rename variable
[MI] mi replace0 . . . . Replace original data
[MI] mi reset . . . . Reset imputed or passive variables
[MI] mi reshape . . . . Reshape mi data
[MI] mi set . . . . Declare multiple-imputation data
[MI] mi stsplit . . . . Stsplit and stjoin mi data
[MI] mi update . . . . Ensure that mi data are consistent
[MI] mi varying . . . . Identify variables that vary across imputations
[MI] mi xeq . . . . Execute command(s) on individual imputations
[MI] mi XXXset . . . . Declare mi data to be svy, st, ts, xt, etc.

[MI]   noupdate option . . . . The noupdate option
[MI]   styles . . . . Dataset styles
[MI]   workflow . . . . Suggested workflow

Utilities

Basic utilities

[GS]   Chapter 13 (GSM, GSU, GSW) . . . . Using the Do-file Editor—automating Stata
[U]   Chapter 4 . . . . Stata's help and search facilities
[U]   Chapter 15 . . . . Saving and printing output—log files
[U]   Chapter 16 . . . . Do-files
[R]   about . . . . Display information about your Stata
[D]   by . . . . Repeat Stata command on subsets of the data
[R]   copyright . . . . Display copyright information
[R]   do . . . . Execute commands from a file
[R]   doedit . . . . Edit do-files and other text files
[R]   exit . . . . Exit Stata
[R]   help . . . . Display online help
[R]   hsearch . . . . Search help files
[R]   level . . . . Set default confidence level
[R]   log . . . . Echo copy of session to file
[D]   obs . . . . Increase the number of observations in a dataset
[R]   #review . . . . Review previous commands
[R]   search . . . . Search Stata documentation
[R]   translate . . . . Print and translate logs
[R]   view . . . . View files and logs
[D]   zipfile . . . . Compress and uncompress files and directories in zip archive format

Error messages

[U]   Chapter 8 . . . . Error messages and return codes
[P]   error . . . . Display generic error message and exit
[R]   error messages . . . . Error messages and return codes
[P]   rmsg . . . . Return messages

Saved results

[U]   Section 13.5 . . . . Accessing coefficients and standard errors
[U]   Section 18.8 . . . . Accessing results calculated by other programs
[U]   Section 18.9 . . . . Accessing results calculated by estimation commands
[U]   Section 18.10 . . . . Saving results
[P]   creturn . . . . Return c-class values
[P]   ereturn . . . . Post the estimation results
[R]   estimates . . . . Save and manipulate estimation results
[R]   estimates describe . . . . Describe estimation results
[R]   estimates for . . . . Repeat postestimation command across models
[R]   estimates notes . . . . Add notes to estimation results
[R]   estimates replay . . . . Redisplay estimation results
[R]   estimates save . . . . Save and use estimation results
[R]   estimates stats . . . . Model statistics
[R]   estimates store . . . . Store and restore estimation results
[R]   estimates table . . . . Compare estimation results

[R] estimates title . . . . Set title for estimation results
[P] _return . . . . Preserve saved results
[P] return . . . . Return saved results
[R] saved results . . . . Saved results

Internet

[U] Chapter 28 . . . . Using the Internet to keep up to date
[R] adoupdate . . . . Update user-written ado-files
[D] checksum . . . . Calculate checksum of file
[D] copy . . . . Copy file from disk or URL
[R] net . . . . Install and manage user-written additions from the Internet
[R] net search . . . . Search the Internet for installable packages
[R] netio . . . . Control Internet connections
[R] news . . . . Report Stata news
[R] sj . . . . Stata Journal and STB installation instructions
[R] ssc . . . . Install and uninstall packages from SSC
[R] update . . . . Update Stata
[D] use . . . . Use Stata dataset

Data types and memory

[U] Chapter 6 . . . . Setting the size of memory
[U] Section 12.2.2 . . . . Numeric storage types
[U] Section 12.4.4 . . . . String storage types
[U] Section 13.11 . . . . Precision and problems therein
[U] Chapter 23 . . . . Working with strings
[D] compress . . . . Compress data in memory
[D] data types . . . . Quick reference for data types
[R] matsize . . . . Set the maximum number of variables in a model
[D] memory . . . . Memory size considerations
[D] missing values . . . . Quick reference for missing values
[D] recast . . . . Change storage type of variable

Advanced utilities

[D] assert . . . . Verify truth of claim
[D] cd . . . . Change directory
[D] changeeol . . . . Convert end-of-line characters of text file
[D] checksum . . . . Calculate checksum of file
[D] copy . . . . Copy file from disk or URL
[P] _datasignature . . . . Determine whether data have changed
[D] datasignature . . . . Determine whether data have changed
[R] db . . . . Launch dialog
[P] dialog programming . . . . Dialog programming
[D] dir . . . . Display filenames
[P] discard . . . . Drop automatically loaded programs
[D] erase . . . . Erase a disk file
[P] file . . . . Read and write ASCII text and binary files
[D] filefilter . . . . Convert ASCII text or binary patterns in a file
[D] hexdump . . . . Display hexadecimal report on file
[D] mkdir . . . . Create directory
[R] more . . . . The —more— message
[R] query . . . . Display system parameters

[P]   quietly .... Quietly and noisily perform Stata command
[D]   rmdir .... Remove directory
[R]   set .... Overview of system parameters
[R]   set_defaults .... Reset system parameters to original Stata defaults
[R]   set emptycells .... Set what to do with empty cells in interactions
[D]   shell .... Temporarily invoke operating system
[P]   signestimationsample .... Determine whether the estimation sample has changed
[P]   smcl .... Stata Markup and Control Language
[P]   sysdir .... Query and set system directories
[D]   type .... Display contents of a file
[R]   which .... Display location and version for an ado-file

Graphics

Common graphs

[G]   graph .... The graph command
[G]   graph bar .... Bar charts
[G]   graph box .... Box plots
[G]   graph combine .... Combine multiple graphs
[G]   graph copy .... Copy graph in memory
[G]   graph describe .... Describe contents of graph in memory or on disk
[G]   graph dir .... List names of graphs in memory and on disk
[G]   graph display .... Display graph stored in memory
[G]   graph dot .... Dot charts (summary statistics)
[G]   graph drop .... Drop graphs from memory
[G]   graph export .... Export current graph
[G]   graph manipulation .... Graph manipulation commands
[G]   graph matrix .... Matrix graphs
[G]   graph other .... Other graphics commands
[G]   graph pie .... Pie charts
[G]   graph play .... Apply edits from a recording on current graph
[G]   graph print .... Print a graph
[G]   graph query .... List available schemes and styles
[G]   graph rename .... Rename graph in memory
[G]   graph save .... Save graph to disk
[G]   graph set .... Set graphics options
[G]   graph twoway .... Twoway graphs
[G]   graph twoway area .... Twoway line plot with area shading
[G]   graph twoway bar .... Twoway bar plots
[G]   graph twoway connected .... Twoway connected plots
[G]   graph twoway dot .... Twoway dot plots
[G]   graph twoway dropline .... Twoway dropped-line plots
[G]   graph twoway fpfit .... Twoway fractional-polynomial prediction plots
[G]   graph twoway fpfitci ...... Twoway fractional-polynomial prediction plots with CIs
[G]   graph twoway function .... Twoway line plot of function
[G]   graph twoway histogram .... Histogram plots
[G]   graph twoway kdensity .... Kernel density plots
[G]   graph twoway lfit .... Twoway linear prediction plots
[G]   graph twoway lfitci .... Twoway linear prediction plots with CIs
[G]   graph twoway line .... Twoway line plots

[G]   graph twoway lowess .... Local linear smooth plots
[G]   graph twoway lpoly .... Local polynomial smooth plots
[G]   graph twoway lpolyci .... Local polynomial smooth plots with CIs
[G]   graph twoway mband .... Twoway median-band plots
[G]   graph twoway mspline .... Twoway median-spline plots
[G]   graph twoway pcarrow .... Paired-coordinate plot with arrows
[G]   graph twoway pcarrowi .... Twoway pcarrow with immediate arguments
[G]   graph twoway pccapsym .... Paired-coordinate plot with spikes and marker symbols
[G]   graph twoway pci .... Twoway paired-coordinate plot with immediate arguments
[G]   graph twoway pcscatter .... Paired-coordinate plot with markers
[G]   graph twoway pcspike .... Paired-coordinate plot with spikes
[G]   graph twoway qfit .... Twoway quadratic prediction plots
[G]   graph twoway qfitci .... Twoway quadratic prediction plots with CIs
[G]   graph twoway rarea .... Range plot with area shading
[G]   graph twoway rbar .... Range plot with bars
[G]   graph twoway rcap .... Range plot with capped spikes
[G]   graph twoway rcapsym .... Range plot with spikes capped with marker symbols
[G]   graph twoway rconnected .... Range plot with connected lines
[G]   graph twoway rline .... Range plot with lines
[G]   graph twoway rscatter .... Range plot with markers
[G]   graph twoway rspike .... Range plot with spikes
[G]   graph twoway scatter .... Twoway scatterplots
[G]   graph twoway scatteri .... Scatter with immediate arguments
[G]   graph twoway spike .... Twoway spike plots
[G]   graph twoway tsline .... Twoway line plots
[G]   graph use .... Display graph stored on disk
[R]   histogram .... Histograms for continuous and categorical variables
[G]   palette .... Display palettes of available selections

Distributional graphs

[R]   diagnostic plots .... Distributional diagnostic plots
[R]   ladder .... Ladder of powers
[R]   spikeplot .... Spike plots and rootograms

Multivariate graphs

[MV]   biplot .... Biplots
[MV]   ca postestimation .... Postestimation tools for ca and camat
[MV]   cluster dendrogram .... Dendrograms for hierarchical cluster analysis
[MV]   mca postestimation .... Postestimation tools for mca
[MV]   mds postestimation .... Postestimation tools for mds, mdsmat, and mdslong
[MV]   scoreplot .... Score and loading plots
[MV]   screeplot .... Scree plot

Quality control

[R]   cusum .... Graph cumulative spectral distribution
[R]   qc .... Quality control charts
[R]   serrbar .... Graph standard error bar chart

Regression diagnostic plots

[R]   regress postestimation .... Postestimation tools for regress

ROC analysis

[R]   logistic postestimation .... Postestimation tools for logistic
[R]   roc .... Receiver operating characteristic (ROC) analysis
[R]   rocfit postestimation .... Postestimation tools for rocfit

Smoothing and densities

[R]   kdensity .... Univariate kernel density estimation
[R]   lowess .... Lowess smoothing
[R]   lpoly .... Kernel-weighted local polynomial smoothing
[R]   sunflower .... Density-distribution sunflower plots

Survival-analysis graphs

[ST]   ltable .... Life tables for survival data
[ST]   stci .... Confidence intervals for means and percentiles of survival time
[ST]   stcox PH-assumption tests .... Tests of proportional-hazards assumption
[ST]   stcurve ... Plot survivor, hazard, cumulative hazard, or cumulative incidence function
[ST]   strate .... Tabulate failure rates and rate ratios
[ST]   sts graph .... Graph the survivor and cumulative hazard functions

Time-series graphs

[TS]   corrgram .... Tabulate and graph autocorrelations
[TS]   cumsp .... Cumulative spectral distribution
[TS]   fcast graph .... Graph forecasts of variables computed by fcast compute
[TS]   irf cgraph ..... Combine graphs of IRFs, dynamic-multiplier functions, and FEVDs
[TS]   irf graph .... Graph IRFs, dynamic-multiplier functions, and FEVDs
[TS]   irf ograph .... Graph overlaid IRFs, dynamic-multiplier functions, and FEVDs
[TS]   pergram .... Periodogram
[TS]   tsline .... Plot time-series data
[TS]   wntestb .... Bartlett's periodogram-based test for white noise
[TS]   xcorr .... Cross-correlogram for bivariate time series

More statistical graphs

[R]   dotplot .... Comparative scatterplots
[ST]   epitab .... Tables for epidemiologists
[R]   fracpoly postestimation .... Postestimation tools for fracpoly
[R]   grmeanby .... Graph means and medians by categorical variables
[R]   pkexamine .... Calculate pharmacokinetic measures
[R]   pksumm .... Summarize pharmacokinetic data
[R]   stem .... Stem-and-leaf displays
[XT]   xtline .... Panel-data line plots

Editing

[G]   graph editor .... Graph Editor

Graph utilities

[G]   set graphics .... Set whether graphs are displayed
[G]   set printcolor .... Set how colors are treated when graphs are printed
[G]   set scheme .... Set default scheme

Graph schemes

[G]	schemes intro	Introduction to schemes
[G]	scheme economist	Scheme description: economist
[G]	scheme s1	Scheme description: s1 family
[G]	scheme s2	Scheme description: s2 family
[G]	scheme sj	Scheme description: sj

Graph concepts

[G]	concept: gph files	Using gph files
[G]	concept: lines	Using lines
[G]	concept: repeated options	Interpretation of repeated options
[G]	text	Text in graphs

Statistics

ANOVA and related

[U]	Chapter 26	Overview of Stata estimation commands
[R]	anova	Analysis of variance and covariance
[R]	loneway	Large one-way ANOVA, random effects, and reliability
[MV]	manova	Multivariate analysis of variance and covariance
[R]	oneway	One-way analysis of variance
[R]	pkcross	Analyze crossover experiments
[R]	pkshape	Reshape (pharmacokinetic) Latin-square data
[R]	regress	Linear regression
[XT]	xtmixed	Multilevel mixed-effects linear regression
[XT]	xtreg	Fixed-, between-, and random-effects, and population-averaged linear models

Basic statistics

[R]	anova	Analysis of variance and covariance
[R]	bitest	Binomial probability test
[R]	ci	Confidence intervals for means, proportions, and counts
[R]	correlate	Correlations (covariances) of variables or coefficients
[D]	egen	Extensions to generate
[R]	mean	Estimate means
[R]	misstable	Tabulate missing values
[MV]	mvtest	Multivariate tests
[R]	oneway	One-way analysis of variance
[R]	proportion	Estimate proportions
[R]	prtest	One- and two-sample tests of proportions
[R]	ranksum	Equality tests on unmatched data
[R]	ratio	Estimate ratios
[R]	regress	Linear regression
[R]	sampsi	Sample size and power determination
[R]	sdtest	Variance-comparison tests
[R]	signrank	Equality tests on matched data
[D]	statsby	Collect statistics for a command across a by list
[R]	summarize	Summary statistics
[R]	table	Tables of summary statistics
[R]	tabstat	Display table of summary statistics

[R]   tabulate oneway .... One-way tables of frequencies
[R]   tabulate twoway .... Two-way tables of frequencies
[R]   tabulate, summarize() .... One- and two-way tables of summary statistics
[R]   total .... Estimate totals
[R]   ttest .... Mean-comparison tests

Binary outcomes

[U]   Chapter 20 .... Estimation and postestimation commands
[U]   Section 26.5 .... Binary-outcome qualitative dependent-variable models
[R]   binreg .... Generalized linear models: Extensions to the binomial family
[R]   biprobit .... Bivariate probit regression
[R]   cloglog .... Complementary log-log regression
[R]   exlogistic .... Exact logistic regression
[R]   glogit .... Logit and probit regression for grouped data
[R]   heckprob .... Probit model with sample selection
[R]   hetprob .... Heteroskedastic probit model
[R]   ivprobit .... Probit model with continuous endogenous regressors
[R]   logistic .... Logistic regression, reporting odds ratios
[R]   logit .... Logistic regression, reporting coefficients
[R]   probit .... Probit regression
[R]   scobit .... Skewed logistic regression
[XT]   xtcloglog .... Random-effects and population-averaged cloglog models
[XT]   xtlogit .... Fixed-effects, random-effects, and population-averaged logit models
[XT]   xtmelogit .... Multilevel mixed-effects logistic regression
[XT]   xtprobit .... Random-effects and population-averaged probit models

Categorical outcomes

[U]   Chapter 20 .... Estimation and postestimation commands
[U]   Section 26.7 .... Multiple-outcome qualitative dependent-variable models
[R]   asclogit .... Alternative-specific conditional logit (McFadden's choice) model
[R]   asmprobit .... Alternative-specific multinomial probit regression
[R]   clogit .... Conditional (fixed-effects) logistic regression
[R]   mlogit .... Multinomial (polytomous) logistic regression
[R]   mprobit .... Multinomial probit regression
[R]   nlogit .... Nested logit regression
[R]   slogit .... Stereotype logistic regression

Cluster analysis

[U]   Section 26.21 .... Multivariate and cluster analysis
[MV]   cluster .... Introduction to cluster-analysis commands
[MV]   cluster dendrogram .... Dendrograms for hierarchical cluster analysis
[MV]   cluster generate   ... Generate summary or grouping variables from a cluster analysis
[MV]   cluster kmeans and kmedians .... Kmeans and kmedians cluster analysis
[MV]   cluster linkage .... Hierarchical cluster analysis
[MV]   cluster notes .... Place notes in cluster analysis
[MV]   cluster programming subroutines .... Add cluster-analysis routines
[MV]   cluster programming utilities .... Cluster-analysis programming utilities
[MV]   cluster stop .... Cluster-analysis stopping rules
[MV]   cluster utility .... List, rename, use, and drop cluster analyses

[MV]   clustermat . . . .   Introduction to clustermat commands
[MV]   matrix dissimilarity . . . .   Compute similarity or dissimilarity measures
[MV]   *measure_option* . . . .   Option for similarity and dissimilarity measures
[MV]   multivariate . . . .   Introduction to multivariate commands

Correspondence analysis

[MV]   ca . . . .   Simple correspondence analysis
[MV]   mca . . . .   Multiple and joint correspondence analysis

Count outcomes

[U]   Chapter 20 . . . .   Estimation and postestimation commands
[U]   Section 26.8 . . . .   Count dependent-variable models
[U]   Section 26.15.5 . . . .   Count dependent-variable models with panel data
[R]   expoisson . . . .   Exact Poisson regression
[R]   nbreg . . . .   Negative binomial regression
[R]   poisson . . . .   Poisson regression
[XT]   xtmepoisson . . . .   Multilevel mixed-effects Poisson regression
[XT]   xtnbreg Fixed-effects, random-effects, & population-averaged negative binomial models
[XT]   xtpoisson . . . .   Fixed-effects, random-effects, and population-averaged Poisson models
[R]   zinb . . . .   Zero-inflated negative binomial regression
[R]   zip . . . .   Zero-inflated Poisson regression
[R]   ztnb . . . .   Zero-truncated negative binomial regression
[R]   ztp . . . .   Zero-truncated Poisson regression

Discriminant analysis

[MV]   candisc . . . .   Canonical linear discriminant analysis
[MV]   discrim . . . .   Discriminant analysis
[MV]   discrim estat . . . .   Postestimation tools for discrim
[MV]   discrim knn . . . .   kth-nearest-neighbor discriminant analysis
[MV]   discrim lda . . . .   Linear discriminant analysis
[MV]   discrim logistic . . . .   Logistic discriminant analysis
[MV]   discrim qda . . . .   Quadratic discriminant analysis
[MV]   scoreplot . . . .   Score and loading plots
[MV]   screeplot . . . .   Scree plot

Do-it-yourself generalized method of moments

[R]   gmm . . . .   Generalized method of moments estimation
[P]   matrix . . . .   Introduction to matrix commands

Do-it-yourself maximum likelihood estimation

[P]   matrix . . . .   Introduction to matrix commands
[R]   ml . . . .   Maximum likelihood estimation

Endogenous covariates

[U]   Chapter 20 . . . .   Estimation and postestimation commands
[U]   Chapter 26 . . . .   Overview of Stata estimation commands
[R]   gmm . . . .   Generalized method of moments estimation
[R]   ivprobit . . . .   Probit model with continuous endogenous regressors
[R]   ivregress . . . .   Single-equation instrumental-variables regression
[R]   ivtobit . . . .   Tobit model with continuous endogenous regressors

[R] reg3 .... Three-stage estimation for systems of simultaneous equations
[XT] xthtaylor .... Hausman–Taylor estimator for error-components models
[XT] xtivreg ..... Instrumental variables and two-stage least squares for panel-data models

Epidemiology and related
[R] binreg .... Generalized linear models: Extensions to the binomial family
[R] dstdize .... Direct and indirect standardization
[ST] epitab .... Tables for epidemiologists
[R] exlogistic .... Exact logistic regression
[D] icd9 .... ICD-9-CM diagnostic and procedure codes
[R] logistic .... Logistic regression, reporting odds ratios
[R] roc .... Receiver operating characteristic (ROC) analysis
[R] rocfit .... Fit ROC models
[ST] st .... Survival-time data
[R] symmetry .... Symmetry and marginal homogeneity tests
[R] tabulate oneway .... One-way tables of frequencies
[R] tabulate twoway .... Two-way tables of frequencies

Estimation related
[R] BIC note .... Calculating and interpreting BIC
[R] constraint .... Define and list constraints
[R] *eform_option* .... Displaying exponentiated coefficients
[R] estimation options .... Estimation options
[R] fracpoly .... Fractional polynomial regression
[R] maximize .... Details of iterative maximization
[R] mfp .... Multivariable fractional polynomial models
[R] mkspline .... Linear and restricted cubic spline construction
[R] stepwise .... Stepwise estimation

Exact statistics
[R] bitest .... Binomial probability test
[R] centile .... Report centile and confidence interval
[R] dstdize .... Direct and indirect standardization
[ST] epitab .... Tables for epidemiologists
[R] exlogistic .... Exact logistic regression
[R] expoisson .... Exact Poisson regression
[R] ksmirnov .... Kolmogorov–Smirnov equality-of-distributions test
[R] loneway .... Large one-way ANOVA, random effects, and reliability
[R] ranksum .... Equality tests on unmatched data
[R] symmetry .... Symmetry and marginal homogeneity tests
[R] tabulate twoway .... Two-way tables of frequencies
[R] tetrachoric .... Tetrachoric correlations for binary variables

Factor analysis and principal components
[R] alpha .... Compute interitem correlations (covariances) and Cronbach's alpha
[MV] canon .... Canonical correlations
[MV] factor .... Factor analysis
[MV] pca .... Principal component analysis
[MV] rotate .... Orthogonal and oblique rotations after factor and pca
[MV] rotatemat .... Orthogonal and oblique rotations of a Stata matrix

[MV]   scoreplot .... Score and loading plots
[MV]   screeplot .... Scree plot
[R]   tetrachoric .... Tetrachoric correlations for binary variables

Generalized linear models

[U]   Chapter 20 .... Estimation and postestimation commands
[U]   Section 26.4 .... Generalized linear models
[R]   binreg .... Generalized linear models: Extensions to the binomial family
[R]   glm .... Generalized linear models
[XT]   xtgee .... Fit population-averaged panel-data models by using GEE

Indicator and categorical variables

[U]   Section 11.4.3 .... Factor variables
[U]   Chapter 25 .... Working with categorical data and factor variables
[R]   fvset .... Declare factor-variable settings

Linear regression and related

[U]   Chapter 20 .... Estimation and postestimation commands
[U]   Chapter 26 .... Overview of Stata estimation commands
[R]   areg .... Linear regression with a large dummy-variable set
[R]   cnsreg .... Constrained linear regression
[R]   constraint .... Define and list constraints
[R]   eivreg .... Errors-in-variables regression
[I]   estimation commands .... Quick reference for estimation commands
[R]   fracpoly .... Fractional polynomial regression
[R]   frontier .... Stochastic frontier models
[R]   glm .... Generalized linear models
[R]   gmm .... Generalized method of moments estimation
[R]   heckman .... Heckman selection model
[R]   intreg .... Interval regression
[R]   ivregress .... Single-equation instrumental-variables regression
[R]   ivtobit .... Tobit model with continuous endogenous regressors
[R]   logit .... Logistic regression, reporting coefficients
[R]   lpoly .... Kernel-weighted local polynomial smoothing
[R]   mfp .... Multivariable fractional polynomial models
[R]   mvreg .... Multivariate regression
[R]   nbreg .... Negative binomial regression
[R]   nestreg .... Nested model statistics
[TS]   newey .... Regression with Newey–West standard errors
[R]   nl .... Nonlinear least-squares estimation
[R]   nlsur .... Estimation of nonlinear systems of equations
[R]   orthog .... Orthogonalize variables and compute orthogonal polynomials
[R]   poisson .... Poisson regression
[TS]   prais .... Prais–Winsten and Cochrane–Orcutt regression
[R]   qreg .... Quantile regression
[R]   reg3 .... Three-stage estimation for systems of simultaneous equations
[R]   regress .... Linear regression
[R]   rocfit .... Fit ROC models
[R]   rreg .... Robust regression
[ST]   stcox .... Cox proportional hazards model
[ST]   stcrreg .... Competing-risks regression

[R]   stepwise . . . . Stepwise estimation
[ST]   streg . . . . Parametric survival models
[R]   sureg . . . . Zellner's seemingly unrelated regression
[R]   tobit . . . . Tobit regression
[R]   treatreg . . . . Treatment-effects model
[R]   truncreg . . . . Truncated regression
[R]   vwls . . . . Variance-weighted least squares
[XT]   xtabond . . . . Arellano–Bond linear dynamic panel-data estimation
[XT]   xtcloglog . . . . Random-effects and population-averaged cloglog models
[XT]   xtdpd . . . . Linear dynamic panel-data estimation
[XT]   xtdpdsys . . . . . . Arellano–Bover/Blundell–Bond linear dynamic panel-data estimation
[XT]   xtfrontier . . . . Stochastic frontier models for panel data
[XT]   xtgee . . . . Fit population-averaged panel-data models by using GEE
[XT]   xtgls . . . . Fit panel-data models by using GLS
[XT]   xthtaylor . . . . Hausman–Taylor estimator for error-components models
[XT]   xtintreg . . . . Random-effects interval-data regression models
[XT]   xtivreg . . . . . Instrumental variables and two-stage least squares for panel-data models
[XT]   xtlogit . . . . Fixed-effects, random-effects, and population-averaged logit models
[XT]   xtmelogit . . . . Multilevel mixed-effects logistic regression
[XT]   xtmepoisson . . . . Multilevel mixed-effects Poisson regression
[XT]   xtmixed . . . . Multilevel mixed-effects linear regression
[XT]   xtnbreg Fixed-effects, random-effects, & population-averaged negative binomial models
[XT]   xtpcse . . . . Linear regression with panel-corrected standard errors
[XT]   xtpoisson . . . . Fixed-effects, random-effects, and population-averaged Poisson models
[XT]   xtprobit . . . . Random-effects and population-averaged probit models
[XT]   xtrc . . . . Random-coefficients model
[XT]   xtreg . . Fixed-, between-, and random-effects, and population-averaged linear models
[XT]   xtregar . . . . Fixed- and random-effects linear models with an AR(1) disturbance
[XT]   xttobit . . . . Random-effects tobit models
[R]   zinb . . . . Zero-inflated negative binomial regression
[R]   zip . . . . Zero-inflated Poisson regression
[R]   ztnb . . . . Zero-truncated negative binomial regression
[R]   ztp . . . . Zero-truncated Poisson regression

Logistic and probit regression

[U]   Chapter 20 . . . . Estimation and postestimation commands
[U]   Chapter 26 . . . . Overview of Stata estimation commands
[R]   asclogit . . . . Alternative-specific conditional logit (McFadden's choice) model
[R]   asmprobit . . . . Alternative-specific multinomial probit regression
[R]   asroprobit . . . . Alternative-specific rank-ordered probit regression
[R]   biprobit . . . . Bivariate probit regression
[R]   clogit . . . . Conditional (fixed-effects) logistic regression
[R]   cloglog . . . . Complementary log-log regression
[R]   exlogistic . . . . Exact logistic regression
[R]   glogit . . . . Logit and probit regression for grouped data
[R]   heckprob . . . . Probit model with sample selection
[R]   hetprob . . . . Heteroskedastic probit model
[R]   ivprobit . . . . Probit model with continuous endogenous regressors
[R]   logistic . . . . Logistic regression, reporting odds ratios
[R]   logit . . . . Logistic regression, reporting coefficients
[R]   mlogit . . . . Multinomial (polytomous) logistic regression

[R]   mprobit . . . . Multinomial probit regression
[R]   nlogit . . . . Nested logit regression
[R]   ologit . . . . Ordered logistic regression
[R]   oprobit . . . . Ordered probit regression
[R]   probit . . . . Probit regression
[R]   rologit . . . . Rank-ordered logistic regression
[R]   scobit . . . . Skewed logistic regression
[R]   slogit . . . . Stereotype logistic regression
[XT]   xtcloglog . . . . Random-effects and population-averaged cloglog models
[XT]   xtgee . . . . Fit population-averaged panel-data models by using GEE
[XT]   xtlogit . . . . Fixed-effects, random-effects, and population-averaged logit models
[XT]   xtmelogit . . . . Multilevel mixed-effects logistic regression
[XT]   xtprobit . . . . Random-effects and population-averaged probit models

Longitudinal data/panel data

[U]   Chapter 20 . . . . Estimation and postestimation commands
[U]   Section 26.15 . . . . Panel-data models
[XT]   quadchk . . . . Check sensitivity of quadrature approximation
[XT]   xt . . . . Introduction to xt commands
[XT]   xtabond . . . . Arellano–Bond linear dynamic panel-data estimation
[XT]   xtcloglog . . . . Random-effects and population-averaged cloglog models
[XT]   xtdata . . . . Faster specification searches with xt data
[XT]   xtdescribe . . . . Describe pattern of xt data
[XT]   xtdpd . . . . Linear dynamic panel-data estimation
[XT]   xtdpdsys . . . . . . Arellano–Bover/Blundell–Bond linear dynamic panel-data estimation
[XT]   xtfrontier . . . . Stochastic frontier models for panel data
[XT]   xtgee . . . . Fit population-averaged panel-data models by using GEE
[XT]   xtgls . . . . Fit panel-data models by using GLS
[XT]   xthtaylor . . . . Hausman–Taylor estimator for error-components models
[XT]   xtintreg . . . . Random-effects interval-data regression models
[XT]   xtivreg . . . . . Instrumental variables and two-stage least squares for panel-data models
[XT]   xtline . . . . Panel-data line plots
[XT]   xtlogit . . . . Fixed-effects, random-effects, and population-averaged logit models
[XT]   xtmelogit . . . . Multilevel mixed-effects logistic regression
[XT]   xtmepoisson . . . . Multilevel mixed-effects Poisson regression
[XT]   xtmixed . . . . Multilevel mixed-effects linear regression
[XT]   xtnbreg Fixed-effects, random-effects, & population-averaged negative binomial models
[XT]   xtpcse . . . . Linear regression with panel-corrected standard errors
[XT]   xtpoisson . . . . Fixed-effects, random-effects, and population-averaged Poisson models
[XT]   xtprobit . . . . Random-effects and population-averaged probit models
[XT]   xtrc . . . . Random-coefficients model
[XT]   xtreg .. Fixed-, between-, and random-effects, and population-averaged linear models
[XT]   xtregar . . . . Fixed- and random-effects linear models with an AR(1) disturbance
[XT]   xtset . . . . Declare data to be panel data
[XT]   xtsum . . . . Summarize xt data
[XT]   xttab . . . . Tabulate xt data
[XT]   xttobit . . . . Random-effects tobit models
[XT]   xtunitroot . . . . Panel-data unit-root tests

Mixed models

[U]   Chapter 20 . . . . Estimation and postestimation commands

[R]   anova . . . . Analysis of variance and covariance
[MV] manova . . . . Multivariate analysis of variance and covariance
[XT] xtcloglog . . . . Random-effects and population-averaged cloglog models
[XT] xtintreg . . . . Random-effects interval-data regression models
[XT] xtlogit . . . . Fixed-effects, random-effects, and population-averaged logit models
[XT] xtmelogit . . . . Multilevel mixed-effects logistic regression
[XT] xtmepoisson . . . . Multilevel mixed-effects Poisson regression
[XT] xtmixed . . . . Multilevel mixed-effects linear regression
[XT] xtprobit . . . . Random-effects and population-averaged probit models
[XT] xtrc . . . . Random-coefficients model
[XT] xtreg .. Fixed-, between-, and random-effects, and population-averaged linear models
[XT] xttobit . . . . Random-effects tobit models

Multidimensional scaling and biplots

[MV] biplot . . . . Biplots
[MV] mds . . . . Multidimensional scaling for two-way data
[MV] mdslong . . . . Multidimensional scaling of proximity data in long format
[MV] mdsmat . . . . Multidimensional scaling of proximity data in a matrix
[MV] *measure_option* . . . . Option for similarity and dissimilarity measures

Multilevel/hierarchical models

[XT] xtmelogit . . . . Multilevel mixed-effects logistic regression
[XT] xtmepoisson . . . . Multilevel mixed-effects Poisson regression
[XT] xtmixed . . . . Multilevel mixed-effects linear regression

Multiple imputation

[U]   Section 26.20 . . . . Multiple imputation
[MI] estimation . . . . Estimation commands for use with mi estimate
[MI] intro substantive . . . . Introduction to multiple-imputation analysis
[MI] mi estimate . . . . Estimation using multiple imputations
[MI] mi estimate using . . . . Estimation using previously saved estimation results
[MI] mi estimate postestimation . . . . Postestimation tools for mi estimate
[MI] mi impute . . . . Impute missing values
[MI] mi impute logit . . . . Impute using logistic regression
[MI] mi impute mlogit . . . . Impute using multinomial logistic regression
[MI] mi impute monotone . . . . Impute missing values in monotone data
[MI] mi impute mvn . . . . Impute using multivariate normal regression
[MI] mi impute ologit . . . . Impute using ordered logistic regression
[MI] mi impute pmm . . . . Impute using predictive mean matching
[MI] mi impute regress . . . . Impute using linear regression

Multivariate analysis of variance and related techniques

[U]   Section 26.21 . . . . Multivariate and cluster analysis
[MV] canon . . . . Canonical correlations
[MV] hotelling . . . . Hotelling's T-squared generalized means test
[MV] manova . . . . Multivariate analysis of variance and covariance
[R]   mvreg . . . . Multivariate regression
[MV] mvtest covariances . . . . Multivariate tests of covariances
[MV] mvtest means . . . . Multivariate tests of means

Nonparametric statistics

[R]   kdensity .... Univariate kernel density estimation
[R]   ksmirnov .... Kolmogorov–Smirnov equality-of-distributions test
[R]   kwallis .... Kruskal–Wallis equality-of-populations rank test
[R]   lowess .... Lowess smoothing
[R]   lpoly .... Kernel-weighted local polynomial smoothing
[R]   lv .... Letter-value displays
[R]   nptrend .... Test for trend across ordered groups
[R]   qreg .... Quantile regression
[R]   ranksum .... Equality tests on unmatched data
[R]   roc .... Receiver operating characteristic (ROC) analysis
[R]   runtest .... Test for random order
[R]   signrank .... Equality tests on matched data
[R]   smooth .... Robust nonlinear smoother
[R]   spearman .... Spearman's and Kendall's correlations
[R]   symmetry .... Symmetry and marginal homogeneity tests

Ordinal outcomes

[U]   Chapter 20 .... Estimation and postestimation commands
[R]   asroprobit .... Alternative-specific rank-ordered probit regression
[R]   ologit .... Ordered logistic regression
[R]   oprobit .... Ordered probit regression
[R]   rologit .... Rank-ordered logistic regression

Other statistics

[R]   alpha .... Compute interitem correlations (covariances) and Cronbach's alpha
[R]   ameans .... Arithmetic, geometric, and harmonic means
[R]   brier .... Brier score decomposition
[R]   centile .... Report centile and confidence interval
[R]   kappa .... Interrater agreement
[MV]   mvtest correlations .... Multivariate tests of correlations
[R]   pcorr .... Partial and semipartial correlation coefficients
[D]   pctile .... Create variable containing percentiles
[D]   range .... Generate numerical range

Pharmacokinetic statistics

[U]   Section 26.22 .... Pharmacokinetic data
[R]   pk .... Pharmacokinetic (biopharmaceutical) data
[R]   pkcollapse .... Generate pharmacokinetic measurement dataset
[R]   pkcross .... Analyze crossover experiments
[R]   pkequiv .... Perform bioequivalence tests
[R]   pkexamine .... Calculate pharmacokinetic measures
[R]   pkshape .... Reshape (pharmacokinetic) Latin-square data
[R]   pksumm .... Summarize pharmacokinetic data

Power and sample size

[R]   sampsi .... Sample size and power determination
[ST]   stpower ..... Sample-size, power, and effect-size determination for survival analysis

[ST]   stpower cox Sample size, power, and effect size for the Cox proportional hazards model
[ST]   stpower exponential .... Sample size and power for the exponential test
[ST]   stpower logrank .... Sample size, power, and effect size for the log-rank test

Quality control

[R]   cusum .... Graph cumulative spectral distribution
[R]   qc .... Quality control charts
[R]   serrbar .... Graph standard error bar chart

Rotation

[MV]   procrustes .... Procrustes transformation
[MV]   rotate .... Orthogonal and oblique rotations after factor and pca
[MV]   rotatemat .... Orthogonal and oblique rotations of a Stata matrix

Sample selection models

[U]   Chapter 20 .... Estimation and postestimation commands
[U]   Section 26.13 .... Models with endogenous sample selection
[R]   heckman .... Heckman selection model
[R]   heckprob .... Probit model with sample selection
[R]   treatreg .... Treatment-effects model

Simulation/resampling

[R]   bootstrap .... Bootstrap sampling and estimation
[R]   bsample .... Sampling with replacement
[R]   jackknife .... Jackknife estimation
[R]   permute .... Monte Carlo permutation tests
[R]   simulate .... Monte Carlo simulations

Standard postestimation tests, tables, and other analyses

[U]   Section 13.5 .... Accessing coefficients and standard errors
[U]   Chapter 20 .... Estimation and postestimation commands
[R]   correlate .... Correlations (covariances) of variables or coefficients
[R]   estat .... Postestimation statistics
[R]   estimates .... Save and manipulate estimation results
[R]   estimates describe .... Describe estimation results
[R]   estimates for .... Repeat postestimation command across models
[R]   estimates notes .... Add notes to estimation results
[R]   estimates replay .... Redisplay estimation results
[R]   estimates save .... Save and use estimation results
[R]   estimates stats .... Model statistics
[R]   estimates store .... Store and restore estimation results
[R]   estimates table .... Compare estimation results
[R]   estimates title .... Set title for estimation results
[R]   hausman .... Hausman specification test
[R]   lincom .... Linear combinations of estimators
[R]   linktest .... Specification link test for single-equation models
[R]   lrtest .... Likelihood-ratio test after estimation
[R]   margins .... Marginal means, predictive margins, and marginal effects
[MV]   mvtest .... Multivariate tests
[R]   nlcom .... Nonlinear combinations of estimators

[R]   predict .... Obtain predictions, residuals, etc., after estimation
[R]   predictnl .... Obtain nonlinear predictions, standard errors, etc., after estimation
[R]   suest .... Seemingly unrelated estimation
[R]   test .... Test linear hypotheses after estimation
[R]   testnl .... Test nonlinear hypotheses after estimation

Survey data

[U]   Chapter 20 .... Estimation and postestimation commands
[U]   Section 26.19 .... Survey data
[SVY]   *brr_options* .... More options for BRR variance estimation
[SVY]   direct standardization .... Direct standardization of means, proportions, and ratios
[SVY]   estat .... Postestimation statistics for survey data
[SVY]   *jackknife_options* .... More options for jackknife variance estimation
[SVY]   ml for svy .... Maximum pseudolikelihood estimation for survey data
[SVY]   poststratification .... Poststratification for survey data
[P]   _robust .... Robust variance estimates
[SVY]   subpopulation estimation .... Subpopulation estimation for survey data
[SVY]   survey .... Introduction to survey commands
[SVY]   svy .... The survey prefix command
[SVY]   svy brr .... Balanced repeated replication for survey data
[SVY]   svy estimation .... Estimation commands for survey data
[SVY]   svy jackknife .... Jackknife estimation for survey data
[SVY]   svy postestimation .... Postestimation tools for svy
[SVY]   svy: tabulate oneway .... One-way tables for survey data
[SVY]   svy: tabulate twoway .... Two-way tables for survey data
[SVY]   svydescribe .... Describe survey data
[SVY]   svymarkout ... Mark observations for exclusion on the basis of survey characteristics
[SVY]   svyset .... Declare survey design for dataset
[MI]   mi XXXset .... Declare mi data to be svy, st, ts, xt, etc.
[SVY]   variance estimation .... Variance estimation for survey data

Survival analysis

[U]   Chapter 20 .... Estimation and postestimation commands
[U]   Section 26.16 .... Survival-time (failure-time) models
[ST]   ct .... Count-time data
[ST]   ctset .... Declare data to be count-time data
[ST]   cttost .... Convert count-time data to survival-time data
[ST]   discrete .... Discrete-time survival analysis
[ST]   ltable .... Life tables for survival data
[ST]   snapspan .... Convert snapshot data to time-span data
[ST]   st .... Survival-time data
[ST]   st_is .... Survival analysis subroutines for programmers
[ST]   stbase .... Form baseline dataset
[ST]   stci .... Confidence intervals for means and percentiles of survival time
[ST]   stcox .... Cox proportional hazards model
[ST]   stcox PH-assumption tests .... Tests of proportional-hazards assumption
[ST]   stcrreg .... Competing-risks regression
[ST]   stcurve ... Plot survivor, hazard, cumulative hazard, or cumulative incidence function
[ST]   stdescribe .... Describe survival-time data
[R]   stepwise .... Stepwise estimation
[ST]   stfill .... Fill in by carrying forward values of covariates

[ST]   stgen . . . . Generate variables reflecting entire histories
[ST]   stir . . . . Report incidence-rate comparison
[ST]   stpower . . . . . Sample-size, power, and effect-size determination for survival analysis
[ST]   stpower cox Sample size, power, and effect size for the Cox proportional hazards model
[ST]   stpower exponential . . . . Sample size and power for the exponential test
[ST]   stpower logrank . . . . Sample size, power, and effect size for the log-rank test
[ST]   stptime . . . . Calculate person-time, incidence rates, and SMR
[ST]   strate . . . . Tabulate failure rates and rate ratios
[ST]   streg . . . . Parametric survival models
[ST]   sts . . . . . Generate, graph, list, and test the survivor and cumulative hazard functions
[ST]   sts generate . . . . Create variables containing survivor and related functions
[ST]   sts graph . . . . Graph the survivor and cumulative hazard functions
[ST]   sts list . . . . List the survivor or cumulative hazard function
[ST]   sts test . . . . Test equality of survivor functions
[ST]   stset . . . . Declare data to be survival-time data
[MI]   mi XXXset . . . . Declare mi data to be svy, st, ts, xt, etc.
[ST]   stsplit . . . . Split and join time-span records
[MI]   mi stsplit . . . . Stsplit and stjoin mi data
[ST]   stsum . . . . Summarize survival-time data
[ST]   sttocc . . . . Convert survival-time data to case–control data
[ST]   sttoct . . . . Convert survival-time data to count-time data
[ST]   stvary . . . . Report whether variables vary over time
[ST]   survival analysis Introduction to survival analysis & epidemiological tables commands

Time series, multivariate

[U]   Section 11.4.4 . . . . Time-series varlists
[U]   Section 13.9 . . . . Time-series operators
[U]   Section 26.14 . . . . Models with time-series data
[TS]   dfactor . . . . Dynamic-factor models
[TS]   dvech . . . . Diagonal vech multivariate GARCH models
[TS]   fcast compute Compute dynamic forecasts of dependent variables after var, svar, or vec
[TS]   fcast graph . . . . Graph forecasts of variables computed by fcast compute
[TS]   haver . . . . Load data from Haver Analytics database
[TS]   irf . . . . Create and analyze IRFs, dynamic-multiplier functions, and FEVDs
[TS]   irf add . . . . Add results from an IRF file to the active IRF file
[TS]   irf cgraph . . . . . Combine graphs of IRFs, dynamic-multiplier functions, and FEVDs
[TS]   irf create . . . . Obtain IRFs, dynamic-multiplier functions, and FEVDs
[TS]   irf ctable . . . . Combine tables of IRFs, dynamic-multiplier functions, and FEVDs
[TS]   irf describe . . . . Describe an IRF file
[TS]   irf drop . . . . Drop IRF results from the active IRF file
[TS]   irf graph . . . . Graph IRFs, dynamic-multiplier functions, and FEVDs
[TS]   irf ograph . . . . Graph overlaid IRFs, dynamic-multiplier functions, and FEVDs
[TS]   irf rename . . . . Rename an IRF result in an IRF file
[TS]   irf set . . . . Set the active IRF file
[TS]   irf table . . . . Create tables of IRFs, dynamic-multiplier functions, and FEVDs
[TS]   rolling . . . . Rolling-window and recursive estimation
[TS]   sspace . . . . State-space models
[TS]   time series . . . . Introduction to time-series commands
[TS]   tsappend . . . . Add observations to a time-series dataset
[TS]   tsfill . . . . Fill in gaps in time variable
[TS]   tsline . . . . Plot time-series data

[TS]   tsreport .... Report time-series aspects of a dataset or estimation sample
[TS]   tsrevar .... Time-series operator programming command
[TS]   tsset .... Declare data to be time-series data
[TS]   var intro .... Introduction to vector autoregressive models
[TS]   var svar .... Structural vector autoregressive models
[TS]   var .... Vector autoregressive models
[TS]   varbasic .... Fit a simple VAR and graph IRFS or FEVDs
[TS]   vargranger .... Perform pairwise Granger causality tests after var or svar
[TS]   varlmar .... Perform LM test for residual autocorrelation after var or svar
[TS]   varnorm .... Test for normally distributed disturbances after var or svar
[TS]   varsoc .... Obtain lag-order selection statistics for VARs and VECMs
[TS]   varstable .... Check the stability condition of VAR or SVAR estimates
[TS]   varwle .... Obtain Wald lag-exclusion statistics after var or svar
[TS]   vec intro .... Introduction to vector error-correction models
[TS]   vec .... Vector error-correction models
[TS]   veclmar .... Perform LM test for residual autocorrelation after vec
[TS]   vecnorm .... Test for normally distributed disturbances after vec
[TS]   vecrank .... Estimate the cointegrating rank of a VECM
[TS]   vecstable .... Check the stability condition of VECM estimates
[TS]   xcorr .... Cross-correlogram for bivariate time series

Time series, univariate

[U]   Section 11.4.4 .... Time-series varlists
[U]   Section 13.9 .... Time-series operators
[U]   Section 26.14 .... Models with time-series data
[TS]   arch ..... Autoregressive conditional heteroskedasticity (ARCH) family of estimators
[TS]   arima .... ARIMA, ARMAX, and other dynamic regression models
[TS]   corrgram .... Tabulate and graph autocorrelations
[TS]   cumsp .... Cumulative spectral distribution
[TS]   dfgls .... DF-GLS unit-root test
[TS]   dfuller .... Augmented Dickey–Fuller unit-root test
[TS]   haver .... Load data from Haver Analytics database
[TS]   newey .... Regression with Newey–West standard errors
[TS]   pergram .... Periodogram
[TS]   pperron .... Phillips–Perron unit-root test
[TS]   prais .... Prais–Winsten and Cochrane–Orcutt regression
[TS]   rolling .... Rolling-window and recursive estimation
[TS]   time series .... Introduction to time-series commands
[TS]   tsappend .... Add observations to a time-series dataset
[TS]   tsfill .... Fill in gaps in time variable
[TS]   tsline .... Plot time-series data
[TS]   tsreport .... Report time-series aspects of a dataset or estimation sample
[TS]   tsrevar .... Time-series operator programming command
[TS]   tsset .... Declare data to be time-series data
[TS]   tssmooth .... Smooth and forecast univariate time-series data
[TS]   tssmooth dexponential .... Double-exponential smoothing
[TS]   tssmooth exponential .... Single-exponential smoothing
[TS]   tssmooth hwinters .... Holt–Winters nonseasonal smoothing
[TS]   tssmooth ma .... Moving-average filter
[TS]   tssmooth nl .... Nonlinear filter
[TS]   tssmooth shwinters .... Holt–Winters seasonal smoothing

[TS]   wntestb .... Bartlett's periodogram-based test for white noise
[TS]   wntestq .... Portmanteau (Q) test for white noise
[TS]   xcorr .... Cross-correlogram for bivariate time series

Transforms and normality tests

[R]   boxcox .... Box–Cox regression models
[R]   fracpoly .... Fractional polynomial regression
[R]   ladder .... Ladder of powers
[R]   lnskew0 .... Find zero-skewness log or Box–Cox transform
[R]   mfp .... Multivariable fractional polynomial models
[MV]   mvtest normality .... Multivariate normality tests
[R]   sktest .... Skewness and kurtosis test for normality
[R]   swilk .... Shapiro–Wilk and Shapiro–Francia tests for normality

Matrix commands

Basics

[U]   Chapter 14 .... Matrix expressions
[P]   matlist .... Display a matrix and control its format
[P]   matrix .... Introduction to matrix commands
[P]   matrix define .... Matrix definition, operators, and functions
[P]   matrix utility .... List, rename, and drop matrices

Programming

[P]   ereturn .... Post the estimation results
[P]   matrix accum .... Form cross-product matrices
[P]   matrix rownames .... Name rows and columns
[P]   matrix score .... Score data from coefficient vectors
[R]   ml .... Maximum likelihood estimation

Other

[P]   makecns .... Constrained estimation
[P]   matrix dissimilarity .... Compute similarity or dissimilarity measures
[P]   matrix eigenvalues .... Eigenvalues of nonsymmetric matrices
[P]   matrix get .... Access system matrices
[P]   matrix mkmat .... Convert variables to matrix and vice versa
[P]   matrix svd .... Singular value decomposition
[P]   matrix symeigen .... Eigenvalues and eigenvectors of symmetric matrices

Mata

[M]   *Mata Reference Manual* ....

Programming

Basics

[U]   Chapter 18 .... Programming Stata
[U]   Section 18.3 .... Macros
[U]   Section 18.11 .... Ado-files
[P]   comments .... Add comments to programs

[P]   fvexpand .... Expand factor varlists
[P]   macro .... Macro definition and manipulation
[P]   program .... Define and manipulate programs
[P]   return .... Return saved results

Program control

[U]   Section 18.11.1 .... Version
[P]   capture .... Capture return code
[P]   continue .... Break out of loops
[P]   error .... Display generic error message and exit
[P]   foreach .... Loop over items
[P]   forvalues .... Loop over consecutive values
[P]   if .... if programming command
[P]   version .... Version control
[P]   while .... Looping

Parsing and program arguments

[U]   Section 18.4 .... Program arguments
[P]   confirm .... Argument verification
[P]   gettoken .... Low-level parsing
[P]   levelsof .... Levels of variable
[P]   numlist .... Parse numeric lists
[P]   syntax .... Parse Stata syntax
[P]   tokenize .... Divide strings into tokens

Console output

[P]   dialog programming .... Dialog programming
[P]   display .... Display strings and values of scalar expressions
[P]   smcl .... Stata Markup and Control Language
[P]   tabdisp .... Display tables

Commonly used programming commands

[P]   byable .... Make programs byable
[P]   #delimit .... Change delimiter
[P]   exit .... Exit from a program or do-file
[R]   fvrevar .... Factor-variables operator programming command
[P]   mark .... Mark observations for inclusion
[P]   matrix .... Introduction to matrix commands
[P]   more .... Pause until key is pressed
[P]   nopreserve option .... nopreserve option
[P]   preserve .... Preserve and restore data
[P]   quietly .... Quietly and noisily perform Stata command
[P]   scalar .... Scalar variables
[P]   smcl .... Stata Markup and Control Language
[P]   sortpreserve .... Sort within programs
[P]   timer .... Time sections of code by recording and reporting time spent
[TS] tsrevar .... Time-series operator programming command

Debugging

[P]	pause	.....	Program debugging command
[P]	timer	.....	Time sections of code by recording and reporting time spent
[P]	trace	.....	Debug Stata programs

Advanced programming commands

[P]	automation	.....	Automation
[P]	break	.....	Suppress Break key
[P]	char	.....	Characteristics
[M-2]	class	.....	Object-oriented programming (classes)
[P]	class	.....	Class programming
[P]	class exit	.....	Exit class-member program and return result
[P]	classutil	.....	Class programming utility
[P]	estat programming	.....	Controlling estat after user-written commands
[P]	_estimates	.....	Manage estimation results
[P]	file	.....	Read and write ASCII text and binary files
[P]	findfile	.....	Find file in path
[P]	include	.....	Include commands from file
[P]	macro	.....	Macro definition and manipulation
[P]	macro lists	.....	Manipulate lists
[R]	ml	.....	Maximum likelihood estimation
[M-5]	moptimize( )	.....	Model optimization
[M-5]	optimize( )	.....	Function optimization
[P]	plugin	.....	Load a plugin
[P]	postfile	.....	Save results in Stata dataset
[P]	_predict	.....	Obtain predictions, residuals, etc., after estimation programming command
[P]	program properties	.....	Properties of user-defined programs
[P]	_return	.....	Preserve saved results
[P]	_rmcoll	.....	Remove collinear variables
[P]	_robust	.....	Robust variance estimates
[P]	serset	.....	Create and manipulate sersets
[D]	snapshot	.....	Save and restore data snapshots
[P]	unab	.....	Unabbreviate variable list
[P]	unabcmd	.....	Unabbreviate command name
[P]	varabbrev	.....	Control variable abbreviation
[P]	viewsource	.....	View source code

Special-interest programming commands

[R]	bstat	.....	Report bootstrap results
[MV]	cluster programming subroutines	.....	Add cluster-analysis routines
[MV]	cluster programming utilities	.....	Cluster-analysis programming utilities
[R]	fvrevar	.....	Factor-variables operator programming command
[P]	matrix dissimilarity	.....	Compute similarity or dissimilarity measures
[MI]	mi select	.....	Programmer's alternative to mi extract
[ST]	st_is	.....	Survival analysis subroutines for programmers
[SVY]	svymarkout	.....	Mark observations for exclusion on the basis of survey characteristics
[MI]	technical	.....	Details for programmers
[TS]	tsrevar	.....	Time-series operator programming command

File formats

| [P] | file formats .dta | ..... | Description of .dta file format |

Mata

[M]   *Mata Reference Manual* ....

Interface features

[GS]   Chapter 1 (GSM, GSU, GSW) .... Introducing Stata—sample session
[GS]   Chapter 2 (GSM, GSU, GSW) .... The Stata user interface
[GS]   Chapter 3 (GSM, GSU, GSW) .... Using the Viewer
[GS]   Chapter 6 (GSM, GSU, GSW) .... Using the Data Editor
[GS]   Chapter 7 (GSM, GSU, GSW) .... Using the Variables Manager
[GS]   Chapter 13 (GSM, GSU, GSW) .... Using the Do-file Editor—automating Stata
[GS]   Chapter 15 (GSM, GSU, GSW) .... Editing graphs
[P]   dialog programming .... Dialog programming
[R]   doedit .... Edit do-files and other text files
[D]   edit .... Browse or edit data with Data Editor
[P]   sleep .... Pause for a specified time
[P]   smcl .... Stata Markup and Control Language
[D]   varmanage .... Manage variable labels, formats, and other properties
[P]   viewsource .... View source code
[P]   window programming .... Programming menus and windows

Title

data types — Quick reference for data types

Description

This entry provides a quick reference for data types allowed by Stata. See [U] **12 Data** for details.

Storage type	Minimum	Maximum	Closest to 0 without being 0	bytes
byte	-127	100	± 1	1
int	$-32{,}767$	$32{,}740$	± 1	2
long	$-2{,}147{,}483{,}647$	$2{,}147{,}483{,}620$	± 1	4
float	$-1.70141173319 \times 10^{38}$	$1.70141173319 \times 10^{38}$	$\pm 10^{-38}$	4
double	$-8.9884656743 \times 10^{307}$	$8.9884656743 \times 10^{307}$	$\pm 10^{-323}$	8

Precision for float is 3.795×10^{-8}

Precision for double is 1.414×10^{-16}

String storage type	Maximum length	Bytes
str1	1	1
str2	2	2
...	.	.
...	.	.
...	.	.
str244	244	244

Also see

[D] **compress** — Compress data in memory

[D] **destring** — Convert string variables to numeric variables and vice versa

[D] **encode** — Encode string into numeric and vice versa

[D] **format** — Set variables' output format

[D] **recast** — Change storage type of variable

[U] **12.2.2 Numeric storage types**

[U] **12.4.4 String storage types**

[U] **12.5 Formats: Controlling how data are displayed**

[U] **13.11 Precision and problems therein**

Title

estimation commands — Quick reference for estimation commands

Description

This entry provides a quick reference for Stata's estimation commands. Because enhancements to Stata are continually being made, type `search estimation commands` for possible additions to this list; see [R] **search**.

For a discussion of properties shared by all estimation commands, see [U] **20 Estimation and postestimation commands**.

For a list of prefix commands that can be used with many of these estimation commands, see [U] **11.1.10 Prefix commands**.

Command	Description	See
anova	Analysis of variance and covariance	[R] **anova**
arch	ARCH family of estimators	[TS] **arch**
areg	Linear regression with a large dummy-variable set	[R] **areg**
arima	ARIMA, ARMAX, and other dynamic regression models	[TS] **arima**
asclogit	Alternative-specific conditional logit (McFadden's choice) model	[R] **asclogit**
asmprobit	Alternative-specific multinomial probit regression	[R] **asmprobit**
asroprobit	Alternative-specific rank-ordered probit regression	[R] **asroprobit**
binreg	Generalized linear models: Extensions to the binomial family	[R] **binreg**
biprobit	Bivariate probit regression	[R] **biprobit**
blogit	Logistic regression for grouped data	[R] **glogit**
boxcox	Box–Cox regression models	[R] **boxcox**
bprobit	Probit regression for grouped data	[R] **glogit**
bsqreg	Quantile regression with bootstrap standard errors	[R] **qreg**
ca	Simple correspondence analysis	[MV] **ca**
camat	Simple correspondence analysis of a matrix	[MV] **ca**
candisc	Canonical linear discriminant analysis	[MV] **candisc**
canon	Canonical correlations	[MV] **canon**
clogit	Conditional (fixed-effects) logistic regression	[R] **clogit**
cloglog	Complementary log-log regression	[R] **cloglog**
cnsreg	Constrained linear regression	[R] **cnsreg**
dfactor	Dynamic-factor models	[TS] **dfactor**
discrim knn	kth-nearest-neighbor discriminant analysis	[MV] **discrim knn**
discrim lda	Linear discriminant analysis	[MV] **discrim lda**
discrim logistic	Logistic discriminant analysis	[MV] **discrim logistic**
discrim qda	Quadratic discriminant analysis	[MV] **discrim qda**
dvech	Diagonal vech GARCH models	[TS] **dvech**

Command	Description	See
eivreg	Errors-in-variables regression	[R] **eivreg**
exlogistic	Exact logistic regression	[R] **exlogistic**
expoisson	Exact Poisson regression	[R] **exlogistic**
factor	Factor analysis	[MV] **factor**
factormat	Factor analysis of a correlation matrix	[MV] **factor**
frontier	Stochastic frontier models	[R] **frontier**
glm	Generalized linear models	[R] **glm**
glogit	Weighted least-squares logistic regression for grouped data	[R] **glogit**
gmm	Generalized method of moments estimation	[R] **gmm**
gnbreg	Generalized negative binomial model	[R] **nbreg**
gprobit	Weighted least-squares probit regression for grouped data	[R] **glogit**
heckman	Heckman selection model	[R] **heckman**
heckprob	Probit model with selection	[R] **heckprob**
hetprob	Heteroskedastic probit model	[R] **hetprob**
intreg	Interval regression	[R] **intreg**
iqreg	Interquantile range regression	[R] **qreg**
ivprobit	Probit model with endogenous regressors	[R] **ivprobit**
ivregress	Single-equation instrumental-variables estimation	[R] **ivregress**
ivtobit	Tobit model with endogenous regressors	[R] **ivtobit**
logistic	Logistic regression, reporting odds ratios	[R] **logistic**
logit	Logistic regression, reporting coefficients	[R] **logit**
manova	Multivariate analysis of variance and covariance	[MV] **manova**
mca	Multiple and joint correspondence analysis	[MV] **mca**
mds	Multidimensional scaling for two-way data	[MV] **mds**
mdslong	Multidimensional scaling of proximity data in long format	[MV] **mdslong**
mdsmat	Multidimensional scaling of proximity data in a matrix	[MV] **mdsmat**
mean	Estimate means	[R] **mean**
mlogit	Multinomial (polytomous) logistic regression	[R] **mlogit**
mprobit	Multinomial probit regression	[R] **mprobit**
mvreg	Multivariate regression	[R] **mvreg**
nbreg	Negative binomial regression	[R] **nbreg**
newey	Regression with Newey–West standard errors	[TS] **newey**
nl	Nonlinear least-squares estimation	[R] **nl**
nlogit	Nested logit regression (RUM-consistent and nonnormalized)	[R] **nlogit**
nlsur	Systems of nonlinear equations	[R] **nlsur**
ologit	Ordered logistic regression	[R] **ologit**
oprobit	Ordered probit regression	[R] **oprobit**
pca	Principal component analysis	[MV] **pca**
pcamat	Principal component analysis of a correlation or covariance matrix	[MV] **pca**

Command	Description	See
poisson	Poisson regression	[R] **poisson**
prais	Prais–Winsten and Cochrane–Orcutt regression	[TS] **prais**
probit	Probit regression	[R] **probit**
procrustes	Procrustes transformation	[MV] **procrustes**
proportion	Estimate proportions	[R] **proportion**
_qreg	Internal estimation command for quantile regression	[R] **qreg**
qreg	Quantile regression	[R] **qreg**
ratio	Estimate ratios	[R] **ratio**
reg3	Three-stage estimation for systems of simultaneous equations	[R] **reg3**
regress	Linear regression	[R] **regress**
rocfit	Fit ROC models	[R] **rocfit**
rologit	Rank-ordered logistic regression	[R] **rologit**
rreg	Robust regression	[R] **rreg**
scobit	Skewed logistic regression	[R] **scobit**
slogit	Stereotype logistic regression	[R] **slogit**
sqreg	Simultaneous-quantile regression	[R] **qreg**
sspace	State-space models	[TS] **sspace**
stcox	Cox proportional hazards model	[ST] **stcox**
stcrreg	Competing-risks regression	[ST] **stcrreg**
streg	Parametric survival models	[ST] **streg**
sureg	Zellner's seemingly unrelated regression	[R] **sureg**
svy: *command*[*]	Estimation commands for survey data	[SVY] **svy estimation**
svy: tabulate oneway	One-way tables for survey data	[SVY] **svy: tabulate oneway**
svy: tabulate twoway	Two-way tables for survey data	[SVY] **svy: tabulate twoway**
tobit	Tobit regression	[R] **tobit**
total	Estimate totals	[R] **total**
treatreg	Treatment-effects model	[R] **treatreg**
truncreg	Truncated regression	[R] **truncreg**
var	Vector autoregressive models	[TS] **var**
var svar	Structural vector autoregressive models	[TS] **var svar**
varbasic	Fit a simple VAR and graph IRFs or FEVDs	[TS] **varbasic**
vec	Vector error-correction models	[TS] **vec**
vwls	Variance-weighted least squares	[R] **vwls**

[*]See the table below for a list of commands that support the svy prefix.

Command	Description	See
xtabond	Arellano–Bond linear dynamic panel-data estimation	[XT] **xtabond**
xtcloglog	Random-effects and population-averaged cloglog models	[XT] **xtcloglog**
xtdpd	Linear dynamic panel-data estimation	[XT] **xtdpd**
xtdpdsys	Arellano–Bond/Blundell–Bond estimation	[XT] **xtdpdsys**
xtfrontier	Stochastic frontier models for panel data	[XT] **xtfrontier**
xtgee	Fit population-averaged panel-data models using GEE	[XT] **xtgee**
xtgls	Fit panel-data models using GLS	[XT] **xtgls**
xthtaylor	Hausman–Taylor estimator for error-components models	[XT] **xthtaylor**
xtintreg	Random-effects interval data regression models	[XT] **xtintreg**
xtivreg	Instrumental variables and two-stage least squares for panel-data models	[XT] **xtivreg**
xtlogit	Fixed-effects, random-effects, and population-averaged logit models	[XT] **xtlogit**
xtmelogit	Multilevel mixed-effects logistic regression	[XT] **xtmelogit**
xtmepoisson	Multilevel mixed-effects Poisson regression	[XT] **xtmepoisson**
xtmixed	Multilevel mixed-effects linear regression	[XT] **xtmixed**
xtnbreg	Fixed-effects, random-effects, and population-averaged negative binomial models	[XT] **xtnbreg**
xtpcse	OLS or Prais–Winsten models with panel-corrected standard errors	[XT] **xtpcse**
xtpoisson	Fixed-effects, random-effects, and population-averaged Poisson models	[XT] **xtpoisson**
xtprobit	Random-effects and population-averaged probit models	[XT] **xtprobit**
xtrc	Random-coefficients models	[XT] **xtrc**
xtreg	Fixed-, between-, and random-effects, and population-averaged linear models	[XT] **xtreg**
xtregar	Fixed- and random-effects linear models with an AR(1) disturbance	[XT] **xtregar**
xttobit	Random-effects tobit models	[XT] **xttobit**
zinb	Zero-inflated negative binomial regression	[R] **zinb**
zip	Zero-inflated Poisson regression	[R] **zip**
ztnb	Zero-truncated negative binomial regression	[R] **ztnb**
ztp	Zero-truncated Poisson regression	[R] **ztp**

The following estimation commands support the mi prefix.

Linear regression

regress	[R] **regress** — Linear regression
cnsreg	[R] **cnsreg** — Constrained linear regression
mvreg	[R] **mvreg** — Multivariate regression

Binary-response regression models

logistic	[R] **logistic** — Logistic regression, reporting odds ratios	
logit	[R] **logit** — Logistic regression, reporting coefficients	
probit	[R] **probit** — Probit regression	
cloglog	[R] **cloglog** — Complementary log-log regression	
binreg	[R] **binreg** — Generalized linear models: Extensions to the binomial family	

Count-response regression models

poisson	[R] **poisson** — Poisson regression
nbreg	[R] **nbreg** — Negative binomial regression
gnbreg	[R] **nbreg** — Negative binomial regression

Ordinal-response regression models

ologit	[R] **ologit** — Ordered logistic regression
oprobit	[R] **oprobit** — Ordered probit regression

Categorical-response regression models

mlogit	[R] **mlogit** — Multinomial (polytomous) logistic regression
mprobit	[R] **mprobit** — Multinomial probit regression
clogit	[R] **clogit** — Conditional (fixed-effects) logistic regression

Quantile regression models

qreg	[R] **qreg** — Quantile regression
iqreg	[R] **qreg** — Quantile regression
sqreg	[R] **qreg** — Quantile regression
bsqreg	[R] **qreg** — Quantile regression

Survival regression models

stcox	[ST] **stcox** — Cox proportional hazards model
streg	[ST] **streg** — Parametric survival models
stcrreg	[ST] **stcrreg** — Competing-risks regression

Other regression models

glm	[R] **glm** — Generalized linear models
rreg	[R] **rreg** — Robust regression
truncreg	[R] **truncreg** — Truncated regression
areg	[R] **areg** — Linear regression with a large dummy-variable set

Basic statistics

mean	[R] **mean** — Estimate means
proportion	[R] **proportion** — Estimate proportions
ratio	[R] **ratio** — Estimate ratios

The following estimation commands support the svy prefix.

Descriptive statistics

mean	[R] **mean** — Estimate means
proportion	[R] **proportion** — Estimate proportions
ratio	[R] **ratio** — Estimate ratios
total	[R] **total** — Estimate totals

Linear regression models

cnsreg	[R] **cnsreg** — Constrained linear regression
glm	[R] **glm** — Generalized linear models
intreg	[R] **intreg** — Interval regression
nl	[R] **nl** — Nonlinear least-squares estimation
regress	[R] **regress** — Linear regression
tobit	[R] **tobit** — Tobit regression
treatreg	[R] **treatreg** — Treatment-effects model
truncreg	[R] **truncreg** — Truncated regression

Survival-data regression models

stcox	[ST] **stcox** — Cox proportional hazards model
streg	[ST] **streg** — Parametric survival models

Binary-response regression models

biprobit	[R] **biprobit** — Bivariate probit regression
cloglog	[R] **cloglog** — Complementary log-log regression
hetprob	[R] **hetprob** — Heteroskedastic probit model
logistic	[R] **logistic** — Logistic regression, reporting odds ratios
logit	[R] **logit** — Logistic regression, reporting coefficients
probit	[R] **probit** — Probit regression
scobit	[R] **scobit** — Skewed logistic regression

Discrete-response regression models

clogit	[R] **clogit** — Conditional (fixed-effects) logistic regression
mlogit	[R] **mlogit** — Multinomial (polytomous) logistic regression
mprobit	[R] **mprobit** — Multinomial probit regression
ologit	[R] **ologit** — Ordered logistic regression
oprobit	[R] **oprobit** — Ordered probit regression
slogit	[R] **slogit** — Stereotype logistic regression

(Continued on next page)

Poisson regression models

gnbreg	Generalized negative binomial regression in [R] **nbreg**
nbreg	[R] **nbreg** — Negative binomial regression
poisson	[R] **poisson** — Poisson regression
zinb	[R] **zinb** — Zero-inflated negative binomial regression
zip	[R] **zip** — Zero-inflated Poisson regression
ztnb	[R] **ztnb** — Zero-truncated negative binomial regression
ztp	[R] **ztp** — Zero-truncated Poisson regression

Instrumental-variables regression models

ivprobit	[R] **ivprobit** — Probit model with continuous endogenous regressors
ivregress	[R] **ivregress** — Single-equation instrumental-variables regression
ivtobit	[R] **ivtobit** — Tobit model with continuous endogenous regressors

Regression models with selection

heckman	[R] **heckman** — Heckman selection model
heckprob	[R] **heckprob** — Probit model with sample selection

Also see

[U] **20 Estimation and postestimation commands**

Title

> **file extensions** — Quick reference for default file extensions

Description

This entry provides a quick reference for default file extensions that are used by various commands.

Extension	Reference	Description
.ado	[U] **17 Ado-files**	automatically loaded do-files
.dct	[D] **infile (fixed format)**	ASCII data dictionary
.do	[U] **16 Do-files**	do-file
.dta	[D] **save**, [D] **use**	Stata-format dataset
.dtasig	[D] **datasignature**	datasignature file
.gph	[G] **graph save**, [G] **graph use**	graph
.grec	[G] **graph editor**	Graph Editor recording (ASCII format)
.irf	[TS] **irf set**	impulse–response function datasets
.log	[R] **log**	log file in text format
.mata	[M-1] **source**	Mata source code
.mlib	[M-3] **mata mlib**	Mata library
.mmat	[M-3] **mata matsave**	Mata matrix
.mo	[M-3] **mata mosave**	Mata object file
.out	[D] **outsheet**	file saved by outsheet
.raw	[D] **infile (free format)**, [D] **insheet**	ASCII-format dataset
.smcl	[R] **log**	log file in SMCL format
.ster	[R] **estimates save**	saved estimates
.sthlp	[P] **smcl**	help files
.stptrace	[MI] **mi ptrace**	parameter-trace file
.sum	[D] **checksum**	checksum files to verify network transfer
.zip	[D] **zipfile**	zip file

The following files are of interest only to advanced programmers or are for Stata's internal use.

Extension	Reference	Description
.class	[P] **class**	class file for object-oriented programming
.dlg	[P] **dialog programming**	dialog resource file
.idlg	[P] **dialog programming**	dialog resource include file
.ihlp	[P] **smcl**	help include file
.key	[R] **search**	search's keyword database file
.maint		maintenance file (for Stata's internal use only)
.mnu		menu file (for Stata's internal use only)
.pkg	[R] **net**	user-site package file
.plugin	[P] **plugin**	compiled addition (DLL)
.scheme	[G] **schemes intro**	control file for a graph scheme
.style	[G] **graph query**	graph style file
.toc	[U] **28.5 Making your own download site**	user-site description file

Title

> **format** — Quick reference for numeric and string display formats

Description

This entry provides a quick reference for display formats.

Remarks

The default formats for each of the numeric variable types are

```
byte    %8.0g
int     %8.0g
long    %12.0g
float   %9.0g
double  %10.0g
```

To change the display format for variable myvar to %9.2f, type

```
format myvar %9.2f
```

or

```
format %9.2f myvar
```

Stata will understand either statement.

Four values displayed in different numeric display formats

%9.0g	%9.0gc	%9.2f	%9.2fc	%-9.0g	%09.2f	%9.2e
12345	12,345	12345.00	12,345.00	12345	012345.00	1.23e+04
37.916	37.916	37.92	37.92	37.916	000037.92	3.79e+01
3567890	3567890	3.57e+06	3.57e+06	3567890	3.57e+06	3.57e+06
.9165	.9165	0.92	0.92	.9165	000000.92	9.16e-01

Left-aligned and right-aligned string display formats

%-17s	%17s
AMC Concord	AMC Concord
AMC Pacer	AMC Pacer
AMC Spirit	AMC Spirit
Buick Century	Buick Century
Buick Opel	Buick Opel

Also see

[U] **12.5 Formats: Controlling how data are displayed**

Title

immediate commands — Quick reference for immediate commands

Description

An *immediate* command is a command that obtains data not from the data stored in memory, but from numbers types as arguments.

Command	Reference	Description
bitesti	[R] **bitest**	Binomial probability test
cci	[ST] **epitab**	Tables for epidemiologists
csi		
iri		
mcci		
cii	[R] **ci**	Confidence intervals for means, proportions, and counts
prtesti	[R] **prtest**	One- and two-sample tests of proportions
sampsi	[R] **sampsi**	Sample size and power determination
sdtesti	[R] **sdtest**	Variance comparison tests
symmi	[R] **symmetry**	Symmetry and marginal homogeneity tests
tabi	[R] **tabulate twoway**	Two-way tables of frequencies
ttesti	[R] **ttest**	Mean comparison tests
twoway pci	[G] **graph twoway pci**	Paired-coordinate plot with spikes or lines
twoway pcarrowi	[G] **graph twoway pcarrowi**	Paired-coordinate plot with arrows
twoway scatteri	[G] **graph twoway scatteri**	Twoway scatterplot

Also see

[U] **19 Immediate commands**

Title

> **missing values** — Quick reference for missing values

Description

This entry provides a quick reference for Stata's missing values.

Remarks

Stata has 27 numeric missing values:

., the default, which is called the *system missing value* or `sysmiss`

and

.a, .b, .c, ..., .z, which are called the *extended missing values.*

Numeric missing values are represented by "large positive values". The ordering is

$$\text{all nonmissing numbers} < \text{ . } < \text{ .a } < \text{ .b } < \cdots < \text{ .z}$$

Thus the expression

$$age > 60$$

is true if variable `age` is greater than 60 or missing.

To exclude missing values, ask whether the value is less than '.'.

    . list if age > 60 & age < .

To specify missing values, ask whether the value is greater than or equal to '.'. For instance,

    . list if age >=.

Stata has one string missing value, which is denoted by "" (blank).

Also see

[U] **12.2.1 Missing values**

Title

postestimation commands — Quick reference for postestimation commands

Description

This entry provides a quick reference for Stata's postestimation commands. Because enhancements to Stata are continually being made, type `search postestimation commands` for possible additions to this list; see [R] **search**.

Available after most estimation commands

Command	Description
estat ic	AIC and BIC
estat summarize	estimation sample summary
estat vce	VCE
estimates	cataloging estimation results
hausman	Hausman's specification test
lincom	point estimates, standard errors, testing, and inference for linear combinations of coefficients
linktest	link test for model specification for single-equation models
lrtest	likelihood-ratio test
margins	marginal means, predictive margins, marginal effects, and average marginal effects
	marginal means, predictive margins, marginal effects, and average marginal effects
nlcom	point estimates, standard errors, testing, and inference for nonlinear combinations of coefficients
predict	predictions, residuals, influence statistics, and other diagnostic measures
predictnl	point estimates, standard errors, testing, and inference for generalized predictions
suest	seemingly unrelated estimation
test	Wald tests of simple and composite linear hypotheses
testnl	Wald tests of nonlinear hypotheses

Special-interest postestimation commands

Command	Description
anova	
acprplot	augmented component-plus-residual plot
avplot	added-variable plot
avplots	all added-variables plots in a single image
cprplot	component-plus-residual plot
dfbeta	DFBETA influence statistics
estat hettest	tests for heteroskedasticity
estat imtest	information matrix test
estat ovtest	Ramsey regression specification-error test for omitted variables
estat szroeter	Szroeter's rank test for heteroskedasticity
estat vif	variance inflation factors for the independent variables
lvr2plot	leverage-versus-squared-residual plot
rvfplot	residual-versus-fitted plot
rvpplot	residual-versus-predictor plot
asclogit	
estat alternatives	alternative summary statistics
estat mfx	marginal effects
asmprobit and asroprobit	
estat alternatives	alternative summary statistics
estat covariance	variance–covariance matrix of the alternatives
estat correlation	correlation matrix of the alternatives
estat facweights	covariance factor weights matrix
estat mfx	marginal effects
bootstrap	
estat bootstrap	table of confidence intervals for each statistic
ca and camat	
cabiplot	biplot of row and column points
caprojection	CA dimension projection plot
estat coordinates	display row and column coordinates
estat distances	display χ^2 distances between row and column profiles
estat inertia	display inertia contributions of the individual cells
estat loadings	display correlations of profiles and axes (*loadings*)
estat profiles	display row and column profiles
† estat summarize	estimation sample summary
estat table	display fitted correspondence table
screeplot	plot singular values

† estat summarize is not available after camat.

Command	Description
candisc	
estat anova	ANOVA summaries table
estat canontest	tests of the canonical discriminant functions
estat classfunctions	classification functions
estat classtable	classification table
estat correlations	correlation matrices and p-values
estat covariance	covariance matrices
estat errorrate	classification error-rate estimation
estat grdistances	Mahalanobis and generalized squared distances between the group means
estat grmeans	group means and variously standardized or transformed means
estat grsummarize	group summaries
estat list	classification listing
estat loadings	canonical discriminant-function coefficients (loadings)
estat manova	MANOVA table
estat structure	canonical structure matrix
estat summarize	estimation sample summary
loadingplot	plot standardized discriminant-function loadings
scoreplot	plot discriminant-function scores
screeplot	plot eigenvalues
canon	
estat correlations	show correlation matrices
estat loadings	show loading matrices
estat rotate	rotate raw coefficients, standard coefficients, or loading matrices
estat rotatecompare	compare rotated and unrotated coefficients or loadings
screeplot	plot canonical correlations
discrim knn and discrim logistic	
estat classtable	classification table
estat errorrate	classification error-rate estimation
estat grsummarize	group summaries
estat list	classification listing
estat summarize	estimation sample summary

Command	Description
discrim lda	
estat anova	ANOVA summaries table
estat canontest	tests of the canonical discriminant functions
estat classfunctions	classification functions
estat classtable	classification table
estat correlations	correlation matrices and p-values
estat covariance	covariance matrices
estat errorrate	classification error-rate estimation
estat grdistances	Mahalanobis and generalized squared distances between the group means
estat grmeans	group means and variously standardized or transformed means
estat grsummarize	group summaries
estat list	classification listing
estat loadings	canonical discriminant-function coefficients (loadings)
estat manova	MANOVA table
estat structure	canonical structure matrix
estat summarize	estimation sample summary
loadingplot	plot standardized discriminant-function loadings
scoreplot	plot discriminant-function scores
screeplot	plot eigenvalues
discrim qda	
estat classtable	classification table
estat correlations	correlation matrices and p-values
estat covariance	covariance matrices
estat errorrate	classification error-rate estimation
estat grdistances	Mahalanobis and generalized squared distances between the group means
estat grsummarize	group summaries
estat list	classification listing
estat summarize	estimation sample summary
exlogistic	
estat predict	single-observation prediction
estat se	report odds ratio or coefficient asymptotic standard errors
expoisson	
estat se	report coefficient asymptotic standard errors

Command	Description
factor and factormat	
estat anti	anti-image correlation and covariance matrices
estat common	correlation matrix of the common factors
estat factors	AIC and BIC model-selection criteria for different numbers of factors
estat kmo	Kaiser–Meyer–Olkin measure of sampling adequacy
estat residuals	matrix of correlation residuals
estat rotatecompare	compare rotated and unrotated loadings
estat smc	squared multiple correlations between each variable and the rest
estat structure	correlations between variables and common factors
[†] estat summarize	estimation sample summary
loadingplot	plot factor loadings
rotate	rotate factor loadings
scoreplot	plot score variables
screeplot	plot eigenvalues
fracpoly	
fracplot	plot data and fit from most recently fitted fractional polynomial model
fracpred	create variable containing prediction, deviance residuals, or SEs of fitted values
gmm	
estat overid	perform test of overidentifying restrictions
ivprobit	
estat classification	reports various summary statistics, including the classification table
lroc	graphs the ROC curve and calculates the area under the curve
lsens	graphs sensitivity and specificity versus probability cutoff
ivregress	
estat endogenous	perform tests of endogeneity
estat firststage	report "first-stage" regression statistics
estat overid	perform tests of overidentifying restrictions
logistic and logit	
estat classification	reports various summary statistics, including the classification table
estat gof	Pearson or Hosmer–Lemeshow goodness-of-fit test
lroc	graphs the ROC curve and calculates the area under the curve
lsens	graphs sensitivity and specificity versus probability cutoff
manova	
manovatest	multivariate tests after manova
screeplot	plot eigenvalues

[†] estat summarize is not available after factormat.

Command	Description
mca	
mcaplot	plot of category coordinates
mcaprojection	MCA dimension projection plot
estat coordinates	display of category coordinates
estat subinertia	matrix of inertias of the active variables (after JCA only)
estat summarize	estimation sample summary
screeplot	plot principal inertias (eigenvalues)
mds, mdslong, and mdsmat	
estat config	coordinates of the approximating configuration
estat correlations	correlations between disparities and distances
estat pairwise	pairwise disparities, approximating distances, and residuals
estat quantiles	quantiles of the residuals per object
estat stress	Kruskal stress (loss) measure (only after classical MDS)
†estat summarize	estimation sample summary
mdsconfig	plot of approximating configuration
mdsshepard	Shepard diagram
screeplot	plot eigenvalues (only after classical MDS)
mfp	
fracplot	plot data and fit from most recently fitted fractional polynomial model
fracpred	create variable containing prediction, deviance residuals, or SEs of fitted values
nlogit	
estat alternatives	alternative summary statistics
pca and pcamat	
estat anti	anti-image correlation and covariance matrices
estat kmo	Kaiser–Meyer–Olkin measure of sampling adequacy
estat loadings	component-loading matrix in one of several normalizations
estat residuals	matrix of correlation or covariance residuals
estat rotatecompare	compare rotated and unrotated components
estat smc	squared multiple correlations between each variable and the rest
†estat summarize	estimation sample summary
loadingplot	plot component loadings
rotate	rotate component loadings
scoreplot	plot score variables
screeplot	plot eigenvalues
poisson	
estat gof	goodness-of-fit test

† estat summarize is not available after mdsmat or pcamat.

Command	Description
probit	
estat classification	reports various summary statistics, including the classification table
estat gof	Pearson or Hosmer–Lemeshow goodness-of-fit test
lroc	graphs the ROC curve and calculates the area under the curve
lsens	graphs sensitivity and specificity versus probability cutoff
procrustes	
estat compare	fit statistics for orthogonal, oblique, and unrestricted transformations
estat mvreg	display multivariate regression resembling unrestricted transformation
estat summarize	display summary statistics over the estimation sample
procoverlay	produce a Procrustes overlay graph
regress	
dfbeta	DFBETA influence statistics
estat hettest	tests for heteroskedasticity
estat imtest	information matrix test
estat ovtest	Ramsey regression specification-error test for omitted variables
estat szroeter	Szroeter's rank test for heteroskedasticity
estat vif	variance inflation factors for the independent variables
acprplot	augmented component-plus-residual plot
avplot	added-variable plot
avplots	all added-variables plots in a single image
cprplot	component-plus-residual plot
lvr2plot	leverage-versus-squared-residual plot
rvfplot	residual-versus-fitted plot
rvpplot	residual-versus-predictor plot
regress postestimation time series	
estat archlm	test for ARCH effects in the residuals
estat bgodfrey	Breusch–Godfrey test for higher-order serial correlation
estat durbinalt	Durbin's alternative test for serial correlation
estat dwatson	Durbin–Watson d statistic to test for first-order serial correlation
rocfit	
rocplot	plot the fitted ROC curve and simultaneous confidence bands
stcox	
estat concordance	Harrell's C
estat phtest	test proportional-hazards assumption based on Schoenfeld residuals
stcurve	plot the survivor, hazard, and cumulative hazard functions

Command	Description
stcrreg	
stcurve	plot the cumulative subhazard and cumulative incidence functions
streg	
stcurve	plot the survivor, hazard, and cumulative hazard functions
svar, var, and varbasic	
fcast compute	obtain dynamic forecasts
fcast graph	graph dynamic forecasts obtained from fcast compute
irf	create and analyze IRFs and FEVDs
vargranger	Granger causality tests
varlmar	LM test for autocorrelation in residuals
varnorm	test for normally distributed residuals
varsoc	lag-order selection criteria
varstable	check stability condition of estimates
varwle	Wald lag-exclusion statistics
vec	
fcast compute	obtain dynamic forecasts
fcast graph	graph dynamic forecasts obtained from fcast compute
irf	create and analyze IRFs and FEVDs
veclmar	LM test for autocorrelation in residuals
vecnorm	test for normally distributed residuals
vecstable	check stability condition of estimates
xtabond, xtdpd, and xtdpdsys	
estat abond	test for autocorrelation
estat sargan	Sargan test of overidentifying restrictions
xtgee	
estat wcorrelation	estimated matrix of the within-group correlations
xtmelogit, xtmepoisson, and xtmixed	
estat group	summarize the composition of the nested groups
estat recovariance	display the estimated random-effects covariance matrix (or matrices)
xtreg	
xttest0	Breusch and Pagan LM test for random effects

Also see

[R] **estat** — Postestimation statistics

[R] **estimates** — Save and manipulate estimation results

[R] **hausman** — Hausman specification test

[R] **lincom** — Linear combinations of estimators

[R] **linktest** — Specification link test for single-equation models

[R] **lrtest** — Likelihood-ratio test after estimation

[R] **margins** — Marginal means, predictive margins, and marginal effects

[R] **nlcom** — Nonlinear combinations of estimators

[R] **predict** — Obtain predictions, residuals, etc., after estimation

[R] **predictnl** — Obtain nonlinear predictions, standard errors, etc., after estimation

[R] **suest** — Seemingly unrelated estimation

[R] **test** — Test linear hypotheses after estimation

[R] **testnl** — Test nonlinear hypotheses after estimation

[U] **20 Estimation and postestimation commands**

Title

prefix commands — Quick reference for prefix commands

Description

Prefix commands operate on other Stata commands. They modify the input, modify the output, and repeat execution of the other Stata command.

Command	Reference	Description
by	[D] **by**	run command on subsets of data
statsby	[D] **statsby**	same as by, but collect statistics from each run
rolling	[TS] **rolling**	run command on moving subsets and collect statistics
bootstrap	[R] **bootstrap**	run command on bootstrap samples
jackknife	[R] **jackknife**	run command on jackknife subsets of data
permute	[R] **permute**	run command on random permutations
simulate	[R] **simulate**	run command on manufactured data
svy	[SVY] **svy**	run command and adjust results for survey sampling
mi estimate	[MI] **mi estimate**	run command on multiply imputed data and adjust results for multiple imputation (MI)
nestreg	[R] **nestreg**	run command with accumulated blocks of regressors, and report nested model comparison tests
stepwise	[R] **stepwise**	run command with stepwise variable inclusion/exclusion
xi	[R] **xi**	run command after expanding factor variables and interactions; for most commands, using factor variables is preferred to using xi (see [U] **11.4.3 Factor variables**)
fracpoly	[R] **fracpoly**	run command with fractional polynomials of one regressor
mfp	[R] **mfp**	run command with multiple fractional polynomial regressors
capture	[P] **capture**	run command and capture its return code
noisily	[P] **quietly**	run command and show the output
quietly	[P] **quietly**	run command and suppress the output
version	[P] **version**	run command under specified version

The last group—capture, noisily, quietly, and version—have to do with programming Stata, and for historical reasons, capture, noisily, and quietly allow you to omit the colon.

Also see

[U] **11.1.10 Prefix commands**

Title

> **reading data** — Quick reference for reading non-Stata data into memory

Description

This entry provides a quick reference for determining which method to use for reading non-Stata data into memory.

Remarks

insheet

- ○ insheet reads text (ASCII) files created by a spreadsheet or a database program.
- ○ The data must be tab separated or comma separated, but not both simultaneously, and cannot be space separated.
- ○ An observation must be on only one line.
- ○ The first line in the file can optionally contain the names of the variables.

See [D] **insheet** for additional information.

infile (free format)—infile without a dictionary

- ○ The data can be space separated, tab separated, or comma separated.
- ○ Strings with embedded spaces or commas must be enclosed in quotes (even if tab- or comma separated).
- ○ An observation can be on more than one line, or there can even be multiple observations per line.

See [D] **infile (free format)** for additional information.

infix (fixed format)

- ○ The data must be in fixed-column format.
- ○ An observation can be on more than one line.
- ○ infix has simpler syntax than infile (fixed format).

See [D] **infix (fixed format)** for additional information.

infile (fixed format)—infile with a dictionary

- ○ The data may be in fixed-column format.
- ○ An observation can be on more than one line.
- ○ infile (fixed format) has the most capabilities for reading data.

See [D] **infile (fixed format)** for additional information.

fdause

○ `fdause` reads SAS XPORT Transport format files—the file format required by the U.S. Food and Drug Administration (FDA).

○ `fdause` will also read value label information from a `formats.xpf` XPORT file, if available.

See [D] **fdasave** for additional information.

haver (Windows only)

○ `haver` reads Haver Analytics (http://www.haver.com/) database files.

○ `haver` is available only for Windows and requires a corresponding DLL (`DLXAPI32.DLL`) available from Haver Analytics.

See [TS] **haver** for additional information.

odbc

○ ODBC, an acronym for Open DataBase Connectivity, is a standard for exchanging data between programs. Stata supports the ODBC standard for importing data via the `odbc` command and can read from any ODBC data source on your computer.

See [D] **odbc**.

xmluse

○ `xmluse` reads extensible markup language (XML) files—highly adaptable text-format files derived from ground station markup language (GSML).

○ `xmluse` can read either an Excel-format XML or a Stata-format XML file into Stata.

See [D] **xmlsave** for additional information.

Also see

[D] **insheet** — Read ASCII (text) data created by a spreadsheet

[D] **infile (free format)** — Read unformatted ASCII (text) data

[D] **infile (fixed format)** — Read ASCII (text) data in fixed format with a dictionary

[D] **infix (fixed format)** — Read ASCII (text) data in fixed format

[D] **infile (fixed format)** — Read ASCII (text) data in fixed format with a dictionary

[D] **fdasave** — Save and use datasets in FDA (SAS XPORT) format

[TS] **haver** — Load data from Haver Analytics database

[D] **odbc** — Load, write, or view data from ODBC sources

[D] **xmlsave** — Save and use datasets in XML format

[U] **21 Inputting data**

Title

2SIV	two-step instrumental variables
2SLS	two-stage least squares
3SLS	three-stage least squares
AF	attributable fraction for the population
AFE	attributable fraction among the exposed
AFT	accelerated failure time
AIC	Akaike information criterion
AIDS	almost ideal demand system
ANCOVA	analysis of covariance
ANOVA	analysis of variance
APE	average partial effects
AR	autoregressive
AR(1)	first-order autoregressive
ARCH	autoregressive conditional heteroskedasticity
ARIMA	autoregressive integrated moving-average
ARMA	autoregressive moving-average
ARMAX	autoregressive moving-average exogenous
ASE	asymptotic standard error
ASL	achieved significance level
AUC	area under the time-versus-concentration curve
BC	bias corrected
BCa	bias-corrected and accelerated
BE	between effects
BFGS	Broyden–Fletcher–Goldfarb–Shanno
BHHH	Berndt–Hall–Hall–Hausman
BIC	Bayesian information criterion
BLUP	best linear unbiased prediction
BRR	balanced repeated replication
CA	correspondence analysis
CCI	conservative confidence interval
CDC	Centers for Disease Control and Prevention
CDF	cumulative distribution function
CES	constant elasticity of substitution
CI	confidence interval
CIF	cumulative incidence function
CMLE	conditional maximum likelihood estimates
ct	count time
cusum	cumulative sum
c.v.	coefficient of variation

DA	data augmentation
DEFF	design effect
DEFT	design effect (standard deviation metric)
DF	dynamic factor
df / d.f.	degree(s) of freedom
DFAR	dynamic factors with vector autoregressive errors
DFP	Davidon–Fletcher–Powell
DPD	dynamic panel data
EGARCH	exponential GARCH
EGLS	estimated generalized least squares
EIM	expected information matrix
EM	expectation maximization
EPS	Encapsulated PostScript
ESS	error sum of squares
FD	first-differenced estimator
FDA	Food and Drug Administration
FE	fixed effects
FEVD	forecast-error variance decomposition
FGLS	feasible generalized least squares
FGNLS	feasible generalized nonlinear least squares
FIVE estimator	full-information instrumental-variables efficient estimator
flong	full long
flongsep	full long and separate
FMI	fraction of missing information
FP	fractional polynomial
FPC	finite population correction
GARCH	generalized autoregressive conditional heteroskedasticity
GEE	generalized estimating equations
GEV	generalized extreme value
GHK	Geweke–Hajivassiliou–Keane
GLIM	generalized linear interactive modeling
GLLAMM	generalized linear latent and mixed models
GLM	generalized linear models
GLS	generalized least squares
GMM	generalized method of moments
GUI	graphical user interface
HAC	heteroskedasticity- and autocorrelation-consistent
HR	hazard ratio

IC	information criteria
ICD-9	International Classification of Diseases, Ninth Revision
IIA	independence of irrelevant alternatives
i.i.d.	independent and identically distributed
IQR	interquartile range
IR	incidence rate
IRF	impulse–response function
IRLS	iterated, reweighted least squares
IRR	incidence-rate ratio
IV	instrumental variables
JCA	joint correspondence analysis
LAPACK	linear algebra package
LAV	least absolute value
LDA	linear discriminant analysis
LIML	limited-information maximum likelihood
LM	Lagrange multiplier
LOO	leave one out
LOWESS	locally weighted scatterplot smoothing
LR	likelihood ratio
LSB	least-significant byte
MA	moving average
MAD	median absolute deviation
MANCOVA	multivariate analysis of covariance
MANOVA	multivariate analysis of variance
MAR	missing at random
MCA	multiple correspondence analysis
MCAR	missing completely at random
MCMC	Markov chain Monte Carlo
MDS	multidimensional scaling
ME	multiple equation
MEFF	misspecification effect
MEFT	misspecification effect (standard deviation metric)
MFP	multivariable fractional polynomial
MI	multiple imputation
mi	multiple imputation
midp	mid-p-value

MINQUE	minimum norm quadratic unbiased estimation
MIVQUE	minimum variance quadratic unbiased estimation
ML	maximum likelihood
MLE	maximum likelihood estimate
mlong	marginal long
MM	method of moments
MNAR	missing not at random
MNP	multinomial probit
MS	mean square
MSB	most-significant byte
MSE	mean squared error
MSL	maximum simulated likelihood
MSS	model sum of squares
MUE	median unbiased estimates
MVN	multivariate normal
MVREG	multivariate regression
NARCH	nonlinear ARCH
NHANES	National Health and Nutrition Examination Survey
NLS	nonlinear least squares
NPARCH	nonlinear power ARCH
NR	Newton–Raphson
ODBC	Open DataBase Connectivity
OIM	observed information matrix
OIRF	orthogonalized impulse–response function
OLE	Object Linking and Embedding (Microsoft product)
OLS	ordinary least squares
OPG	outer product of the gradient
OR	odds ratio
PA	population averaged
PARCH	power ARCH
PCA	principal component analysis
PCSE	panel-corrected standard error
p.d.f.	probability density function
PF	prevented fraction for the population
PFE	prevented fraction among the exposed
PH	proportional hazards
pk	pharmacokinetic data
PMM	predictive mean matching
PNG	Portable Network Graphics
PSU	primary sampling unit

QDA	quadratic discriminant analysis
QML	quasi maximum likelihood
rc	return code
RCT	randomized controlled trial
RE	random effects
REML	restricted (or residual) maximum likelihood
RESET	regression specification-error test
ROC	receiver operating characteristic
ROP	rank-ordered probit
ROT	rule of thumb
RR	relative risk
RRR	relative-risk ratio
RSS	residual sum of squares
RUM	random utility maximization
RVI	relative variance increase
SAARCH	simple asymmetric ARCH
SARIMA	seasonal ARIMA
s.d.	standard deviation
SE / s.e.	standard error
SF	static factor
SFAR	static factors with vector autoregressive errors
SIR	standardized incidence ratio
SJ	Stata Journal
SMCL	Stata Markup and Control Language
SMR	standardized mortality/morbidity ratio
SMSA	standard metropolitan statistical area
SOR	standardized odds ratio
SQL	Structured Query Language
SRD	standardized rate difference
SRR	standardized risk ratio
SRS	simple random sample/sampling
SRSWR	SRS with replacement
SSC	Statistical Software Components
SSCP	sum of squares and cross products
SSU	secondary sampling unit
st	survival time
STB	Stata Technical Bulletin
STS	structural time series
SUR	seemingly unrelated regression
SURE	seemingly unrelated regression estimation
SVAR	structural vector autoregressive model
SVD	singular value decomposition

TARCH	threshold ARCH
TDT	transmission/disequilibrium test
TIFF	tagged image file format
TSS	total sum of squares
VAR	vector autoregressive model
VAR(1)	first-order vector autoregressive
VARMA	vector autoregressive moving-average
VARMA(1,1)	first-order vector autoregressive moving-average
VCE	variance–covariance estimate
VECM	vector error-correction model
VIF	variance inflation factor
WLC	worst linear combination
WLF	worst linear function
WLS	weighted least squares
XML	Extensible Markup Language
ZINB	zero-inflated negative binomial
ZIP	zero-inflated Poisson
ZTNB	zero-truncated negative binomial
ZTP	zero-truncated Poisson

Vignettes index

Aalen, O. O. (1947–), [ST] **sts**
Akaike, H. (1927–), [R] **estat**
Arellano, M. (1957–), [XT] **xtabond**

Bartlett, M. S. (1910–2002), [TS] **wntestb**
Berkson, J. (1899–1982), [R] **logit**
Bliss, C. I. (1899–1979), [R] **probit**
Bond, S. R. (1963–), [XT] **xtabond**
Bonferroni, C. E. (1892–1960), [R] **correlate**
Box, G. E. P. (1919–), [TS] **arima**
Breusch, T. S. (1949–), [R] **regress postestimation time series**
Brier, G. W. (1913–1998), [R] **brier**

Cholesky, A.-L. (1875–1918), [M-5] **cholesky( )**
Cleveland, W. S. (1943–), [R] **lowess**
Cochran, W. G. (1909–1980), [SVY] **survey**
Cochrane, D. (1917–1983), [TS] **prais**
Cohen, J. (1923–1998), [R] **kappa**
Cornfield, J. (1912–1979), [ST] **epitab**
Cox, D. R. (1924–), [ST] **stcox**
Cox, G. M. (1900–1978), [R] **anova**
Cronbach, L. J. (1916–2001), [R] **alpha**

Dickey, D. A. (1945–), [TS] **dfuller**
Durbin, J. (1923–), [R] **regress postestimation time series**

Efron, B. (1938–), [R] **bootstrap**
Engle, R. F. (1942–), [TS] **arch**

Fisher, R. A. (1890–1962), [R] **anova**
Fourier, J. B. J. (1768–1830), [R] **cumul**
Fuller, W. A. (1931–), [TS] **dfuller**

Gabriel, K. R. (1929–2003), [MV] **biplot**
Galton, F. (1822–1911), [R] **regress**
Gauss, J. C. F. (1777–1855), [R] **regress**
Gnanadesikan, R. (1932–), [R] **diagnostic plots**
Godfrey, L. G. (1946–), [R] **regress postestimation time series**
Gompertz, B. (1779–1865), [ST] **streg**
Gosset, W. S. (1876–1937), [R] **ttest**
Granger, C. W. J. (1934–2009), [TS] **vargranger**
Greenwood, M. (1880–1949), [ST] **sts**

Hadamard, J. S. (1865–1963), [D] **functions**
Haenszel, W. M. (1910–1998), [ST] **strate**
Hartley, H. O. (1912–1980), [MI] **mi impute**
Hausman, J. A. (1946–), [R] **hausman**
Heckman, J. J. (1944–), [R] **heckman**
Henderson, C. R. (1911–1989), [XT] **xtmixed**
Hermite, C. (1822–1901), [M-5] **issymmetric( )**

Hilbert, D. (1862–1943), [M-5] **Hilbert( )**
Hotelling, H. (1895–1973), [MV] **hotelling**
Householder, A. S. (1904–1993), [M-5] **qrd( )**
Huber, P. J. (1934–), [U] **20 Estimation and postestimation commands**

Jaccard, P. (1868–1944), [MV] *measure_option*
Jeffreys, H. (1891–1989), [R] **ci**
Jenkins, G. M. (1933–1982), [TS] **arima**
Johansen, S. (1939–), [TS] **vecrank**

Kaiser, H. F. (1927–1992), [MV] **rotate**
Kaplan, E. L. (1920–2006), [ST] **sts**
Kendall, M. G. (1907–1983), [R] **spearman**
Kish, L. (1910–2000), [SVY] **survey**
Kolmogorov, A. N. (1903–1987), [R] **ksmirnov**
Kronecker, L. (1823–1891), [M-2] **op_kronecker**
Kruskal, J. B. (1928–), [MV] **mds**
Kruskal, W. H. (1919–2005), [R] **kwallis**

Laplace, P.-S. (1749–1827), [R] **regress**
Legendre, A.-M. (1752–1833), [R] **regress**
Lexis, W. (1837–1914), [ST] **stsplit**
Lorenz, M. O. (1876–1959), [R] **inequality**

Mahalanobis, P. C. (1893–1972), [MV] **hotelling**
Mann, H. B. (1905–2000), [R] **ranksum**
Mantel, N. (1919–2002), [ST] **strate**
McFadden, D. L. (1937–), [R] **asclogit**
McNemar, Q. (1900–1986), [ST] **epitab**
Meier, P. (1924–), [ST] **sts**
Moore, E. H. (1862–1932), [M-5] **pinv( )**
Murrill, W. A. (1867–1957), [MV] **discrim knn**

Nelder, J. A. (1924–), [R] **glm**
Nelson, W. B. (1936–), [ST] **sts**
Newey, W. K. (1954–), [TS] **newey**
Newton, I. (1643–1727), [M-5] **optimize( )**
Neyman, J. (1894–1981), [R] **ci**

Orcutt, G. H. (1917–), [TS] **prais**

Pearson, K. (1857–1936), [R] **correlate**
Penrose, R. (1931–), [M-5] **pinv( )**
Perron, P. (1959–), [TS] **pperron**
Phillips, P. C. B. (1948–), [TS] **pperron**
Playfair, W. (1759–1823), [G] **graph pie**
Poisson, S.-D. (1781–1840), [R] **poisson**
Prais, S. J. (1928–), [TS] **prais**

Raphson, J. (1648–1715), [M-5] **optimize( )**
Rubin, D. B. (1943–), [MI] **intro substantive**

Scheffé, H. (1907–1977), [R] **oneway**
Schwarz, G. E. (1933–), [R] **estat**

Shapiro, S. S. (1930–), [R] **swilk**
Shepard, R. N. (1929–), [MV] **mds postestimation**
Shewhart, W. A. (1891–1967), [R] **qc**
Šidák, Z. (1933–1999), [R] **correlate**
Simpson, T. (1710–1761), [M-5] **optimize( )**
Smirnov, N. V. (1900–1966), [R] **ksmirnov**
Sneath, P. H. A. (1923–), [MV] *measure_option*
Sokal, R. R. (1926–), [MV] *measure_option*
Spearman, C. E. (1863–1945), [R] **spearman**

Theil, H. (1924–2000), [R] **reg3**
Thiele, T. N. (1838–1910), [R] **summarize**
Tobin, J. (1918–2002), [R] **tobit**
Toeplitz, O. (1881–1940), [M-5] **Toeplitz( )**
Tukey, J. W. (1915–2000), [R] **jackknife**

Vandermonde, A.-T. (1735–1796),
    [M-5] **Vandermonde( )**

Wald, A. (1902–1950), [TS] **varwle**
Wallis, W. A. (1912–1998), [R] **kwallis**
Ward, J. H. (1926–), [MV] **cluster linkage**
Watson, G. S. (1921–1998), [R] **regress postestimation**
    time series
Wedderburn, R. W. M. (1947–1975), [R] **glm**
Weibull, E. H. W. (1887–1979), [ST] **streg**
West, K. D. (1953–), [TS] **newey**
White, H. L. Jr. (1950–), [U] **20 Estimation and**
    postestimation commands
Whitney, D. R. (1915–2007), [R] **ranksum**
Wilcoxon, F. (1892–1965), [R] **signrank**
Wilk, M. B. (1922–), [R] **diagnostic plots**
Wilks, S. S. (1906–1964), [MV] **manova**
Wilson, E. B. (1879–1964), [R] **ci**
Winsten, C. B. (1923–2005), [TS] **prais**
Woolf, B. (1902–1983), [ST] **epitab**

Zellner, A. (1927–), [R] **sureg**

Author index

A

Aalen, O. O., [ST] **stcrreg postestimation**, [ST] **sts**

Abraham, B., [TS] **tssmooth**, [TS] **tssmooth dexponential**, [TS] **tssmooth exponential**, [TS] **tssmooth hwinters**, [TS] **tssmooth shwinters**

Abraira-García, L., [ST] **epitab**

Abramowitz, M., [D] **functions**, [R] **orthog**, [XT] **xtmelogit**, [XT] **xtmepoisson**

Abrams, K. R., [R] **meta**

Abramson, J. H., [R] **kappa**, [ST] **epitab**

Abramson, Z. H., [R] **kappa**, [ST] **epitab**

Achen, C. H., [R] **scobit**

Achenback, T. M., [MV] **mvtest**

Acock, A. C., [R] **alpha**, [R] **anova**, [R] **correlate**, [R] **nestreg**, [R] **oneway**, [R] **prtest**, [R] **ranksum**, [R] **ttest**

Adkins, L. C., [R] **heckman**, [R] **regress**, [R] **regress postestimation**, [TS] **arch**

Afifi, A. A., [R] **anova**, [R] **stepwise**, [U] **20.20 References**

Agresti, A., [R] **ci**, [R] **expoisson**, [R] **tabulate twoway**, [ST] **epitab**

Ahn, S. K., [TS] **vec intro**

Ahrens, J. H., [D] **functions**

Aigner, D., [R] **frontier**, [XT] **xtfrontier**

Aiken, L. S., [R] **pcorr**

Aisbett, C. W., [ST] **stcox**, [ST] **streg**

Aitchison, J., [R] **ologit**, [R] **oprobit**

Aitken, A. C., [R] **reg3**

Aitkin, M. A., [MV] **mvtest correlations**

Aivazian, S. A., [R] **ksmirnov**

Akaike, H., [MV] **factor postestimation**, [R] **BIC note**, [R] **estat**, [R] **glm**, [ST] **streg**, [TS] **varsoc**

Albert, A., [MV] **discrim logistic**, [MV] **discrim**

Albert, P. S., [XT] **xtgee**

Aldenderfer, M. S., [MV] **cluster**

Aldrich, J. H., [R] **logit**, [R] **probit**

Alexandersson, A., [R] **regress**

Alf Jr., E., [R] **rocfit**

Alldredge, J. R., [R] **pk**, [R] **pkcross**

Allen, M. J., [R] **alpha**

Allison, M. J., [MV] **manova**

Allison, P. D., [MI] **intro substantive**, [MI] **mi impute**, [R] **rologit**, [R] **testnl**, [ST] **discrete**, [XT] **xtlogit**, [XT] **xtpoisson**, [XT] **xtreg**

Altman, D. G., [R] **anova**, [R] **fracpoly**, [R] **kappa**, [R] **kwallis**, [R] **meta**, [R] **mfp**, [R] **nptrend**, [R] **oneway**

Alvarez, J., [XT] **xtabond**

Ambler, G., [R] **fracpoly**, [R] **mfp**, [R] **regress**

Amemiya, T., [R] **glogit**, [R] **intreg**, [R] **ivprobit**, [R] **nlogit**, [R] **tobit**, [TS] **varsoc**, [XT] **xthtaylor**, [XT] **xtivreg**

Amisano, G., [TS] **irf create**, [TS] **var intro**, [TS] **var svar**, [TS] **vargranger**, [TS] **varwle**

Anderberg, M. R., [MV] **cluster**, [MV] *measure_option*

Andersen, E. B., [R] **clogit**

Andersen, P. K., [ST] **stcrreg**

Anderson, B. D. O., [TS] **sspace**

Anderson, E., [M-1] **LAPACK**, [M-5] **lapack( )**, [MV] **clustermat**, [MV] **discrim estat**, [MV] **discrim lda**, [MV] **discrim lda postestimation**, [MV] **mvtest**, [MV] **mvtest normality**, [P] **matrix eigenvalues**

Anderson, J. A., [R] **ologit**, [R] **slogit**

Anderson, M. L., [ST] **stcrreg**

Anderson, R. E., [R] **rologit**

Anderson, R. L., [R] **anova**

Anderson, S., [R] **pkequiv**

Anderson, T. W., [MI] **intro substantive**, [MV] **discrim**, [MV] **manova**, [MV] **pca**, [R] **ivregress postestimation**, [TS] **vec**, [TS] **vecrank**, [XT] **xtabond**, [XT] **xtdpd**, [XT] **xtdpdsys**, [XT] **xtivreg**

Andrews, D. F., [D] **egen**, [MV] **discrim lda postestimation**, [MV] **discrim qda**, [MV] **discrim qda postestimation**, [MV] **manova**, [R] **rreg**

Andrews, D. W. K., [R] **ivregress**

Andrews, M., [XT] **xtmelogit**, [XT] **xtmepoisson**, [XT] **xtmixed**, [XT] **xtreg**

Angrist, J. D., [R] **ivregress**, [R] **ivregress postestimation**, [R] **qreg**, [R] **regress**, [U] **20.20 References**

Anscombe, F. J., [R] **binreg postestimation**, [R] **glm**, [R] **glm postestimation**

Ansley, C. F., [TS] **arima**

Arbuthnott, J., [R] **signrank**

Archer, K. J., [R] **logistic**, [R] **logistic postestimation**, [R] **logit**, [R] **logit postestimation**

Arellano, M., [R] **gmm**, [XT] **xtabond**, [XT] **xtdpd**, [XT] **xtdpd postestimation**, [XT] **xtdpdsys**, [XT] **xtdpdsys postestimation**, [XT] **xtreg**

Arminger, G., [R] **suest**

Armitage, P., [R] **ameans**, [R] **expoisson**, [R] **pkcross**, [R] **sdtest**

Armstrong, R. D., [R] **qreg**

Arnold, S. F., [MV] **manova**

Arora, S. S., [XT] **xtivreg**, [XT] **xtreg**

Arseven, E., [MV] **discrim lda**

Arthur, M., [R] **symmetry**

Aten, B., [XT] **xtunitroot**

Atkinson, A. C., [D] **functions**, [R] **boxcox**, [R] **nl**

Azen, S. P., [R] **anova**, [U] **20.20 References**

Aznar, A., [TS] **vecrank**

B

Babiker, A., [R] **sampsi**, [ST] **epitab**, [ST] **stpower**, [ST] **stpower cox**, [ST] **sts test**

Babin, B. J., [R] **rologit**

Babu, A. J. G., [D] **functions**
Bai, Z., [M-1] **LAPACK**, [M-5] **lapack( )**, [P] **matrix eigenvalues**
Baker, R. J., [R] **glm**
Baker, R. M., [R] **ivregress postestimation**
Bakker, A., [R] **mean**
Balaam, L. N., [R] **pkcross**
Balakrishnan, N., [D] **functions**
Baldus, W. P., [ST] **stcrreg**
Balestra, P., [XT] **xtivreg**
Baltagi, B. H., [R] **hausman**, [XT] **xt**, [XT] **xtabond**,
   [XT] **xtdpd**, [XT] **xtdpdsys**, [XT] **xthtaylor**,
   [XT] **xtivreg**, [XT] **xtmixed**, [XT] **xtreg**,
   [XT] **xtreg postestimation**, [XT] **xtregar**,
   [XT] **xtunitroot**
Bamber, D., [R] **roc**, [R] **rocfit**
Bancroft, T. A., [R] **stepwise**
Banerjee, A., [XT] **xtunitroot**
Barbin, É., [M-5] **cholesky( )**
Barnard, G. A., [R] **spearman**, [R] **ttest**
Barnard, J., [MI] **intro substantive**, [MI] **mi estimate**,
   [MI] **mi estimate postestimation**, [MI] **mi
   estimate using**
Barnett, A. G., [R] **glm**
Barnow, B. S., [R] **treatreg**
Barrison, I. G., [R] **binreg**
Barthel, F. M.-S., [ST] **stcox PH-assumption tests**,
   [ST] **stpower**, [ST] **stpower cox**
Bartlett, M. S., [MV] **factor**, [MV] **factor
   postestimation**, [MV] **Glossary**, [R] **oneway**,
   [TS] **wntestb**
Bartus, T., [R] **margins**
Basford, K. E., [G] **graph matrix**, [XT] **xtmelogit**,
   [XT] **xtmepoisson**, [XT] **xtmixed**
Basilevsky, A. T., [MV] **factor**, [MV] **pca**
Basmann, R. L., [R] **ivregress**, [R] **ivregress
   postestimation**
Bassett Jr., G., [R] **qreg**
Basu, A., [R] **glm**
Bates, D. M., [XT] **xtmelogit**, [XT] **xtmepoisson**,
   [XT] **xtmixed**, [XT] **xtmixed postestimation**
Battese, G. E., [XT] **xtfrontier**
Baum, C. F., [D] **cross**, [D] **fillin**, [D] **joinby**,
   [D] **reshape**, [D] **separate**, [D] **stack**, [D] **xpose**,
   [M-1] **intro**, [MV] **mvtest**, [MV] **mvtest
   normality**, [P] **intro**, [P] **levelsof**, [R] **gmm**,
   [R] **heckman**, [R] **heckprob**, [R] **ivregress**,
   [R] **ivregress postestimation**, [R] **net**, [R] **net
   search**, [R] **regress postestimation**, [R] **regress
   postestimation time series**, [R] **ssc**, [TS] **arch**,
   [TS] **arima**, [TS] **dfgls**, [TS] **rolling**, [TS] **time
   series**, [TS] **tsset**, [TS] **var**, [TS] **wntestq**,
   [XT] **xtgls**, [XT] **xtreg**, [U] **11.7 References**,
   [U] **16.5 References**, [U] **18.13 References**,
   [U] **20.20 References**
Bauwens, L., [TS] **dvech**
Bayart, D., [R] **qc**
Beale, E. M. L., [R] **stepwise**, [R] **test**
Beall, G., [MV] **mvtest**, [MV] **mvtest covariances**

Beaton, A. E., [R] **rreg**
Beck, N., [XT] **xtgls**, [XT] **xtpcse**
Becker, R. A., [G] **graph matrix**
Becketti, S., [P] **pause**, [R] **fracpoly**, [R] **runtest**,
   [R] **spearman**, [TS] **corrgram**
Beerstecher, E., [MV] **manova**
Beggs, S., [R] **rologit**
Belanger, A. J., [R] **sktest**
Bellman, R. E., [MV] **Glossary**
Bellocco, R., [ST] **epitab**
Belsley, D. A., [R] **estat**, [R] **regress postestimation**,
   [U] **18.13 References**
Bendel, R. B., [R] **stepwise**
Benedetti, J. K., [R] **tetrachoric**
Beniger, J. R., [G] **graph bar**, [G] **graph pie**,
   [G] **graph twoway histogram**, [R] **cumul**
Bentham, G., [XT] **xtmepoisson**
Bentler, P. M., [MV] **Glossary**, [MV] **rotate**,
   [MV] **rotatemat**
Bera, A. K., [R] **sktest**, [TS] **arch**, [TS] **varnorm**,
   [TS] **vecnorm**, [XT] **xtreg**, [XT] **xtreg
   postestimation**, [XT] **xtregar**
Beran, R. J., [R] **regress postestimation time series**
Berglund, P., [SVY] **subpopulation estimation**
Berk, K. N., [R] **stepwise**
Berk, R. A., [R] **rreg**
Berkes, I., [TS] **dvech**
Berkson, J., [R] **logit**, [R] **probit**
Bern, P. H., [R] **nestreg**
Bernaards, C. A., [MV] **rotatemat**
Bernasco, W., [R] **tetrachoric**
Berndt, E. K., [M-5] **optimize( )**, [R] **glm**, [TS] **arch**,
   [TS] **arima**
Berndt, E. R., [R] **treatreg**, [R] **truncreg**
Bernstein, I. H., [R] **alpha**
Berry, G., [R] **ameans**, [R] **expoisson**, [R] **sdtest**
Best, D. J., [D] **functions**
Bewley, R., [R] **reg3**
Beyer, W. H., [R] **qc**
Beyersman, J., [ST] **stcrreg**
Bhargava, A., [XT] **xtregar**
Bibby, J. M., [MI] **mi impute mvn**, [MV] **discrim**,
   [MV] **discrim lda**, [MV] **factor**, [MV] **manova**,
   [MV] **matrix dissimilarity**, [MV] **mds**,
   [MV] **mds postestimation**, [MV] **mdslong**,
   [MV] **mdsmat**, [MV] **mvtest**, [MV] **mvtest
   means**, [MV] **mvtest normality**, [MV] **pca**,
   [MV] **procrustes**, [P] **matrix dissimilarity**
Bickeböller, H., [R] **symmetry**
Bickel, P. J., [D] **egen**, [R] **rreg**
Binder, D. A., [MI] **intro substantive**, [P] **_robust**,
   [SVY] **svy estimation**, [SVY] **variance
   estimation**, [U] **20.20 References**
Birdsall, T. G., [R] **logistic postestimation**
Bischof, C., [M-1] **LAPACK**, [M-5] **lapack( )**,
   [P] **matrix eigenvalues**
Black, F., [TS] **arch**
Black, W. C., [R] **rologit**

Blackburne III, E. F., [XT] **xtabond**, [XT] **xtdpd**,
  [XT] **xtdpdsys**
Blackford, S., [M-1] **LAPACK**, [M-5] **lapack( )**,
  [P] **matrix eigenvalues**
Blackwell III, J. L., [R] **areg**, [XT] **xtgls**, [XT] **xtpcse**,
  [XT] **xtreg**
Bland, M., [R] **ranksum**, [R] **sdtest**, [R] **signrank**,
  [R] **spearman**
Blashfield, R. K., [MV] **cluster**
Blasius, J., [MV] **ca**, [MV] **mca**
Blasnik, M., [D] **clonevar**, [D] **split**, [D] **statsby**
Bleda, M.-J., [R] **alpha**
Bliss, C. I., [R] **probit**
Bloch, D. A., [R] **brier**
Bloomfield, P., [R] **qreg**
Blundell, R., [R] **gmm**, [R] **ivprobit**, [XT] **xtdpd**,
  [XT] **xtdpdsys**
BMDP, [R] **symmetry**
Boggess, M., [ST] **stcrreg**, [ST] **stcrreg postestimation**
Boice Jr., J. D., [R] **bitest**, [ST] **epitab**
Boland, P. J., [R] **ttest**
Bolduc, D., [R] **asmprobit**
Bollen, K. A., [MV] **factor postestimation**, [R] **regress
  postestimation**
Bollerslev, T., [TS] **arch**, [TS] **arima**, [TS] **dvech**
Bond, S., [R] **gmm**, [XT] **xtabond**, [XT] **xtdpd**,
  [XT] **xtdpd postestimation**, [XT] **xtdpdsys**,
  [XT] **xtdpdsys postestimation**
Borenstein, M., [R] **meta**
Borg, I., [MV] **mds**, [MV] **mds postestimation**,
  [MV] **mdslong**, [MV] **mdsmat**
Borgan, Ø., [ST] **stcrreg**
Borowczyk, J., [M-5] **cholesky( )**
Boshuizen, H. C., [MI] **intro substantive**, [MI] **mi
  impute monotone**
Boswijk, H. P., [TS] **vec**
Bottai, M., [ST] **epitab**, [XT] **xtreg**
Bound, J., [R] **ivregress postestimation**
Bover, O., [XT] **xtdpd**, [XT] **xtdpdsys**
Bowerman, B. L., [TS] **tssmooth**, [TS] **tssmooth
  dexponential**, [TS] **tssmooth exponential**,
  [TS] **tssmooth hwinters**, [TS] **tssmooth
  shwinters**
Bowker, A. H., [R] **symmetry**
Box, G. E. P., [MV] **manova**, [MV] **mvtest
  covariances**, [R] **anova**, [R] **boxcox**,
  [R] **lnskew0**, [TS] **arima**, [TS] **corrgram**,
  [TS] **cumsp**, [TS] **dfuller**, [TS] **pergram**,
  [TS] **pperron**, [TS] **wntestq**, [TS] **xcorr**
Box, J. F., [R] **anova**
Box-Steffensmeier, J. M., [ST] **stcox**, [ST] **streg**
Boyd, N. F., [R] **kappa**
Boyle, J. M., [P] **matrix symeigen**
Boyle, P., [XT] **xtmepoisson**
Bradley, R. A., [R] **signrank**
Brady, A. R., [R] **logistic**, [R] **spikeplot**
Brady, T., [D] **edit**
Brant, R., [R] **ologit**

Bray, R. J., [MV] **clustermat**
Bray, T. A., [D] **functions**
Breitung, J., [XT] **xtunitroot**
Brent, R. P., [MV] **mdsmat**, [MV] **mvtest means**
Breslow, N. E., [R] **clogit**, [R] **dstdize**, [R] **symmetry**,
  [ST] **epitab**, [ST] **stcox**, [ST] **stcox PH-
  assumption tests**, [ST] **sts**, [ST] **sts test**,
  [XT] **xtmelogit**, [XT] **xtmepoisson**
Breusch, T. S., [R] **mvreg**, [R] **regress postestimation**,
  [R] **regress postestimation time series**,
  [R] **sureg**, [TS] **Glossary**, [XT] **xtreg
  postestimation**
Brier, G. W., [R] **brier**
Brillinger, D. R., [R] **jackknife**
Brockwell, P. J., [TS] **corrgram**, [TS] **sspace**
Brook, R. H., [R] **brier**
Brown, B. W., [ST] **sts graph**
Brown, C. C., [ST] **epitab**
Brown, D. R., [R] **anova**, [R] **loneway**, [R] **oneway**
Brown, J. D., [MV] **manova**
Brown, L. D., [R] **ci**
Brown, M. B., [R] **sdtest**, [R] **tetrachoric**
Brown, S. E., [R] **symmetry**
Browne, M. W., [MV] **procrustes**
Bru, B., [R] **poisson**
Bruno, G. S. F., [XT] **xtabond**, [XT] **xtdpd**,
  [XT] **xtdpdsys**, [XT] **xtreg**
Bryk, A. S., [XT] **xtmelogit**, [XT] **xtmepoisson**,
  [XT] **xtmixed**
Brzinsky-Fay, C., [G] **graph twoway rbar**
Buchholz, A., [ST] **stcrreg**
Buchner, D. M., [R] **ladder**
Bunch, D. S., [R] **asmprobit**
Buot, M.-L. G., [MV] **mvtest means**
Burnam, M. A., [R] **lincom**, [R] **mlogit**, [R] **mprobit**,
  [R] **mprobit postestimation**, [R] **predictnl**,
  [R] **slogit**
Burr, I. W., [R] **qc**
Buskens, V., [R] **tabstat**

C

Cai, T. T., [R] **ci**
Cailliez, F., [MV] **mdsmat**
Cain, G. G., [R] **treatreg**
Caines, P. E., [TS] **sspace**
Califf, R. M., [ST] **stcox postestimation**
Caliński, T., [MV] **cluster**, [MV] **cluster stop**
Cameron, A. C., [R] **asclogit**, [R] **asmprobit**,
  [R] **bootstrap**, [R] **gmm**, [R] **heckman**,
  [R] **intreg**, [R] **ivregress**, [R] **ivregress
  postestimation**, [R] **logit**, [R] **mprobit**,
  [R] **nbreg**, [R] **ologit**, [R] **oprobit**, [R] **poisson**,
  [R] **probit**, [R] **qreg**, [R] **regress**, [R] **regress
  postestimation**, [R] **simulate**, [R] **sureg**,
  [R] **tobit**, [R] **ztnb**, [R] **ztp**, [XT] **xt**,
  [XT] **xtmixed**, [XT] **xtnbreg**, [XT] **xtpoisson**

Campbell, M. J., [R] ci, [R] kappa, [R] logistic
postestimation, [R] poisson, [R] tabulate
twoway, [ST] stpower, [ST] stpower cox,
[ST] stpower logrank

Cappellari, L., [D] corr2data, [D] egen, [R] asmprobit

Cardell, S., [R] rologit

Carlile, T., [R] kappa

Carlin, J. B., [MI] intro, [MI] intro substantive,
[MI] mi estimate, [MI] mi impute mvn, [MI] mi
impute regress, [R] ameans, [ST] epitab

Carnes, B. A., [ST] streg

Carpenter, J. R., [MI] intro, [MI] intro substantive,
[XT] xtmelogit

Carroll, J. B., [MV] rotatemat

Carroll, R. J., [R] boxcox, [R] rreg, [R] sdtest,
[XT] xtmixed

Carson, R. T., [R] ztnb, [R] ztp

Carter, S., [R] frontier, [R] lrtest, [R] nbreg, [R] ztnb,
[ST] stcox, [ST] streg, [XT] xt, [XT] xtmelogit,
[XT] xtmepoisson, [XT] xtmixed

Casals, J., [TS] sspace

Casella, G., [XT] xtmixed

Cattell, R. B., [MV] factor postestimation,
[MV] pca postestimation, [MV] procrustes,
[MV] screeplot

Caudill, S. B., [R] frontier, [XT] xtfrontier

Caulcutt, R., [R] qc

Center for Human Resource Research, [XT] xt

Chabert, J.-L., [M-5] cholesky( )

Chadwick, J., [R] poisson

Chamberlain, G., [R] clogit, [R] gmm

Chambers, J. M., [G] by_option, [G] graph box,
[G] graph matrix, [R] diagnostic plots,
[R] grmeanby, [R] lowess, [U] 1.4 References

Chang, I. M., [R] margins

Chang, Y., [TS] sspace, [XT] xtivreg, [XT] xtreg

Chao, E. C., [XT] xtmelogit, [XT] xtmelogit
postestimation, [XT] xtmepoisson,
[XT] xtmepoisson postestimation

Charlett, A., [R] fracpoly

Chatfield, C., [TS] arima, [TS] corrgram,
[TS] Glossary, [TS] pergram, [TS] tssmooth,
[TS] tssmooth dexponential, [TS] tssmooth
exponential, [TS] tssmooth hwinters,
[TS] tssmooth ma, [TS] tssmooth shwinters

Chatterjee, S., [R] poisson, [R] regress, [R] regress
postestimation, [TS] prais

Chen, X., [R] logistic, [R] logistic postestimation,
[R] logit

Cheung, Y.-W., [TS] dfgls

Chiang, C. L., [ST] ltable

Chiburis, R., [R] heckman, [R] heckprob, [R] oprobit

Choi, B. C. K., [R] roc, [R] rocfit

Choi, I., [XT] xtunitroot

Choi, M.-D., [M-5] Hilbert( )

Choi, S. C., [MV] discrim knn

Cholesky, A.-L., [M-5] cholesky( )

Chou, R. Y., [TS] arch

Chow, S.-C., [R] pk, [R] pkcross, [R] pkequiv,
[R] pkexamine, [R] pkshape, [ST] stpower,
[ST] stpower exponential

Christakis, N., [R] rologit

Christiano, L. J., [TS] irf create, [TS] var svar

Chu, C.-S. J., [XT] xtunitroot

Chu-Chun-Lin, S., [TS] sspace

Clark, V. A., [ST] ltable

Clarke, M. R. B., [MV] factor

Clarke, R. D., [R] poisson

Clarke-Pearson, D. L., [R] roc

Clarkson, D. B., [R] tabulate twoway

Clayton, D. G., [D] egen, [R] cloglog, [R] cumul,
[ST] epitab, [ST] stptime, [ST] strate,
[ST] stsplit, [ST] sttocc, [XT] xtmelogit,
[XT] xtmepoisson

Cleland, J., [XT] xtmelogit

Clerget-Darpoux, F., [R] symmetry

Cleveland, W. S., [G] by_option, [G] graph
box, [G] graph dot, [G] graph intro,
[G] graph matrix, [G] graph twoway lowess,
[R] diagnostic plots, [R] lowess, [R] lpoly,
[R] sunflower, [U] 1.4 References

Cleves, M. A., [R] binreg, [R] dstdize, [R] logistic,
[R] logit, [R] roc, [R] rocfit, [R] sdtest,
[R] symmetry, [ST] st, [ST] stcox, [ST] stcrreg,
[ST] stcurve, [ST] stdescribe, [ST] streg,
[ST] sts, [ST] stset, [ST] stsplit, [ST] stvary,
[ST] survival analysis

Cliff, N., [MV] canon postestimation

Clogg, C. C., [R] suest

Clopper, C. J., [R] ci

Cobb, G. W., [R] anova

Cochran, W. G., [P] levelsof, [R] ameans, [R] anova,
[R] correlate, [R] dstdize, [R] mean,
[R] oneway, [R] poisson, [R] probit,
[R] proportion, [R] ranksum, [R] ratio,
[R] signrank, [R] total, [SVY] estat,
[SVY] subpopulation estimation, [SVY] survey,
[SVY] svyset, [SVY] variance estimation

Cochrane, D., [TS] prais

Coelli, T. J., [R] frontier, [XT] xtfrontier

Coffey, C., [MI] intro substantive

Cohen, J., [R] kappa, [R] pcorr

Cohen, P., [R] pcorr

Coleman, J. S., [R] poisson

Collett, D., [R] clogit, [R] logistic, [R] logistic
postestimation, [ST] stci, [ST] stcox
postestimation, [ST] stcrreg postestimation,
[ST] stpower, [ST] stpower logrank, [ST] streg
postestimation, [ST] sts test, [ST] stsplit

Collins, E., [SVY] survey, [SVY] svy estimation

Comrey, A. L., [MV] Glossary, [MV] rotate,
[MV] rotatemat

Comte, F., [TS] dvech

Cong, R., [R] tobit, [R] tobit postestimation,
[R] treatreg, [R] truncreg

Conover, W. J., [R] **centile**, [R] **ksmirnov**, [R] **kwallis**,
    [R] **nptrend**, [R] **sdtest**, [R] **spearman**,
    [R] **tabulate twoway**
Conroy, R. M., [R] **intreg**
Conway, M. R., [XT] **xtlogit**, [XT] **xtprobit**
Cook, A., [R] **ci**
Cook, I., [U] **1.4 References**
Cook, R. D., [P] **_predict**, [R] **boxcox**, [R] **regress
    postestimation**
Cooper, M. C., [MV] **cluster**, [MV] **cluster
    programming subroutines**, [MV] **cluster stop**
Cornfield, J., [ST] **epitab**
Coull, B. A., [R] **ci**
Coviello, V., [ST] **stcrreg**, [ST] **stcrreg postestimation**,
    [ST] **sttocc**
Cox, C. S., [SVY] **survey**, [SVY] **svy estimation**
Cox, D. R., [R] **boxcox**, [R] **exlogistic**, [R] **expoisson**,
    [R] **lnskew0**, [ST] **ltable**, [ST] **stcox**,
    [ST] **stcox PH-assumption tests**, [ST] **stcrreg**,
    [ST] **stpower**, [ST] **stpower cox**, [ST] **streg**,
    [ST] **streg postestimation**, [ST] **sts**
Cox, G. M., [P] **levelsof**
Cox, M. A. A., [MV] **biplot**, [MV] **ca**, [MV] **Glossary**,
    [MV] **mds**, [MV] **mds postestimation**,
    [MV] **mdsmat**, [MV] **procrustes**
Cox, N. J., [D] **by**, [D] **clonevar**, [D] **contract**,
    [D] **describe**, [D] **destring**, [D] **drop**,
    [D] **duplicates**, [D] **egen**, [D] **fillin**,
    [D] **functions**, [D] **rename**, [D] **reshape**,
    [D] **sample**, [D] **separate**, [D] **split**, [G] **graph
    bar**, [G] **graph box**, [G] **graph dot**, [G] **graph
    intro**, [G] **graph twoway function**, [G] **graph
    twoway histogram**, [G] **graph twoway kdensity**,
    [G] **graph twoway lowess**, [G] **graph twoway
    lpoly**, [G] **graph twoway pcarrow**, [G] **graph
    twoway scatter**, [MV] **mvtest**, [MV] **mvtest
    normality**, [P] **levelsof**, [P] **matrix define**, [R] **ci**,
    [R] **cumul**, [R] **diagnostic plots**, [R] **histogram**,
    [R] **inequality**, [R] **kappa**, [R] **kdensity**,
    [R] **ladder**, [R] **lowess**, [R] **lpoly**, [R] **net**,
    [R] **net search**, [R] **regress postestimation**,
    [R] **search**, [R] **serrbar**, [R] **sktest**, [R] **smooth**,
    [R] **spikeplot**, [R] **ssc**, [R] **stem**, [R] **sunflower**,
    [R] **tabulate twoway**, [TS] **tssmooth hwinters**,
    [TS] **tssmooth shwinters**, [XT] **xtdescribe**,
    [U] **12.10 References**, [U] **13.12 References**,
    [U] **17.10 Reference**, [U] **23.5 Reference**,
    [U] **24.7 Reference**
Cox, T. F., [MV] **biplot**, [MV] **ca**, [MV] **Glossary**,
    [MV] **mds**, [MV] **mds postestimation**,
    [MV] **mdsmat**, [MV] **procrustes**
Cozad, J. B., [MV] **discrim lda**
Cragg, J. G., [R] **ivregress postestimation**
Cramer, E. M., [MV] **procrustes**
Cramér, H., [R] **tabulate twoway**
Cramer, J. S., [R] **logit**
Crawford, C. B., [MV] **Glossary**, [MV] **rotate**,
    [MV] **rotatemat**
Critchley, F., [MV] **mdsmat**
Cronbach, L. J., [R] **alpha**

Crowder, M. J., [ST] **stcrreg**, [ST] **streg**
Crowe, P. R., [G] **graph box**
Crowley, J., [ST] **stcox**, [ST] **stcrreg**, [ST] **stset**
Cui, J., [R] **symmetry**, [ST] **stcox**, [ST] **streg**,
    [XT] **xtgee**
Curtis, J. T., [MV] **clustermat**
Curts-García, J., [R] **smooth**
Cutler, S. J., [ST] **ltable**
Cuzick, J., [R] **kappa**, [R] **nptrend**
Czekanowski, J., [MV] *measure_option*

D

D'Agostino, R. B., [MV] **mvtest normality**, [R] **sktest**
D'Agostino Jr., R. B., [R] **sktest**
Daniel, C., [R] **diagnostic plots**, [R] **oneway**
Danuso, F., [R] **nl**
Das, S., [XT] **xtunitroot**
DasGupta, A., [R] **ci**
Davey Smith, G., [R] **meta**
David, H. A., [D] **egen**, [R] **spearman**, [R] **summarize**
David, J. S., [TS] **arima**
Davidon, W. C., [M-5] **optimize( )**
Davidson, R., [R] **boxcox**, [R] **cnsreg**, [R] **gmm**,
    [R] **intreg**, [R] **ivregress**, [R] **ivregress
    postestimation**, [R] **mlogit**, [R] **nl**, [R] **nlsur**,
    [R] **reg3**, [R] **regress**, [R] **regress postestimation
    time series**, [R] **tobit**, [R] **truncreg**, [TS] **arch**,
    [TS] **arima**, [TS] **Glossary**, [TS] **prais**,
    [TS] **sspace**, [TS] **varlmar**, [XT] **xtgls**,
    [XT] **xtpcse**
Davis, R. A., [TS] **corrgram**, [TS] **sspace**
Davison, A. C., [R] **bootstrap**
Day, N. E., [R] **clogit**, [R] **dstdize**, [R] **symmetry**,
    [ST] **epitab**
Day, W. H. E., [MV] **cluster**
De Hoyos, R. E., [XT] **xtreg**
de Irala-Estévez, J., [R] **logistic**
De Jong, P., [TS] **dfactor**, [TS] **sspace**, [TS] **sspace
    postestimation**
de Leeuw, J., [MV] **ca postestimation**
De Luca, G., [R] **biprobit**, [R] **heckprob**, [R] **probit**
De Stavola, B. L., [ST] **stcox**, [ST] **stset**
de Wolf, I., [R] **rologit**
Deaton, A., [R] **nlsur**, [U] **20.20 References**
Deb, P., [R] **nbreg**
DeGroot, M. H., [TS] **arima**
Deistler, M., [TS] **sspace**
DeLong, D. M., [R] **roc**
DeLong, E. R., [R] **roc**
DeMaris, A., [R] **regress postestimation**
Demidenko, E., [XT] **xtmixed**
Demmel, J., [M-1] **LAPACK**, [M-5] **lapack( )**,
    [P] **matrix eigenvalues**
Demnati, A., [SVY] **direct standardization**,
    [SVY] **poststratification**, [SVY] **variance
    estimation**

Dempster, A. P., [MI] **intro substantive**, [MI] **mi impute mvn**, [XT] **xtmixed**
Denis, D., [G] **graph twoway scatter**
Desu, M. M., [ST] **stpower**, [ST] **stpower exponential**
Deville, J.-C., [SVY] **direct standardization**, [SVY] **poststratification**, [SVY] **variance estimation**
Devroye, L., [D] **functions**
Dewey, M. E., [R] **correlate**
Dice, L. R., [MV] *measure_option*
Dickens, R., [TS] **prais**
Dickey, D. A., [TS] **dfuller**, [TS] **Glossary**, [TS] **pperron**
Dickson, E. R., [ST] **stcrreg**
Diebold, F. X., [TS] **arch**
Dieter, U., [D] **functions**
Digby, P. G. N., [R] **tetrachoric**
Diggle, P. J., [TS] **arima**, [TS] **wntestq**, [XT] **xtmixed**
Dijksterhuis, G. B., [MV] **procrustes**
DiNardo, J., [XT] **xtrc**
Ding, Z., [TS] **arch**
Dixon, W. J., [R] **ttest**
Dobbin, K., [ST] **stpower**
Dobson, A. J., [R] **glm**
Dohoo, I., [R] **regress**, [ST] **epitab**
Doll, R., [R] **poisson**, [ST] **epitab**
Donald, S. G., [R] **ivregress postestimation**
Dongarra, J. J., [M-1] **LAPACK**, [M-5] **lapack( )**, [P] **matrix eigenvalues**, [P] **matrix symeigen**
Donner, A., [R] **loneway**
Donoho, D. L., [R] **lpoly**
Doornik, J. A., [MV] **mvtest**, [MV] **mvtest normality**, [TS] **vec**
Dore, C. J., [R] **fracpoly**
Dorfman, D. D., [R] **rocfit**
Draper, N., [R] **eivreg**, [R] **oneway**, [R] **regress**, [R] **stepwise**
Driver, H. E., [MV] *measure_option*
Drukker, D. M., [R] **asmprobit**, [R] **boxcox**, [R] **frontier**, [R] **lrtest**, [R] **nbreg**, [R] **tobit**, [R] **treatreg**, [R] **ztnb**, [ST] **stcox**, [ST] **streg**, [TS] **sspace**, [TS] **vec**, [XT] **xt**, [XT] **xtmelogit**, [XT] **xtmepoisson**, [XT] **xtmixed**, [XT] **xtregar**
Du Croz, J., [M-1] **LAPACK**, [M-5] **lapack( )**, [P] **matrix eigenvalues**
Duan, N., [R] **heckman**
Dubes, R. C., [MV] **cluster**
Duda, R. O., [MV] **cluster**, [MV] **cluster stop**
Duncan, A. J., [R] **qc**
Dunlop, D. D., [R] **sampsi**
Dunn, G., [MV] **discrim**, [MV] **discrim qda postestimation**, [MV] **mca**, [R] **kappa**
Dunnett, C. W., [R] **mprobit**
Dunnington, G. W., [R] **regress**
Dupont, W. D., [R] **logistic**, [R] **mkspline**, [R] **sunflower**, [ST] **epitab**, [ST] **stcox**, [ST] **stir**, [ST] **sts**

Durbin, J., [R] **ivregress postestimation**, [R] **regress postestimation time series**, [TS] **Glossary**, [TS] **prais**
Durlauf, S. N., [TS] **vec**, [TS] **vec intro**, [TS] **vecrank**
Duval, R. D., [R] **bootstrap**, [R] **jackknife**
Dwyer, J. H., [XT] **xtreg**

E

Edelsbrunner, H., [MV] **cluster**
Ederer, F., [ST] **ltable**
Edgington, E. S., [R] **runtest**
Edwards, A. L., [R] **anova**
Edwards, A. W. F., [R] **tetrachoric**
Edwards, J. H., [R] **tetrachoric**
Efron, B., [R] **bootstrap**, [R] **qreg**
Efroymson, M. A., [R] **stepwise**
Egger, M., [R] **meta**
Eichenbaum, M., [TS] **irf create**, [TS] **var svar**
Eisenhart, C., [R] **correlate**, [R] **runtest**
Elliott, G., [TS] **dfgls**, [TS] **Glossary**
Ellis, C. D., [R] **poisson**
Eltinge, J. L., [R] **test**, [SVY] **estat**, [SVY] **survey**, [SVY] **svy postestimation**, [SVY] **svydescribe**, [SVY] **variance estimation**
Emerson, J. D., [R] **lv**, [R] **stem**
Enas, G. G., [MV] **discrim knn**
Ender, P. B., [MV] **canon**
Enders, W., [TS] **arch**, [TS] **arima**, [TS] **corrgram**
Engel, A., [SVY] **estat**, [SVY] **subpopulation estimation**, [SVY] **survey**, [SVY] **svy**, [SVY] **svy brr**, [SVY] **svy estimation**, [SVY] **svy jackknife**, [SVY] **svy postestimation**, [SVY] **svy: tabulate oneway**, [SVY] **svy: tabulate twoway**, [SVY] **svydescribe**
Engle, R. F., [R] **regress postestimation time series**, [TS] **arch**, [TS] **arima**, [TS] **dfactor**, [TS] **dvech**, [TS] **vec**, [TS] **vec intro**, [TS] **vecrank**
Erdreich, L. S., [R] **roc**, [R] **rocfit**
Esman, R. M., [D] **egen**
Eubank, R. L., [R] **lpoly**
Evans, C. L., [TS] **irf create**, [TS] **var svar**
Evans, M. A., [R] **pk**, [R] **pkcross**
Everitt, B. S., [MV] **cluster**, [MV] **cluster stop**, [MV] **discrim**, [MV] **discrim qda postestimation**, [MV] **mca**, [MV] **pca**, [MV] **screeplot**, [R] **anova**, [R] **gllamm**, [R] **glm**, [U] **1.4 References**
Ewens, W. J., [R] **symmetry**
Ezekiel, M., [R] **regress postestimation**
Ezzati-Rice, T. M., [MI] **intro substantive**

F

Fan, J., [R] **lpoly**
Fan, Y.-A., [R] **tabulate twoway**
Fang, K.-T., [R] **asmprobit**

Feinleib, M., [XT] **xtreg**
Feiveson, A. H., [R] **nlcom**, [R] **ranksum**,
      [ST] **stpower**
Feldman, J. J., [SVY] **survey**, [SVY] **svy estimation**
Feldt, L. S., [R] **anova**
Feller, W., [TS] **wntestb**
Feltbower, R., [ST] **epitab**
Ferguson, G. A., [MV] **Glossary**, [MV] **rotate**,
      [MV] **rotatemat**
Ferri, H. A., [R] **kappa**
Fibrinogen Studies Collaboration, [ST] **stcox**
      postestimation
Fidell, L. S., [MV] **discrim**, [MV] **discrim lda**
Fieller, E. C., [R] **pkequiv**
Fienberg, S. E., [R] **kwallis**, [R] **tabulate twoway**
Filon, L. N. G., [R] **correlate**
Findley, D. F., [R] **estat**
Findley, T. W., [R] **ladder**
Fine, J. P., [ST] **intro**, [ST] **stcrreg**
Finney, D. J., [R] **probit**, [R] **tabulate twoway**
Fiocco, M., [ST] **stcrreg**, [ST] **stcrreg postestimation**
Fiorio, C. V., [R] **kdensity**
Fiser, D. H., [R] **logistic postestimation**
Fishell, E., [R] **kappa**
Fisher, L. D., [MV] **factor**, [MV] **pca**, [R] **anova**,
      [R] **dstdize**, [R] **oneway**, [ST] **epitab**
Fisher, M. R., [XT] **xtcloglog**, [XT] **xtgee**,
      [XT] **xtintreg**, [XT] **xtlogit**, [XT] **xtprobit**,
      [XT] **xttobit**
Fisher, N. I., [R] **regress postestimation time series**
Fisher, R. A., [MV] **clustermat**, [MV] **discrim**,
      [MV] **discrim estat**, [MV] **discrim lda**,
      [MV] **Glossary**, [P] **levelsof**, [R] **anova**,
      [R] **ranksum**, [R] **signrank**, [R] **tabulate**
      twoway, [ST] **streg**
Fix, E., [MV] **discrim knn**
Flannery, B. P., [D] **functions**, [P] **matrix symeigen**,
      [R] **dydx**, [R] **vwls**, [TS] **arch**, [TS] **arima**
Fleiss, J. L., [R] **dstdize**, [R] **kappa**, [R] **sampsi**,
      [ST] **epitab**
Fleming, T. R., [ST] **stcox**, [ST] **sts test**
Fletcher, R., [M-5] **optimize( )**
Ford, J. M., [R] **frontier**, [XT] **xtfrontier**
Forsythe, A. B., [R] **sdtest**
Forthofer, R. N., [R] **dstdize**
Foster, A., [R] **regress**
Foulkes, M. A., [ST] **stpower**, [ST] **stpower cox**,
      [ST] **stpower exponential**, [U] **1.4 References**
Fourier, J. B. J., [R] **cumul**
Fox, J., [R] **kdensity**, [R] **lv**
Fox, W. C., [R] **logistic postestimation**
Francia, R. S., [R] **swilk**
Frank, M. W., [XT] **xtabond**, [XT] **xtdpd**,
      [XT] **xtdpdsys**
Frankel, M. R., [P] **_robust**, [SVY] **variance**
      estimation, [U] **20.20 References**
Franklin, C. H., [D] **cross**
Franzese, R. J., Jr., [XT] **xtpcse**

Franzini, L., [XT] **xtregar**
Frechette, G. R., [XT] **xtprobit**
Freedman, L. S., [ST] **stpower**, [ST] **stpower cox**,
      [ST] **stpower exponential**, [ST] **stpower**
      logrank, [U] **1.4 References**
Freeman, J. L., [ST] **epitab**, [SVY] **svy: tabulate**
      twoway
Freeman Jr., D. H., [SVY] **svy: tabulate twoway**
Freese, J., [R] **asroprobit**, [R] **clogit**, [R] **cloglog**,
      [R] **logistic**, [R] **logit**, [R] **mlogit**, [R] **mprobit**,
      [R] **nbreg**, [R] **ologit**, [R] **oprobit**,
      [R] **poisson**, [R] **probit**, [R] **regress**,
      [R] **regress postestimation**, [R] **zinb**, [R] **zip**,
      [U] **20.20 References**
Friedman, J., [MV] **discrim knn**
Friedman, M., [TS] **arima**
Friendly, M., [G] **graph twoway scatter**
Frison, L., [R] **sampsi**
Frome, E. L., [R] **qreg**
Fu, V. K., [R] **ologit**
Fuller, W. A., [MV] **factor**, [P] **_robust**, [R] **regress**,
      [R] **spearman**, [SVY] **svy: tabulate**
      twoway, [SVY] **variance estimation**,
      [TS] **dfuller**, [TS] **Glossary**, [TS] **pperron**,
      [U] **20.20 References**
Funkhouser, H. G., [G] **graph pie**
Fyler, D. C., [ST] **epitab**
Fyles, A., [ST] **stcrreg**, [ST] **stcrreg postestimation**

G

Gabriel, K. R., [MV] **biplot**
Gail, M. H., [P] **_robust**, [ST] **stcrreg**, [ST] **stpower**,
      [ST] **stpower exponential**, [ST] **strate**,
      [U] **1.4 References**, [U] **20.20 References**
Galati, J. C., [MI] **intro**, [MI] **intro substantive**
Galecki, A. T., [XT] **xtmixed**
Gall, J.-R. L., [R] **logistic**, [R] **logistic postestimation**
Gallant, A. R., [R] **ivregress**, [R] **nl**
Galton, F., [R] **correlate**, [R] **cumul**, [R] **summarize**
Gan, F. F., [R] **diagnostic plots**
Gange, S. J., [XT] **xtcloglog**, [XT] **xtgee**,
      [XT] **xtintreg**, [XT] **xtlogit**, [XT] **xtprobit**,
      [XT] **xttobit**
Gani, J., [TS] **wntestb**
Garbow, B. S., [P] **matrix symeigen**
Gardiner, J. S., [TS] **tssmooth**, [TS] **tssmooth**
      dexponential, [TS] **tssmooth exponential**,
      [TS] **tssmooth hwinters**, [TS] **tssmooth**
      shwinters
Gardner Jr., E. S., [TS] **tssmooth dexponential**,
      [TS] **tssmooth hwinters**
Garrett, J. M., [R] **fracpoly**, [R] **logistic**, [R] **logistic**
      postestimation, [R] **regress postestimation**,
      [ST] **stcox PH-assumption tests**
Garsd, A., [R] **exlogistic**
Gart, J. J., [ST] **epitab**
Gasser, T., [R] **lpoly**
Gates, R., [R] **asmprobit**

Gauvreau, K., [R] **dstdize**, [R] **logistic**, [R] **sampsi**, [ST] **ltable**, [ST] **sts**

Gehan, E. A., [ST] **sts test**

Geisser, S., [R] **anova**

Gelbach, J., [R] **ivprobit**, [R] **ivtobit**

Gelman, A., [MI] **intro substantive**, [MI] **mi impute mvn**, [MI] **mi impute regress**

Gelman, R., [R] **margins**

Gentle, J. E., [D] **functions**, [R] **anova**, [R] **nl**

Genz, A., [R] **asmprobit**

George, S. L., [ST] **stpower**, [ST] **stpower exponential**

Gerkins, V. R., [R] **symmetry**

Geskus, R. B., [ST] **stcrreg**, [ST] **stcrreg postestimation**

Geweke, J., [R] **asmprobit**, [TS] **dfactor**

Giannini, C., [TS] **irf create**, [TS] **var intro**, [TS] **var svar**, [TS] **vargranger**, [TS] **varwle**

Gibbons, J. D., [R] **ksmirnov**, [R] **spearman**

Gichangi, A., [ST] **stcrreg**

Giesen, D., [R] **tetrachoric**

Gifi, A., [MV] **mds**

Gijbels, I., [R] **lpoly**

Gilbert, G. K., [MV] *measure_option*

Giles, D. E. A., [TS] **prais**

Gill, R. D., [ST] **stcrreg**

Gillham, N. W., [R] **regress**

Gillispie, C. C., [R] **regress**

Gini, R., [R] **vwls**, [ST] **epitab**

Ginther, O. J., [XT] **xtmixed**

Girshick, M. A., [MV] **pca**

Glass, R. I., [ST] **epitab**

Gleason, J. R., [D] **cf**, [D] **describe**, [D] **functions**, [D] **generate**, [D] **infile (fixed format)**, [D] **label**, [D] **notes**, [D] **order**, [R] **anova**, [R] **bootstrap**, [R] **ci**, [R] **correlate**, [R] **loneway**, [R] **summarize**, [R] **ttest**, [ST] **epitab**

Gleick, J., [M-5] **optimize( )**

Gleser, G., [R] **alpha**

Glidden, D. V., [R] **logistic**, [ST] **stcox**

Gloeckler, L., [ST] **discrete**

Glosten, L. R., [TS] **arch**

Gnanadesikan, R., [MV] **manova**, [R] **cumul**, [R] **diagnostic plots**

Godambe, V. P., [SVY] **variance estimation**

Godfrey, L. G., [R] **regress postestimation time series**

Goeden, G. B., [R] **kdensity**

Goldberger, A. S., [R] **intreg**, [R] **tobit**, [R] **treatreg**

Goldblatt, A., [ST] **epitab**

Golden, C. D., [SVY] **survey**, [SVY] **svy estimation**

Goldfarb, D., [M-5] **optimize( )**

Goldman, N., [XT] **xtmelogit**

Goldstein, H., [XT] **xtmelogit**, [XT] **xtmepoisson**, [XT] **xtmixed**

Goldstein, R., [D] **egen**, [R] **brier**, [R] **correlate**, [R] **inequality**, [R] **nl**, [R] **ologit**, [R] **oprobit**, [R] **ranksum**, [R] **regress postestimation**, [XT] **xtreg**

Golub, G. H., [R] **orthog**, [R] **tetrachoric**

Gonzalez Jr., J. F., [SVY] **estat**, [SVY] **subpopulation estimation**, [SVY] **svy estimation**

Gonzalo, J., [TS] **vec intro**, [TS] **vecrank**

Good, P. I., [G] **graph intro**, [R] **permute**, [R] **symmetry**, [R] **tabulate twoway**

Goodall, C., [R] **lowess**, [R] **rreg**

Goodman, L. A., [R] **tabulate twoway**

Gooley, T. A., [ST] **stcrreg**

Gordon, A. D., [MV] **biplot**, [MV] **cluster**, [MV] **cluster stop**, [MV] *measure_option*

Gordon, M. G., [R] **binreg**

Gorman, J. W., [R] **stepwise**

Gorsuch, R. L., [MV] **factor**, [MV] **rotate**, [MV] **rotatemat**

Gosset [Student, pseud.], W. S., [R] **ttest**

Gosset, W. S., [R] **ttest**

Gould, W. W., [D] **datasignature**, [D] **destring**, [D] **egen**, [D] **icd9**, [D] **infile (fixed format)**, [D] **reshape**, [M-1] **how**, [M-1] **interactive**, [M-2] **exp**, [M-2] **goto**, [M-2] **struct**, [M-2] **subscripts**, [M-2] **syntax**, [M-4] **stata**, [M-5] **deriv( )**, [M-5] **inbase( )**, [M-5] **moptimize( )**, [M-5] **st_addvar( )**, [M-5] **st_global( )**, [M-5] **st_local( )**, [M-5] **st_view( )**, [P] **_datasignature**, [P] **_robust**, [P] **intro**, [P] **matrix mkmat**, [P] **postfile**, [R] **bootstrap**, [R] **dydx**, [R] **frontier**, [R] **grmeanby**, [R] **jackknife**, [R] **kappa**, [R] **logistic**, [R] **margins**, [R] **maximize**, [R] **mkspline**, [R] **ml**, [R] **net search**, [R] **nlcom**, [R] **ologit**, [R] **oprobit**, [R] **predictnl**, [R] **qreg**, [R] **rreg**, [R] **simulate**, [R] **sktest**, [R] **smooth**, [R] **swilk**, [R] **testnl**, [ST] **stcox**, [ST] **stcrreg**, [ST] **stdescribe**, [ST] **streg**, [ST] **stset**, [ST] **stsplit**, [ST] **stvary**, [ST] **survival analysis**, [SVY] **ml for svy**, [SVY] **survey**, [XT] **xtfrontier**, [U] **13.12 References**, [U] **18.13 References**, [U] **26.25 Reference**

Gourieroux, C., [R] **hausman**, [R] **suest**, [R] **test**, [TS] **arima**

Govindarajulu, Z., [D] **functions**

Gower, J. C., [MV] **biplot**, [MV] **ca**, [MV] **mca**, [MV] *measure_option*, [MV] **procrustes**

Graham, J. W., [MI] **intro substantive**, [MI] **mi impute**

Grambsch, P. M., [ST] **stcox**, [ST] **stcox PH-assumption tests**, [ST] **stcox postestimation**, [ST] **stcrreg**

Granger, C. W. J., [TS] **arch**, [TS] **vargranger**, [TS] **vec**, [TS] **vec intro**, [TS] **vecrank**

Graubard, B. I., [R] **margins**, [R] **ml**, [R] **test**, [SVY] **direct standardization**, [SVY] **survey**, [SVY] **svy**, [SVY] **svy estimation**, [SVY] **svy postestimation**, [SVY] **svy: tabulate twoway**, [SVY] **variance estimation**

Gray, R. J., [ST] **intro**, [ST] **stcrreg**

Graybill, F. A., [R] **centile**

Green, B. F., [MV] **discrim lda**, [MV] **procrustes**

Green, D. M., [R] **logistic postestimation**

Green, P. E., [MV] cluster
Greenacre, M. J., [MV] ca, [MV] mca, [MV] mca postestimation
Greenbaum, A., [M-1] LAPACK, [M-5] lapack( ), [P] matrix eigenvalues
Greene, W. H., [P] matrix accum, [R] asclogit, [R] asmprobit, [R] biprobit, [R] clogit, [R] cnsreg, [R] frontier, [R] gmm, [R] heckman, [R] heckprob, [R] hetprob, [R] ivregress, [R] logit, [R] lrtest, [R] margins, [R] mkspline, [R] mlogit, [R] nlogit, [R] nlsur, [R] pcorr, [R] probit, [R] reg3, [R] regress, [R] regress postestimation time series, [R] sureg, [R] testnl, [R] treatreg, [R] truncreg, [R] zinb, [R] zip, [TS] arch, [TS] arima, [TS] corrgram, [TS] var, [XT] xt, [XT] xtgls, [XT] xtpcse, [XT] xtpoisson, [XT] xtrc, [XT] xtreg
Greenfield, S., [MV] factor, [MV] factor postestimation, [R] alpha, [R] lincom, [R] mlogit, [R] mprobit, [R] mprobit postestimation, [R] predictnl, [R] slogit
Greenhouse, J. B., [ST] epitab
Greenhouse, S. W., [R] anova, [ST] epitab
Greenland, S., [R] ci, [R] glogit, [R] ologit, [R] poisson, [ST] epitab
Greenwood, M., [ST] ltable, [ST] sts
Greenwood, P., [MI] intro substantive
Gregoire, A., [R] kappa
Griffith, J. L., [R] brier
Griffith, R., [R] gmm
Griffiths, W. E., [R] cnsreg, [R] estat, [R] glogit, [R] ivregress, [R] ivregress postestimation, [R] logit, [R] probit, [R] regress, [R] regress postestimation, [R] test, [TS] arch, [TS] prais, [XT] xtgls, [XT] xtpcse, [XT] xtrc, [XT] xtreg
Griliches, Z., [XT] xtgls, [XT] xtnbreg, [XT] xtpcse, [XT] xtpoisson, [XT] xtrc
Grizzle, J. E., [R] vwls
Groenen, P. J. F., [MV] mds, [MV] mds postestimation, [MV] mdslong, [MV] mdsmat
Grogger, J. T., [R] ztnb, [R] ztp
Gronau, R., [R] heckman
Gropper, D. M., [R] frontier, [XT] xtfrontier
Gross, A. J., [ST] ltable
Grunfeld, Y., [XT] xtgls, [XT] xtpcse, [XT] xtrc
Guan, W., [R] bootstrap
Guerry, A.-M., [G] graph twoway histogram
Guilkey, D. K., [XT] xtprobit
Guillemot, M., [M-5] cholesky( )
Guimarães, P., [XT] xtnbreg
Gutierrez, R. G., [R] frontier, [R] lpoly, [R] lrtest, [R] nbreg, [R] ztnb, [ST] stcox, [ST] stcrreg, [ST] stdescribe, [ST] streg, [ST] stset, [ST] stsplit, [ST] stvary, [ST] survival analysis, [XT] xt, [XT] xtmelogit, [XT] xtmepoisson, [XT] xtmixed

H

Haaland, J.-A., [G] graph intro
Haan, P., [R] asmprobit, [R] mlogit, [R] mprobit
Hadamard, J. S., [D] functions
Hadi, A. S., [R] poisson, [R] regress, [R] regress postestimation, [TS] prais
Hadorn, D. C., [R] brier
Hadri, K., [XT] xtunitroot
Haenszel, W., [ST] epitab, [ST] sts test
Hahn, J., [R] ivregress postestimation
Hair Jr., J. F., [R] rologit
Hajivassiliou, V. A., [R] asmprobit
Hakkio, C. S., [D] egen
Hald, A., [R] qreg, [R] regress, [R] signrank, [R] summarize
Haldane, J. B. S., [ST] epitab
Hall, A. D., [R] frontier
Hall, A. R., [R] gmm, [R] gmm postestimation, [R] ivregress, [R] ivregress postestimation
Hall, B. H., [M-5] optimize( ), [R] glm, [TS] arch, [TS] arima, [XT] xtnbreg, [XT] xtpoisson
Hall, P., [R] bootstrap, [R] regress postestimation time series
Hall, R. E., [M-5] optimize( ), [R] glm, [TS] arch, [TS] arima
Hall, W. J., [MV] biplot, [R] roc, [R] rocfit
Halley, E., [ST] ltable
Hallock, K., [R] qreg
Halvorsen, K. T., [R] tabulate twoway
Hamann, U., [MV] measure_option
Hamer, R. M., [MV] mds, [MV] mdslong, [MV] mdsmat
Hamerle, A., [R] clogit
Hamilton, J. D., [P] matrix eigenvalues, [R] gmm, [TS] arch, [TS] arima, [TS] corrgram, [TS] dfuller, [TS] fcast compute, [TS] Glossary, [TS] irf, [TS] irf create, [TS] pergram, [TS] pperron, [TS] sspace, [TS] sspace postestimation, [TS] time series, [TS] var, [TS] var intro, [TS] var svar, [TS] vargranger, [TS] varnorm, [TS] varsoc, [TS] varstable, [TS] varwle, [TS] vec, [TS] vec intro, [TS] vecnorm, [TS] vecrank, [TS] vecstable, [TS] xcorr
Hamilton, L. C., [D] xpose, [G] graph intro, [MV] factor, [MV] screeplot, [R] bootstrap, [R] diagnostic plots, [R] ladder, [R] lv, [R] mlogit, [R] regress, [R] regress postestimation, [R] rreg, [R] simulate, [R] summarize, [R] ttest
Hammarling, S., [M-1] LAPACK, [M-5] lapack( ), [P] matrix eigenvalues
Hampel, F. R., [D] egen, [R] rreg, [U] 20.20 References
Hand, D. J., [MV] biplot, [MV] ca, [MV] discrim, [MV] mca
Hankey, B., [ST] strate
Hanley, J. A., [R] roc, [R] rocfit

Hannachi, A., [MV] **pca**

Hannan, E. J., [TS] **sspace**

Hansen, H., [MV] **mvtest**, [MV] **mvtest normality**

Hansen, L. P., [R] **gmm**, [R] **ivregress**, [R] **ivregress postestimation**, [XT] **xtabond**, [XT] **xtdpd**, [XT] **xtdpdsys**

Hao, L., [R] **qreg**

Harabasz, J., [MV] **cluster**, [MV] **cluster stop**

Hardin, J. W., [D] **statsby**, [G] **graph intro**, [R] **binreg**, [R] **biprobit**, [R] **estat**, [R] **glm**, [R] **glm postestimation**, [R] **regress postestimation**, [TS] **newey**, [TS] **prais**, [XT] **xtgee**, [XT] **xtmelogit postestimation**, [XT] **xtmepoisson postestimation**, [XT] **xtpoisson**

Haritou, A., [R] **suest**

Harkness, J., [R] **ivprobit**, [R] **ivtobit**

Harley, J. B., [ST] **stpower cox**

Harman, H. H., [MV] **factor**, [MV] **factor postestimation**, [MV] **rotate**, [MV] **rotatemat**

Harrell Jr., F. E., [R] **mkspline**, [R] **ologit**, [ST] **stcox postestimation**

Harrington, D. P., [ST] **stcox**, [ST] **sts test**

Harris, E. K., [MV] **discrim**, [MV] **discrim logistic**

Harris, R. D. F., [XT] **xtunitroot**

Harris, R. J., [MV] **canon postestimation**

Harris, R. L., [R] **qc**

Harris, T., [R] **qreg**

Harrison, D. A., [D] **list**, [G] **graph twoway histogram**, [R] **histogram**, [R] **tabulate oneway**, [R] **tabulate twoway**

Harrison, J. A., [R] **dstdize**

Harrison, J. M., [ST] **stcrreg**

Hart, P. E., [MV] **cluster**, [MV] **cluster stop**

Hartigan, J. A., [G] **graph matrix**

Hartley, H. O., [MI] **intro substantive**, [MI] **mi impute**

Harvey, A. C., [R] **hetprob**, [TS] **arch**, [TS] **arima**, [TS] **prais**, [TS] **sspace**, [TS] **sspace postestimation**, [TS] **var svar**

Harville, D. A., [XT] **xtmixed**

Hassell, J., [ST] **sts**

Hastie, T. J., [MV] **discrim knn**, [R] **grmeanby**, [R] **slogit**

Hastorf, A. H., [ST] **epitab**

Hauck, W. W., [R] **pkequiv**, [XT] **xtcloglog**, [XT] **xtlogit**, [XT] **xtprobit**

Hausman, J. A., [M-5] **optimize( )**, [R] **glm**, [R] **hausman**, [R] **ivregress postestimation**, [R] **nlogit**, [R] **rologit**, [R] **suest**, [TS] **arch**, [TS] **arima**, [XT] **xthtaylor**, [XT] **xtnbreg**, [XT] **xtpoisson**, [XT] **xtreg postestimation**

Hayashi, F., [R] **gmm**, [R] **ivregress**, [R] **ivregress postestimation**

Hayes, R. J., [R] **permute**

Haynam, G. E., [D] **functions**

Hays, R. D., [R] **lincom**, [R] **mlogit**, [R] **mprobit**, [R] **mprobit postestimation**, [R] **predictnl**, [R] **slogit**

Heagerty, P. J., [MV] **factor**, [MV] **pca**, [R] **anova**, [R] **dstdize**, [R] **oneway**, [ST] **epitab**, [XT] **xtmixed**

Heckman, J., [R] **biprobit**, [R] **heckman**, [R] **heckman postestimation**, [R] **heckprob**

Hedges, L. V., [R] **meta**

Hedley, D., [ST] **stcrreg**, [ST] **stcrreg postestimation**

Heeringa, S. G., [SVY] **subpopulation estimation**

Heinecke, K., [P] **matrix mkmat**

Heinonen, O. P., [ST] **epitab**

Heiss, F., [R] **nlogit**

Henderson, B. E., [R] **symmetry**

Henderson, C. R., [XT] **xtmixed**

Hendrickson, A. E., [MV] **Glossary**, [MV] **rotate**, [MV] **rotatemat**

Hendrickx, J., [R] **mlogit**, [R] **xi**

Henry-Amar, M., [ST] **ltable**

Hensher, D. A., [R] **nlogit**

Henze, N., [MV] **mvtest**, [MV] **mvtest normality**

Hermite, C., [M-5] **issymmetric( )**

Herriot, J. G., [M-5] **spline3( )**

Hertz, S., [ST] **stsplit**

Herzberg, A. M., [MV] **discrim lda postestimation**, [MV] **discrim qda**, [MV] **discrim qda postestimation**, [MV] **manova**

Hess, K. R., [ST] **stcox PH-assumption tests**, [ST] **sts graph**

Heston, A., [XT] **xtunitroot**

Heyde, C. C., [U] **1.4 References**

Hickam, D. H., [R] **brier**

Higbee, K. T., [D] **clonevar**

Higgins, J. E., [R] **anova**

Higgins, J. P. T., [R] **meta**

Higgins, M. L., [TS] **arch**

Hilbe, J. M., [D] **functions**, [MV] **discrim lda**, [MV] **manova**, [MV] *measure_option*, [R] **cloglog**, [R] **estat**, [R] **glm**, [R] **glm postestimation**, [R] **logistic**, [R] **logit**, [R] **nbreg**, [R] **poisson**, [R] **probit**, [R] **sampsi**, [R] **ztnb**, [R] **ztp**, [XT] **xtgee**, [XT] **xtmelogit postestimation**, [XT] **xtmepoisson postestimation**, [XT] **xtpoisson**

Hilbert, D., [M-5] **Hilbert( )**

Hildreth, C., [TS] **prais**

Hilferty, M. M., [MV] **mvtest normality**

Hilgard, E. R., [ST] **epitab**

Hill, A. B., [R] **poisson**, [ST] **epitab**

Hill, R. C., [R] **cnsreg**, [R] **estat**, [R] **glogit**, [R] **heckman**, [R] **ivregress**, [R] **ivregress postestimation**, [R] **logit**, [R] **probit**, [R] **regress**, [R] **regress postestimation**, [R] **test**, [TS] **arch**, [TS] **prais**, [XT] **xtgls**, [XT] **xtpcse**, [XT] **xtrc**, [XT] **xtreg**

Hill, R. P., [ST] **stcrreg**, [ST] **stcrreg postestimation**

Hill, W. G., [ST] **epitab**

Hills, M., [D] **egen**, [R] **cloglog**, [R] **cumul**, [ST] **epitab**, [ST] **stcox**, [ST] **stptime**, [ST] **strate**, [ST] **stset**, [ST] **stsplit**, [ST] **sttocc**

Hinkley, D. V., [R] **bootstrap**

Hipel, K. W., [TS] **arima**

Hirji, K. F., [R] **exlogistic**, [R] **expoisson**

Hlouskova, J., [XT] **xtunitroot**

Hoaglin, D. C., [R] **diagnostic plots**, [R] **lv**, [R] **regress postestimation**, [R] **smooth**, [R] **stem**

Hochberg, Y., [R] **oneway**

Hocking, R. R., [MI] **intro substantive**, [R] **stepwise**, [XT] **xtmixed**

Hodges, J. L., [MV] **discrim knn**

Hoechle, D., [XT] **xtgls**, [XT] **xtpcse**, [XT] **xtreg**, [XT] **xtregar**

Hoel, P. G., [R] **bitest**, [R] **ttest**

Hoffmann, J. P., [R] **glm**

Hofman, A. F., [ST] **stcrreg**

Hogben, L., [ST] **sts**

Hole, A. R., [R] **asmprobit**, [R] **clogit**, [R] **mlogit**, [R] **mprobit**

Holloway, L., [R] **brier**

Holm, S., [R] **test**

Holmes, S., [R] **bootstrap**

Holmgren, J., [ST] **epitab**

Holt, C. C., [TS] **tssmooth**, [TS] **tssmooth dexponential**, [TS] **tssmooth exponential**, [TS] **tssmooth hwinters**, [TS] **tssmooth shwinters**

Holt, D., [SVY] **estat**, [SVY] **survey**

Holtz-Eakin, D., [XT] **xtabond**, [XT] **xtdpd**, [XT] **xtdpdsys**

Honoré, B. E., [XT] **xttobit**

Hood, W. C., [R] **ivregress**

Hooker, P. F., [ST] **streg**

Horst, P., [MV] **factor postestimation**, [MV] **rotate**, [MV] **rotatemat**

Horton, N. J., [MI] **intro substantive**

Horváth, L., [TS] **dvech**

Hosmer Jr., D. W., [R] **clogit**, [R] **clogit postestimation**, [R] **glm**, [R] **glogit**, [R] **lincom**, [R] **logistic**, [R] **logistic postestimation**, [R] **logit**, [R] **logit postestimation**, [R] **lrtest**, [R] **mlogit**, [R] **predictnl**, [R] **stepwise**, [ST] **stcox**, [ST] **stpower**, [ST] **stpower cox**, [ST] **streg**, [XT] **xtgee**

Hossain, K. M., [ST] **epitab**

Hotelling, H., [MV] **canon**, [MV] **hotelling**, [MV] **manova**, [MV] **pca**, [R] **roc**, [R] **rocfit**

Hougaard, P., [ST] **streg**

Householder, A. S., [M-5] **qrd( )**, [MV] **mds**, [MV] **mdslong**, [MV] **mdsmat**

Hoşten, S., [MV] **mvtest means**

Hsiao, C., [XT] **xt**, [XT] **xtabond**, [XT] **xtdpd**, [XT] **xtdpdsys**, [XT] **xtivreg**, [XT] **xtregar**

Hsieh, F. Y., [ST] **stpower**, [ST] **stpower cox**, [ST] **stpower logrank**

Hu, M., [ST] **stcox**, [ST] **stset**

Huang, C., [R] **sunflower**

Huang, D. S., [R] **nlsur**, [R] **sureg**

Hubálek, Z., [MV] *measure_option*

Huber, P. J., [D] **egen**, [P] **_robust**, [R] **qreg**, [R] **rreg**, [R] **suest**, [U] **20.20 References**

Huberty, C. J., [MV] **candisc**, [MV] **discrim**, [MV] **discrim estat**, [MV] **discrim lda**, [MV] **discrim lda postestimation**, [MV] **discrim qda**

Hubrich, K., [TS] **vec intro**, [TS] **vecrank**

Hughes, J. B., [MV] **manova**

Hunter, D. R., [R] **qreg**

Huq, M. I., [ST] **epitab**

Huq, N. M., [XT] **xtmelogit**

Hurd, M., [R] **intreg**, [R] **tobit**

Hurley, J. R., [MV] **procrustes**

Hutto, C., [R] **exlogistic**

Huynh, H., [R] **anova**

I

Iglewicz, B., [R] **lv**

Ikebe, Y., [P] **matrix symeigen**

Im, K. S., [XT] **xtunitroot**

Isaacs, D., [R] **fracpoly**

Ishiguro, M., [R] **BIC note**

ISSP, [MV] **ca**, [MV] **mca**, [MV] **mca postestimation**

J

Jaccard, P., [MV] *measure_option*

Jackman, R. W., [R] **regress postestimation**

Jackson, J. E., [MV] **pca**, [MV] **pca postestimation**

Jacobs, K. B., [R] **symmetry**

Jacobs, M., [D] **duplicates**

Jacoby, W. G., [MV] **biplot**

Jaeger, D. A., [R] **ivregress postestimation**

Jagannathan, R., [TS] **arch**

Jain, A. K., [MV] **cluster**

James, G. S., [MV] **mvtest**, [MV] **mvtest means**

James, I., [M-2] **op_kronecker**, [M-5] **issymmetric( )**, [M-5] **pinv( )**

Janes, H., [R] **rocfit**

Jang, D. S., [SVY] **variance estimation**

Jann, B., [P] **mark**, [R] **estimates store**, [R] **ksmirnov**, [R] **saved results**, [R] **tabulate twoway**, [SVY] **svy: tabulate twoway**

Jarque, C. M., [R] **sktest**, [TS] **varnorm**, [TS] **vecnorm**

Jeffreys, H., [R] **ci**, [R] **spearman**

Jenkins, B., [M-5] **hash1( )**

Jenkins, G. M., [TS] **arima**, [TS] **corrgram**, [TS] **cumsp**, [TS] **dfuller**, [TS] **pergram**, [TS] **pperron**, [TS] **xcorr**

Jenkins, S. P., [D] **corr2data**, [D] **egen**, [D] **rename**, [R] **asmprobit**, [R] **do**, [R] **inequality**, [ST] **discrete**, [ST] **stcox**

Jennrich, R. I., [MV] **Glossary**, [MV] **mvtest**, [MV] **mvtest correlations**, [MV] **rotate**, [MV] **rotatemat**

Jensen, A. R., [MV] **rotate**

Jensen, D. R., [MV] **mvtest**, [MV] **mvtest means**

Jerez, M., [TS] **sspace**

Jewell, N. P., [ST] **epitab**

Jick, H., [ST] **epitab**

Joe, H., [R] **tabulate twoway**, [XT] **xtmelogit**, [XT] **xtmepoisson**

Johansen, S., [TS] **irf create**, [TS] **varlmar**, [TS] **vec**, [TS] **vec intro**, [TS] **veclmar**, [TS] **vecnorm**, [TS] **vecrank**, [TS] **vecstable**

Johnson, D. E., [MV] **manova**, [R] **anova**

Johnson, L. A., [TS] **tssmooth**, [TS] **tssmooth dexponential**, [TS] **tssmooth exponential**, [TS] **tssmooth hwinters**, [TS] **tssmooth shwinters**

Johnson, M. E., [R] **sdtest**

Johnson, M. M., [R] **sdtest**

Johnson, N. L., [D] **functions**, [R] **ksmirnov**, [R] **nbreg**, [R] **poisson**, [U] **1.4 References**

Johnson, R. A., [MV] **canon**, [MV] **discrim**, [MV] **discrim estat**, [MV] **discrim lda**, [MV] **discrim lda postestimation**, [MV] **mvtest**, [MV] **mvtest correlations**, [MV] **mvtest covariances**, [MV] **mvtest means**

Johnson, W., [MI] **intro substantive**, [SVY] **survey**

Johnston, J., [XT] **xtrc**

Jolliffe, D., [R] **inequality**, [R] **qreg**, [R] **regress**

Jolliffe, I. T., [MV] **biplot**, [MV] **pca**, [R] **brier**

Jones, A., [R] **heckman**, [R] **logit**, [R] **probit**

Jones, B. S., [ST] **stcox**, [ST] **streg**

Jones, D. R., [R] **meta**

Jones, M. C., [R] **kdensity**, [R] **lpoly**

Jones, P. S., [M-5] **Vandermonde( )**

Jöreskog, K. G., [MV] **factor postestimation**

Jorgensen, R. A., [ST] **stcrreg**

Jorner, U., [G] **graph intro**

Judge, G. G., [R] **estat**, [R] **glogit**, [R] **ivregress**, [R] **ivregress postestimation**, [R] **logit**, [R] **probit**, [R] **regress postestimation**, [R] **test**, [TS] **arch**, [TS] **prais**, [XT] **xtgls**, [XT] **xtpcse**, [XT] **xtrc**, [XT] **xtreg**

Judkins, D. R., [SVY] **svy brr**, [SVY] **variance estimation**

Judson, D. H., [R] **poisson**, [R] **tabulate twoway**, [R] **ztp**

Jung, B. C., [XT] **xtmixed**

Juul, S., [R] **dstdize**, [R] **roc**

K

Kachitvichyanukul, V., [D] **functions**

Kadane, J. B., [XT] **xtmelogit**, [XT] **xtmepoisson**

Kahn, H. A., [R] **dstdize**, [ST] **epitab**, [ST] **ltable**, [ST] **stcox**

Kaiser, H. F., [MV] **factor postestimation**, [MV] **Glossary**, [MV] **pca postestimation**, [MV] **rotate**, [MV] **rotatemat**

Kaiser, J., [R] **permute**, [R] **signrank**

Kalbfleisch, J. D., [ST] **ltable**, [ST] **stcox**, [ST] **stcox PH-assumption tests**, [ST] **stcox postestimation**, [ST] **streg**, [ST] **sts**, [ST] **sts test**, [ST] **stset**, [XT] **xtcloglog**, [XT] **xtlogit**, [XT] **xtprobit**

Kalman, R. E., [TS] **arima**

Kalmijn, M., [R] **tetrachoric**

Kantor, D., [D] **cf**, [D] **functions**

Kaplan, E. L., [ST] **stcrreg**, [ST] **stcrreg postestimation**, [ST] **sts**

Katz, J. N., [XT] **xtgls**, [XT] **xtpcse**

Kaufman, L., [MV] **cluster**, [MV] **clustermat**, [MV] **matrix dissimilarity**, [MV] *measure_option*, [P] **matrix dissimilarity**

Keane, M. P., [R] **asmprobit**

Keeler, E. B., [R] **brier**

Keiding, N., [ST] **stcrreg**, [ST] **stsplit**

Kemp, A. W., [D] **functions**, [R] **nbreg**, [R] **poisson**

Kemp, C. D., [D] **functions**

Kempthorne, P. J., [R] **regress postestimation**

Kendall, D. G., [MV] **mds**

Kendall, M. G., [MV] *measure_option*, [R] **centile**, [R] **spearman**, [R] **tabulate twoway**, [R] **tobit**

Kennedy Jr., W. J., [P] **_robust**, [R] **anova**, [R] **nl**, [R] **regress**, [R] **stepwise**, [SVY] **svy: tabulate twoway**

Kent, J. T., [MI] **mi impute mvn**, [MV] **discrim**, [MV] **discrim lda**, [MV] **factor**, [MV] **manova**, [MV] **matrix dissimilarity**, [MV] **mds**, [MV] **mds postestimation**, [MV] **mdslong**, [MV] **mdsmat**, [MV] **mvtest**, [MV] **mvtest means**, [MV] **mvtest normality**, [MV] **pca**, [MV] **procrustes**, [P] **_robust**, [P] **matrix dissimilarity**, [U] **20.20 References**

Kenward, M. G., [MI] **intro substantive**

Kettenring, J. R., [R] **diagnostic plots**

Keynes, J. M., [R] **ameans**

Khan, M. R., [ST] **epitab**

Khanti-Akom, S., [XT] **xthtaylor**

Khare, M., [MI] **intro substantive**

Kiernan, M., [R] **kappa**

Kim, I.-M., [TS] **vec**, [TS] **vec intro**, [TS] **vecrank**

Kim, J. O., [MV] **factor**

Kimber, A. C., [ST] **streg**

Kimbrough, J. W., [MV] **discrim knn**

Kinderman, A. J., [D] **functions**

King, M. L., [TS] **prais**

King, R. G., [TS] **vecrank**

Kirkwood, B. R., [R] **dstdize**, [R] **summarize**

Kish, L., [P] **_robust**, [R] **loneway**, [SVY] **estat**, [SVY] **survey**, [SVY] **variance estimation**, [U] **20.20 References**

Kitagawa, G., [R] **BIC note**

Kiviet, J. F., [XT] **xtabond**

Klar, J., [R] **logistic postestimation**

Klecka, W. R., [MV] **discrim**, [MV] **discrim lda**

Kleiber, C., [R] **inequality**

Klein, J. P., [ST] **stci**, [ST] **stcox postestimation**, [ST] **stcox**, [ST] **stcrreg**, [ST] **stpower**, [ST] **stpower cox**, [ST] **streg**, [ST] **sts**, [ST] **sts graph**, [ST] **sts test**

Klein, L. R., [R] **reg3**, [R] **regress postestimation time series**

Klein, M., [R] **binreg**, [R] **clogit**, [R] **logistic**, [R] **lrtest**, [R] **mlogit**, [R] **ologit**, [XT] **xtgee**

Kleinbaum, D. G., [R] **binreg**, [R] **clogit**, [R] **logistic**, [R] **lrtest**, [R] **mlogit**, [R] **ologit**, [ST] **epitab**, [XT] **xtgee**

Kleiner, B., [G] *by_option*, [G] **graph box**, [G] **graph matrix**, [R] **diagnostic plots**, [R] **lowess**, [U] **1.4 References**

Kleinman, K. P., [MI] **intro substantive**

Klema, V. C., [P] **matrix symeigen**

Kmenta, J., [R] **eivreg**, [R] **ivregress**, [R] **regress**, [TS] **arch**, [TS] **prais**, [TS] **rolling**, [XT] **xtpcse**

Knook, D. L., [MI] **intro substantive**, [MI] **mi impute monotone**

Knuth, D., [D] **functions**

Koch, G. G., [R] **anova**, [R] **kappa**, [R] **vwls**, [SVY] **svy: tabulate twoway**

Koehler, K. J., [R] **diagnostic plots**

Koenker, R., [R] **qreg**, [R] **regress postestimation**

Kohler, U., [D] **input**, [G] **graph twoway rbar**, [MV] **biplot**, [R] **kdensity**, [R] **regress**, [R] **regress postestimation**

Kohn, R., [TS] **arima**

Kolev, G. I., [P] **scalar**, [U] **11.7 References**

Kolmogorov, A. N., [R] **ksmirnov**

Koopmans, T. C., [R] **ivregress**

Korin, B. P., [MV] **mvtest**

Korn, E. L., [R] **margins**, [R] **ml**, [R] **test**, [SVY] **direct standardization**, [SVY] **survey**, [SVY] **svy**, [SVY] **svy estimation**, [SVY] **svy postestimation**, [SVY] **svy: tabulate twoway**, [SVY] **variance estimation**

Kotz, S., [D] **functions**, [R] **inequality**, [R] **ksmirnov**, [R] **nbreg**, [R] **nlogit**, [R] **poisson**, [U] **1.4 References**

Krakauer, H., [ST] **ltable**

Krall, J. M., [ST] **stpower cox**

Kramer, C. Y., [MV] **mvtest**, [MV] **mvtest means**

Krauss, N., [SVY] **estat**, [SVY] **subpopulation estimation**, [SVY] **svy estimation**

Kreidberg, M. B., [ST] **epitab**

Kreuter, F., [R] **kdensity**, [R] **regress**, [R] **regress postestimation**, [SVY] **survey**

Krishnaiah, P. R., [MV] **mvtest**

Krishnamoorthy, K., [MV] **mvtest**, [MV] **mvtest means**

Kroeber, A. L., [MV] *measure_option*

Kronecker, L., [M-2] **op_kronecker**

Kroner, K. F., [TS] **arch**

Krus, D. J., [MV] **canon postestimation**

Krushelnytskyy, B., [R] **inequality**, [R] **qreg**

Kruskal, J. B., [MV] **Glossary**, [MV] **mds**, [MV] **mds postestimation**, [MV] **mdslong**, [MV] **mdsmat**

Kruskal, W. H., [R] **kwallis**, [R] **ranksum**, [R] **spearman**, [R] **tabulate twoway**

Kshirsagar, A. M., [MV] **discrim lda**, [MV] **pca**

Kuehl, R. O., [R] **anova**, [R] **oneway**

Kuh, E., [R] **estat**, [R] **regress postestimation**, [U] **18.13 References**

Kulczynski, S., [MV] *measure_option*

Kumbhakar, S. C., [R] **frontier**, [XT] **xtfrontier**

Kung, D. S., [R] **qreg**

Kupper, L. L., [ST] **epitab**

Kutner, M. H., [R] **pkcross**, [R] **pkequiv**, [R] **pkshape**, [R] **regress postestimation**

Kwiatkowski, D., [XT] **xtunitroot**

L

Lachenbruch, P. A., [MV] **discrim estat**, [MV] **discrim lda**, [MV] **discrim**, [R] **diagnostic plots**

Lachin, J. M., [ST] **stpower**, [ST] **stpower cox**, [ST] **stpower exponential**, [U] **1.4 References**

Lacy, M. G., [R] **permute**

Lafontaine, F., [R] **boxcox**

Lahiri, K., [R] **tobit**, [XT] **xtgls**

Lai, K. S., [TS] **dfgls**

Lai, S., [R] **exlogistic**

Laird, N. M., [MI] **intro substantive**, [MI] **mi impute mvn**, [R] **expoisson**, [XT] **xtmelogit**, [XT] **xtmepoisson**, [XT] **xtmixed**

Lakatos, E., [ST] **stpower**, [ST] **stpower exponential**, [ST] **stpower logrank**

Lal, R., [D] **functions**

Lambert, D., [R] **zip**

Lambert, P. C., [ST] **stcrreg**

LaMotte, L. R., [XT] **xtmixed**

Lan, K. K. G., [ST] **stpower**, [ST] **stpower exponential**, [ST] **stpower logrank**

Lance, G. N., [MV] **cluster**

Landau, S., [MV] **cluster**, [MV] **cluster stop**

Landis, J. R., [R] **kappa**

Lane, M. A., [SVY] **survey**, [SVY] **svy estimation**

Lane, P. W., [R] **margins**

Lange, K., [R] **qreg**

Lange, S. M., [ST] **stcrreg**

Langford, I. H., [XT] **xtmepoisson**

Langholz, B., [ST] **sttocc**

Larsen, W. A., [R] **regress postestimation**

Lash, T. L., [R] **ci**, [R] **glogit**, [R] **poisson**, [ST] **epitab**

Latouche, A., [ST] **stcrreg**

Laurent, S., [TS] **dvech**

Lauritsen, J. M., [D] **labelbook**, [D] **list**

Lauritzen, S. L., [R] **summarize**

Lavori, P. W., [ST] **stpower**, [ST] **stpower cox**

Lawless, J. F., [ST] **ltable**, [ST] **stpower**

Lawley, D. N., [MV] **canon**, [MV] **factor**, [MV] **factor postestimation**, [MV] **manova**, [MV] **mvtest**, [MV] **mvtest correlations**, [MV] **pca**

Layard, R., [XT] **xtabond**, [XT] **xtdpd**, [XT] **xtdpdsys**

Ledolter, J., [TS] **tssmooth**, [TS] **tssmooth dexponential**, [TS] **tssmooth exponential**, [TS] **tssmooth hwinters**, [TS] **tssmooth shwinters**

Lee, E. S., [R] **dstdize**

Lee, E. T., [R] **roc**, [R] **rocfit**, [ST] **streg**

Lee, J. C., [MV] **mvtest**

Lee, J. W., [XT] **xtmixed**

Lee, K. L., [ST] **stcox postestimation**

Lee, L. F., [XT] **xtreg**

Lee, P., [ST] **streg**

Lee, T.-C., [R] **estat**, [R] **glogit**, [R] **ivregress**, [R] **ivregress postestimation**, [R] **logit**, [R] **probit**, [R] **regress postestimation**, [R] **test**, [TS] **arch**, [TS] **prais**, [XT] **xtgls**, [XT] **xtpcse**, [XT] **xtrc**, [XT] **xtreg**

Lee, W. C., [R] **roc**

Leese, M., [MV] **cluster**, [MV] **cluster stop**

Legendre, A. M., [R] **regress**

Lehmann, E. L., [R] **oneway**

Leisenring, W., [ST] **stcrreg**

Lemeshow, S., [R] **clogit**, [R] **clogit postestimation**, [R] **glm**, [R] **glogit**, [R] **lincom**, [R] **logistic**, [R] **logistic postestimation**, [R] **logit**, [R] **logit postestimation**, [R] **lrtest**, [R] **mlogit**, [R] **predictnl**, [R] **stepwise**, [ST] **stcox**, [ST] **stpower**, [ST] **stpower cox**, [ST] **streg**, [SVY] **poststratification**, [SVY] **survey**, [XT] **xtgee**

Leonard, M., [XT] **xtgee**

Leone, F. C., [D] **functions**

Lepkowski, J. M., [MI] **intro substantive**, [MI] **mi impute**, [MI] **mi impute logit**, [MI] **mi impute mlogit**, [MI] **mi impute monotone**, [MI] **mi impute ologit**

Leroy, A. M., [R] **qreg**, [R] **regress postestimation**, [R] **rreg**

Lesaffre, E., [MV] **discrim logistic**

LeSage, G., [ST] **stcrreg**

Levene, H., [R] **sdtest**

Levin, A., [XT] **xtunitroot**

Levin, B., [R] **dstdize**, [R] **kappa**, [R] **sampsi**, [ST] **epitab**

Levin, W., [ST] **stcrreg**, [ST] **stcrreg postestimation**

Levinsohn, J., [R] **frontier**

Levy, D., [R] **sunflower**

Levy, P. S., [SVY] **poststratification**, [SVY] **survey**

Lewis, H. G., [R] **heckman**

Lewis, I. G., [R] **binreg**

Lewis, J. D., [R] **fracpoly**

Lexis, W. H., [ST] **stsplit**

Leyland, A. H., [XT] **xtmelogit**, [XT] **xtmepoisson**

Li, G., [R] **rreg**

Li, K.-H., [MI] **intro substantive**, [MI] **mi estimate**, [MI] **mi estimate postestimation**, [MI] **mi estimate using**, [MI] **mi impute mvn**

Li, N., [MI] **intro substantive**

Li, Q., [XT] **xtivreg**, [XT] **xtreg postestimation**, [XT] **xtregar**

Li, W., [R] **pkcross**, [R] **pkequiv**, [R] **pkshape**

Liang, K.-Y., [XT] **xtcloglog**, [XT] **xtgee**, [XT] **xtlogit**, [XT] **xtmelogit**, [XT] **xtmepoisson**, [XT] **xtmixed**, [XT] **xtnbreg**, [XT] **xtpoisson**, [XT] **xtprobit**

Lieberman, O., [TS] **dvech**

Likert, R. A., [R] **alpha**

Lilien, D. M., [TS] **arch**

Lilienfeld, D. E., [ST] **epitab**

Lim, G. C., [R] **cnsreg**, [R] **regress**, [R] **regress postestimation**, [TS] **arch**

Lin, C.-F., [XT] **xtunitroot**

Lin, D. Y., [P] **_robust**, [ST] **stcox**, [ST] **stcrreg**, [SVY] **svy estimation**, [U] **20.20 References**

Lin, X., [XT] **xtmelogit**, [XT] **xtmepoisson**

Lincoff, G. H., [MV] **discrim knn**

Linde-Zwirble, W., [D] **functions**

Lindelow, M., [SVY] **svy estimation**, [SVY] **svyset**

Lindley, D. V., [R] **ci**

Lindor, K. D., [ST] **stcrreg**

Lindstrom, M. J., [XT] **xtcloglog**, [XT] **xtgee**, [XT] **xtintreg**, [XT] **xtlogit**, [XT] **xtprobit**, [XT] **xttobit**

Lingoes, J. C., [MV] **mds**, [MV] **mdslong**, [MV] **mdsmat**

Linhart, J. M., [D] **format**, [M-5] **mindouble( )**, [R] **lpoly**, [ST] **sts**, [U] **13.12 References**

Lipset, S. M., [R] **histogram**

Lipsitz, S. R., [MI] **intro substantive**

Littell, R. C., [XT] **xtmelogit**

Little, R. J. A., [MI] **intro substantive**, [MI] **mi impute mvn**, [MI] **mi impute pmm**

Liu, J.-P., [R] **pk**, [R] **pkcross**, [R] **pkequiv**, [R] **pkexamine**, [R] **pkshape**, [ST] **stpower**

Liu, Q., [XT] **xtcloglog**, [XT] **xtintreg**, [XT] **xtlogit**, [XT] **xtmelogit**, [XT] **xtmepoisson**, [XT] **xtpoisson**, [XT] **xtprobit**, [XT] **xttobit**

Ljung, G. M., [TS] **wntestq**

Locke, C. S., [R] **pkequiv**

Loftsgaarden, D. O., [MV] **discrim knn**

Lokshin, M., [R] **heckman**, [R] **heckprob**, [R] **oprobit**

Long, J. S., [D] **codebook**, [D] **label**, [D] **notes**, [R] **asroprobit**, [R] **clogit**, [R] **cloglog**, [R] **intreg**, [R] **logistic**, [R] **logit**, [R] **mlogit**, [R] **mprobit**, [R] **nbreg**, [R] **ologit**, [R] **oprobit**, [R] **poisson**, [R] **probit**, [R] **regress**, [R] **regress postestimation**, [R] **testnl**, [R] **tobit**, [R] **zinb**, [R] **zip**, [R] **ztnb**, [R] **ztp**, [U] **12.10 References**, [U] **16.5 References**, [U] **20.20 References**

Longley, J. D., [R] **kappa**

Longton, G., [R] **rocfit**

López-Feldman, A., [R] **inequality**

López-Vizcaíno, M. E., [ST] **epitab**

Lorenz, M. O., [R] **inequality**

Louis, T. A., [R] **tabulate twoway**

Lovell, C. A. K., [R] **frontier**, [XT] **xtfrontier**

Lovie, A. D., [R] **spearman**

Lovie, P., [R] **spearman**

Lu, J. Y., [TS] **prais**

Lucas, H. L., [R] **pkcross**

Luce, R. D., [R] **rologit**

Luckman, B., [MV] **screeplot**

Ludwig, J., [ST] **stcrreg**

Lumley, T. S., [MV] **factor**, [MV] **pca**, [R] **anova**,
[R] **dstdize**, [R] **oneway**, [ST] **epitab**

Luniak, M., [MV] **biplot**

Lunn, M., [ST] **stcrreg**

Lunt, M., [R] **ologit**, [R] **slogit**

Lurie, M. B., [MV] **manova**

Lütkepohl, H., [M-5] **Dmatrix( )**, [M-5] **Kmatrix( )**,
[M-5] **Lmatrix( )**, [R] **estat**, [R] **glogit**,
[R] **ivregress**, [R] **ivregress postestimation**,
[R] **logit**, [R] **probit**, [R] **regress postestimation**,
[R] **test**, [TS] **arch**, [TS] **dfactor**, [TS] **dvech**,
[TS] **fcast compute**, [TS] **irf**, [TS] **irf
create**, [TS] **prais**, [TS] **sspace**, [TS] **sspace
postestimation**, [TS] **time series**, [TS] **var**,
[TS] **var intro**, [TS] **var svar**, [TS] **varbasic**,
[TS] **vargranger**, [TS] **varnorm**, [TS] **varsoc**,
[TS] **varstable**, [TS] **varwle**, [TS] **vec intro**,
[TS] **vecnorm**, [TS] **vecrank**, [TS] **vecstable**,
[XT] **xtgls**, [XT] **xtpcse**, [XT] **xtrc**, [XT] **xtreg**

M

Ma, G., [R] **roc**, [R] **rocfit**

Machin, D., [R] **ci**, [R] **kappa**, [R] **tabulate twoway**,
[ST] **stpower**, [ST] **stpower cox**, [ST] **stpower
logrank**

Mack, T. M., [R] **symmetry**

MacKinnon, J. G., [P] **_robust**, [R] **boxcox**, [R] **cnsreg**,
[R] **gmm**, [R] **intreg**, [R] **ivregress**, [R] **ivregress
postestimation**, [R] **mlogit**, [R] **nl**, [R] **nlsur**,
[R] **reg3**, [R] **regress**, [R] **regress postestimation
time series**, [R] **tobit**, [R] **truncreg**, [TS] **arch**,
[TS] **arima**, [TS] **dfuller**, [TS] **Glossary**,
[TS] **pperron**, [TS] **prais**, [TS] **sspace**,
[TS] **varlmar**, [XT] **xtgls**, [XT] **xtpcse**,
[U] **20.20 References**

MacLaren, M. D., [D] **functions**

MacMahon, B., [ST] **epitab**

MacRae, K. D., [R] **binreg**

MaCurdy, T. E., [XT] **xthtaylor**

Madans, J. H., [SVY] **survey**, [SVY] **svy estimation**

Madansky, A., [R] **runtest**

Maddala, G. S., [R] **nlogit**, [R] **tobit**, [R] **treatreg**,
[TS] **vec**, [TS] **vec intro**, [TS] **vecrank**,
[XT] **xtgls**, [XT] **xtunitroot**

Magnus, J. R., [TS] **var svar**

Mahalanobis, P. C., [MV] **discrim lda**, [MV] **Glossary**,
[MV] **hotelling**

Mair, C. S., [XT] **xtmepoisson**

Mallows, C. L., [R] **regress postestimation**

Manchul, L., [ST] **stcrreg**, [ST] **stcrreg postestimation**

Mandelbrot, B., [TS] **arch**

Mander, A., [R] **symmetry**, [ST] **stsplit**

Mangel, M., [TS] **varwle**

Manly, B. F. J., [MV] **discrim qda postestimation**

Mann, H. B., [R] **kwallis**, [R] **ranksum**

Manning, W. G., [R] **heckman**

Manski, C. F., [R] **gmm**

Mantel, N., [R] **stepwise**, [ST] **epitab**, [ST] **sts test**

Marcellino, M., [XT] **xtunitroot**

Marchenko, Y., [R] **anova**, [R] **loneway**, [R] **oneway**,
[ST] **stcox**, [ST] **stcrreg**, [ST] **stdescribe**,
[ST] **streg**, [ST] **stset**, [ST] **stsplit**, [ST] **stvary**,
[ST] **survival analysis**, [XT] **xtmelogit**,
[XT] **xtmepoisson**, [XT] **xtmixed**

Marden, J. I., [R] **rologit**

Mardia, K. V., [MI] **mi impute mvn**, [MV] **discrim**,
[MV] **discrim lda**, [MV] **factor**, [MV] **manova**,
[MV] **matrix dissimilarity**, [MV] **mds**,
[MV] **mds postestimation**, [MV] **mdslong**,
[MV] **mdsmat**, [MV] **mvtest**, [MV] **mvtest
means**, [MV] **mvtest normality**, [MV] **pca**,
[MV] **procrustes**, [P] **matrix dissimilarity**

Mark, D. B., [ST] **stcox postestimation**

Markowski, C. A., [R] **sdtest**

Markowski, E. P., [R] **sdtest**

Marquardt, D. W., [M-5] **optimize( )**

Marr, J. W., [ST] **stsplit**

Marsaglia, G., [D] **functions**

Marschak, J., [R] **ivregress**

Martin, W., [R] **regress**, [ST] **epitab**

Martínez, M. A., [R] **logistic**

Marubini, E., [ST] **stcrreg**, [ST] **stpower**, [ST] **stpower
logrank**, [ST] **sts test**

Massey, J. T., [SVY] **estat**, [SVY] **subpopulation
estimation**, [SVY] **survey**, [SVY] **svy**,
[SVY] **svy brr**, [SVY] **svy estimation**,
[SVY] **svy jackknife**, [SVY] **svy
postestimation**, [SVY] **svy: tabulate oneway**,
[SVY] **svy: tabulate twoway**, [SVY] **svydescribe**

Massey Jr., F. J., [R] **ttest**

Master, I. M., [R] **exlogistic**

Mastrucci, M. T., [R] **exlogistic**

Matthews, J. N. S., [R] **ameans**, [R] **expoisson**,
[R] **sdtest**

Maurer, K., [SVY] **estat**, [SVY] **subpopulation
estimation**, [SVY] **survey**, [SVY] **svy**,
[SVY] **svy brr**, [SVY] **svy estimation**,
[SVY] **svy jackknife**, [SVY] **svy
postestimation**, [SVY] **svy: tabulate oneway**,
[SVY] **svy: tabulate twoway**, [SVY] **svydescribe**

Maxwell, A. E., [MV] **factor**, [MV] **factor
postestimation**, [R] **symmetry**

May, S., [ST] **stcox**, [ST] **stpower**, [ST] **stpower cox**,
[ST] **streg**

Mazýa, V., [D] **functions**

McCarthy, P. J., [SVY] **survey**, [SVY] **svy brr**,
[SVY] **variance estimation**

McCleary, S. J., [R] **regress postestimation**

McCullagh, P., [R] **binreg**, [R] **binreg postestimation**,
[R] **glm**, [R] **glm postestimation**, [R] **ologit**,
[R] **rologit**, [XT] **vce_options**, [XT] **xtgee**,
[XT] **xtmelogit postestimation**,

McCullagh, P., *continued*
[XT] **xtmepoisson postestimation**,
[XT] **xtpoisson**
McCulloch, C. E., [R] **logistic**, [ST] **stcox**,
[XT] **xtmelogit**, [XT] **xtmepoisson**,
[XT] **xtmixed**
McCullough, B. D., [TS] **corrgram**
McDonald, A., [XT] **xtmepoisson**
McDonald, J. A., [R] **sunflower**
McDonald, J. F., [R] **tobit**, [R] **tobit postestimation**
McDowell, A., [SVY] **estat**, [SVY] **subpopulation estimation**, [SVY] **survey**, [SVY] **svy**,
[SVY] **svy brr**, [SVY] **svy estimation**,
[SVY] **svy jackknife**, [SVY] **svy**
postestimation, [SVY] **svy: tabulate oneway**,
[SVY] **svy: tabulate twoway**, [SVY] **svydescribe**
McDowell, A. W., [R] **sureg**, [TS] **arima**
McFadden, D. L., [R] **asclogit**, [R] **asmprobit**,
[R] **clogit**, [R] **hausman**, [R] **maximize**,
[R] **nlogit**, [R] **suest**
McGilchrist, C. A., [ST] **stcox**, [ST] **streg**
McGill, R., [R] **sunflower**
McGinnis, R. E., [R] **symmetry**
McGuire, T. J., [R] **dstdize**
McKelvey, R. D., [R] **ologit**
McKenney, A., [M-1] **LAPACK**, [M-5] **lapack( )**,
[P] **matrix eigenvalues**
McLachlan, G. J., [MV] **discrim**, [MV] **discrim**
estat, [MV] **discrim knn**, [MV] **discrim**
lda, [XT] **xtmelogit**, [XT] **xtmepoisson**,
[XT] **xtmixed**
McLeod, A. I., [TS] **arima**
McNeil, B. J., [R] **roc**, [R] **rocfit**
McNeil, D., [R] **poisson**, [ST] **stcrreg**
McNemar, Q., [ST] **epitab**
Mead, R., [M-5] **optimize( )**
Meeusen, W., [R] **frontier**, [XT] **xtfrontier**
Mehta, C. R., [R] **exlogistic**, [R] **exlogistic**
postestimation, [R] **expoisson**, [R] **tabulate**
twoway
Meier, P., [ST] **stcrreg**, [ST] **stcrreg postestimation**,
[ST] **sts**
Meiselman, D., [TS] **arima**
Mendenhall III, W., [SVY] **survey**
Meng, X.-L., [MI] **intro substantive**, [MI] **mi estimate**,
[MI] **mi estimate postestimation**, [MI] **mi**
estimate using
Mensing, R. W., [R] **anova postestimation**
Metz, C. E., [R] **logistic postestimation**
Meyer, B. D., [ST] **discrete**
Michel-Pajus, A., [M-5] **cholesky( )**
Michels, K. M., [R] **anova**, [R] **loneway**, [R] **oneway**
Michener, C. D., [MV] *measure_option*
Mickey, M. R., [MV] **discrim estat**
Midthune, D., [SVY] **svy estimation**
Mielke, P., [R] **brier**
Miettinen, O. S., [ST] **epitab**

Milan, L., [MV] **ca**, [MV] **factor**, [MV] **mca**,
[MV] **pca**
Miller, A. B., [R] **kappa**
Miller, H. W., [SVY] **survey**, [SVY] **svy estimation**
Miller, J. I., [TS] **sspace**
Miller Jr., R. G., [R] **diagnostic plots**, [R] **oneway**
Milligan, G. W., [MV] **cluster**, [MV] **cluster**
programming subroutines, [MV] **cluster stop**
Milliken, G. A., [MV] **manova**, [R] **anova**,
[R] **margins**, [XT] **xtmelogit**
Milosevic, M., [ST] **stcrreg**, [ST] **stcrreg**
postestimation
Miranda, A., [R] **heckprob**, [R] **ivprobit**, [R] **ivtobit**,
[R] **logistic**, [R] **logit**, [R] **nbreg**, [R] **ologit**,
[R] **oprobit**, [R] **poisson**, [R] **probit**
Mitchell, C., [R] **exlogistic**
Mitchell, M. N., [G] **graph intro**, [R] **logistic**,
[R] **logistic postestimation**, [R] **logit**
Moeschberger, M. L., [ST] **stci**, [ST] **stcox**, [ST] **stcox**
postestimation, [ST] **stcrreg**, [ST] **stpower**,
[ST] **stpower cox**, [ST] **streg**, [ST] **sts**, [ST] **sts**
graph, [ST] **sts test**
Moffitt, R. A., [R] **tobit**, [R] **tobit postestimation**
Molenberghs, G., [XT] **xtmixed**
Moler, C. B., [P] **matrix symeigen**
Monahan, J. F., [D] **functions**
Monfort, A., [R] **hausman**, [R] **suest**, [R] **test**,
[TS] **arima**
Monshouwer, K., [MV] **mvtest**
Monson, R. R., [R] **bitest**, [ST] **epitab**
Montgomery, D. C., [TS] **tssmooth**, [TS] **tssmooth**
dexponential, [TS] **tssmooth exponential**,
[TS] **tssmooth hwinters**, [TS] **tssmooth**
shwinters
Mood, A. M., [R] **centile**
Moon, H. R., [XT] **xtunitroot**
Mooney, C. Z., [R] **bootstrap**, [R] **jackknife**
Moore, E. H., [M-5] **pinv( )**
Moore, J. B., [TS] **sspace**
Moore, R. J., [D] **functions**
Moran, J. L., [R] **dstdize**
Morgenstern, H., [ST] **epitab**
Mori, M., [ST] **stcrreg**
Morris, C., [R] **bootstrap**
Morris, J. N., [ST] **stsplit**
Morris, N. F., [R] **binreg**
Morrison, D. F., [MV] **clustermat**, [MV] **discrim lda**,
[MV] **discrim logistic**, [MV] **discrim logistic**
postestimation, [MV] **manova**
Morrow, A., [ST] **epitab**
Mosier, C. I., [MV] **procrustes**
Moskowitz, M., [R] **kappa**
Mosteller, F., [R] **jackknife**, [R] **regress**, [R] **regress**
postestimation, [R] **rreg**
Moulton, L. H., [R] **permute**
Muellbauer, J., [R] **nlsur**
Mueller, C. W., [MV] **factor**
Mueller, R. O., [MV] **discrim lda**

Muirhead, R. J., [MV] **pca**
Mulaik, S. A., [MV] **factor**, [MV] **rotate**
Mullahy, J., [R] **gmm**, [R] **zinb**, [R] **zip**
Müller, H.-G., [R] **lpoly**, [ST] **sts graph**
Mundlak, Y., [XT] **xtregar**
Munnell, A., [XT] **xtmixed**
Murphy, A. H., [R] **brier**
Murphy, J. L., [XT] **xtprobit**
Murphy, R. S., [SVY] **survey**, [SVY] **svy estimation**
Murray, R. M., [XT] **xtmelogit**
Murray-Lyon, I. M., [R] **binreg**
Murrill, W. A., [MV] **discrim knn**
Murtaugh, P. A., [ST] **stcrreg**
Mussolino, M. E., [SVY] **survey**, [SVY] **svy estimation**
Muñoz, J., [R] **exlogistic**
Myland, J. C., [D] **functions**
Mátyás, L., [R] **gmm**

N

Nachtsheim, C. J., [R] **pkcross**, [R] **pkequiv**,
    [R] **pkshape**, [R] **regress postestimation**
Nadarajah, S., [R] **nlogit**
Nadaraya, E. A., [R] **lpoly**
Nagel, R., [MV] **discrim lda**
Nagler, J., [R] **scobit**
Naiman, D. Q., [R] **qreg**
Nannicini, T., [R] **treatreg**
Nardi, G., [ST] **epitab**
Narendranathan, W., [XT] **xtregar**
Narula, S. C., [R] **qreg**
Nash, J. C., [G] **graph box**
Nash, J. D., [D] **infile (fixed format)**, [D] **merge**
Naylor, J. C., [XT] **xtcloglog**, [XT] **xtintreg**,
    [XT] **xtlogit**, [XT] **xtmelogit**, [XT] **xtmepoisson**,
    [XT] **xtpoisson**, [XT] **xtprobit**, [XT] **xttobit**
Nee, J. C. M., [R] **kappa**
Neff, R. K., [ST] **epitab**
Neimann, H., [MV] **mdsmat**
Nel, D. G., [MV] **mvtest**, [MV] **mvtest means**
Nelder, J. A., [M-5] **optimize( )**, [R] **binreg**, [R] **binreg
    postestimation**, [R] **glm**, [R] **glm postestimation**,
    [R] **margins**, [R] **ologit**, [XT] *vce_options*,
    [XT] **xtgee**, [XT] **xtmelogit postestimation**,
    [XT] **xtmepoisson postestimation**,
    [XT] **xtpoisson**
Nelson, C. R., [R] **ivregress postestimation**
Nelson, D. B., [TS] **arch**, [TS] **arima**, [TS] **dvech**
Nelson, E. C., [MV] **factor**, [MV] **factor
    postestimation**, [R] **alpha**, [R] **lincom**,
    [R] **mlogit**, [R] **mprobit**, [R] **mprobit
    postestimation**, [R] **predictnl**, [R] **slogit**
Nelson, F. D., [R] **logit**, [R] **probit**
Nelson, W., [ST] **stcrreg postestimation**, [ST] **sts**
Nelson, W. C., [MV] **mvtest correlations**
Neter, J., [R] **pkcross**, [R] **pkequiv**, [R] **pkshape**,
    [R] **regress postestimation**
Neudecker, H., [TS] **var svar**

Neuhaus, J. M., [XT] **xtcloglog**, [XT] **xtintreg**,
    [XT] **xtlogit**, [XT] **xtmelogit**, [XT] **xtmepoisson**,
    [XT] **xtmixed**, [XT] **xtprobit**
Nevels, K., [MV] **procrustes**
Newbold, P., [TS] **arima**, [TS] **vec intro**
Newey, W. K., [R] **glm**, [R] **gmm**, [R] **ivprobit**,
    [R] **ivregress**, [R] **ivtobit**, [TS] **newey**,
    [TS] **pperron**, [XT] **xtabond**, [XT] **xtdpd**,
    [XT] **xtdpdsys**, [XT] **xtunitroot**
Newman, S. C., [R] **poisson**, [ST] **epitab**, [ST] **stcox**,
    [ST] **sts**
Newson, R., [D] **contract**, [D] **generate**, [D] **statsby**,
    [R] **centile**, [R] **glm**, [R] **inequality**, [R] **kwallis**,
    [R] **mkspline**, [R] **ranksum**, [R] **signrank**,
    [R] **spearman**, [R] **tabulate twoway**, [ST] **stcox
    postestimation**
Newton, H. J., [R] **kdensity**, [TS] **arima**,
    [TS] **corrgram**, [TS] **cumsp**, [TS] **dfuller**,
    [TS] **pergram**, [TS] **wntestb**, [TS] **xcorr**,
    [XT] **xtgee**
Newton, I., [M-5] **optimize( )**
Newton, M. A., [XT] **xtcloglog**, [XT] **xtgee**,
    [XT] **xtintreg**, [XT] **xtlogit**, [XT] **xtprobit**,
    [XT] **xttobit**
Neyman, J., [R] **ci**
Ng, E. S. W., [XT] **xtmelogit**
Ng, S., [TS] **dfgls**
Nicewander, W. A., [R] **correlate**
Nichols, A., [R] **ivregress**, [R] **reg3**, [R] **treatreg**,
    [XT] **xtmixed**, [XT] **xtrc**, [XT] **xtreg**
Nickell, S. J., [R] **gmm**, [XT] **xtabond**, [XT] **xtdpd**,
    [XT] **xtdpdsys**, [XT] **xtunitroot**
Nielsen, B., [TS] **varsoc**, [TS] **vec intro**
Nolan, D., [R] **diagnostic plots**
Nordlund, D. J., [MV] **discrim lda**
Nunnally, J. C., [R] **alpha**

O

O'Connell, P. G. J., [XT] **xtunitroot**
O'Connell, R. T., [TS] **tssmooth**, [TS] **tssmooth
    dexponential**, [TS] **tssmooth exponential**,
    [TS] **tssmooth hwinters**, [TS] **tssmooth
    shwinters**
O'Donnell, C. J., [XT] **xtfrontier**
O'Donnell, O., [SVY] **svy estimation**, [SVY] **svyset**
O'Fallon, W. M., [R] **logit**
Oakes, D., [ST] **ltable**, [ST] **stcox**, [ST] **stcox PH-
    assumption tests**, [ST] **stpower**, [ST] **streg**,
    [ST] **sts**
Obstfeld, M., [XT] **xtunitroot**
Ochiai, A., [MV] *measure_option*
Odum, E. P., [MV] **clustermat**
Oehlert, G. W., [R] **nlcom**
Oh, K.-Y., [XT] **xtunitroot**
Oldham, K. B., [D] **functions**
Olivier, D., [R] **expoisson**
Olkin, I., [MV] **hotelling**, [R] **kwallis**, [TS] **wntestb**
Olsen, M. K., [MI] **intro substantive**

Olshansky, S. J., [ST] **streg**
Olson, J. M., [R] **symmetry**
Orcutt, G. H., [TS] **prais**
Ord, J. K., [R] **centile**, [R] **mean**, [R] **proportion**,
    [R] **qreg**, [R] **ratio**, [R] **summarize**, [R] **total**
Orsini, N., [ST] **epitab**, [XT] **xtreg**
Osbat, C., [XT] **xtunitroot**
Osterwald-Lenum, M., [TS] **vecrank**
Ostle, B., [R] **anova postestimation**
Ott, R. L., [SVY] **survey**
Over, M., [R] **regress**, [XT] **xtivreg**

P

Pacheco, J. M., [R] **dstdize**
Pagan, A. R., [R] **frontier**, [R] **mvreg**, [R] **regress
    postestimation**, [R] **sureg**, [TS] **Glossary**,
    [XT] **xtreg postestimation**
Pagano, M., [R] **dstdize**, [R] **logistic**, [R] **margins**,
    [R] **sampsi**, [R] **tabulate twoway**, [ST] **ltable**,
    [ST] **sts**
Paik, M. C., [R] **dstdize**, [R] **kappa**, [R] **sampsi**,
    [ST] **epitab**
Palta, M., [XT] **xtcloglog**, [XT] **xtgee**, [XT] **xtintreg**,
    [XT] **xtlogit**, [XT] **xtprobit**, [XT] **xttobit**
Pampel, F. C., [R] **logistic**, [R] **logit**, [R] **probit**
Panis, C., [R] **mkspline**
Park, H. J., [P] **_robust**, [R] **regress**,
    [SVY] **svy: tabulate twoway**
Park, J. Y., [R] **boxcox**, [R] **margins**, [R] **nlcom**,
    [R] **predictnl**, [R] **testnl**, [TS] **sspace**, [TS] **vec**,
    [TS] **vec intro**, [TS] **vecrank**
Parks, W. P., [R] **exlogistic**
Parmar, M. K. B., [ST] **stpower**, [ST] **stpower cox**
Parzen, E., [R] **estat**, [R] **kdensity**
Pasquini, J., [R] **vwls**, [ST] **epitab**
Patel, N. R., [R] **exlogistic**, [R] **exlogistic
    postestimation**, [R] **expoisson**, [R] **tabulate
    twoway**
Paterson, L., [XT] **xtmelogit**
Patterson, H. D., [R] **pkcross**
Patterson, K., [XT] **xtunitroot**
Paul, C., [R] **logistic**
Paulsen, J., [TS] **varsoc**, [TS] **vec intro**
Pearce, M. S., [R] **logistic**, [ST] **epitab**
Pearson, E. S., [R] **ci**, [R] **ttest**
Pearson, K., [G] **graph twoway histogram**,
    [MV] **mds**, [MV] *measure_option*, [MV] **pca**,
    [R] **correlate**, [R] **tabulate twoway**
Peen, C., [MV] **procrustes**
Pendergast, J. F., [XT] **xtcloglog**, [XT] **xtgee**,
    [XT] **xtintreg**, [XT] **xtlogit**, [XT] **xtprobit**,
    [XT] **xttobit**
Penrose, R., [M-5] **pinv( )**
Pepe, M. S., [R] **roc**, [R] **rocfit**, [ST] **stcrreg**
Peracchi, F., [R] **regress**, [R] **regress postestimation**
Pérez-Hernández, M. A., [R] **kdensity**
Pérez-Hoyos, S., [R] **lrtest**

Pérez-Santiago, M. I., [ST] **epitab**
Perkins, A. M., [R] **ranksum**
Perrin, E., [MV] **factor**, [MV] **factor postestimation**,
    [R] **alpha**, [R] **lincom**, [R] **mlogit**, [R] **mprobit**,
    [R] **mprobit postestimation**, [R] **predictnl**,
    [R] **slogit**
Perron, P., [TS] **dfgls**, [TS] **Glossary**, [TS] **pperron**
Persson, R., [G] **graph intro**
Pesaran, M. H., [XT] **xtunitroot**
Peterson, B., [R] **ologit**
Peterson, W. W., [R] **logistic postestimation**
Petitclerc, M., [R] **kappa**
Petkova, E., [R] **suest**
Peto, J., [ST] **sts test**
Peto, R., [ST] **stcox**, [ST] **streg**, [ST] **sts test**
Petrin, A., [R] **frontier**
Pfeffer, R. I., [R] **symmetry**
Phillips, P. C. B., [R] **boxcox**, [R] **margins**, [R] **nlcom**,
    [R] **predictnl**, [R] **regress postestimation time
    series**, [R] **testnl**, [TS] **Glossary**, [TS] **pperron**,
    [TS] **vargranger**, [TS] **vec**, [TS] **vec intro**,
    [TS] **vecrank**, [XT] **xtunitroot**
Piantadosi, S., [P] **_robust**, [U] **20.20 References**
Pickles, A., [R] **gllamm**, [R] **glm**, [XT] **xtgee**,
    [XT] **xtmelogit**, [XT] **xtmepoisson**, [XT] **xtreg**
Pierce, D. A., [TS] **wntestq**, [XT] **xtcloglog**,
    [XT] **xtintreg**, [XT] **xtlogit**, [XT] **xtmelogit**,
    [XT] **xtmepoisson**, [XT] **xtpoisson**,
    [XT] **xtprobit**, [XT] **xttobit**
Pierson, R. A., [XT] **xtmixed**
Pike, M. C., [R] **symmetry**, [ST] **ltable**, [ST] **streg**
Pillai, K. C. S., [MV] **canon**, [MV] **manova**
Pindyck, R. S., [R] **biprobit**, [R] **heckprob**
Pinheiro, J. C., [XT] **xtmelogit**, [XT] **xtmelogit
    postestimation**, [XT] **xtmepoisson**,
    [XT] **xtmepoisson postestimation**,
    [XT] **xtmixed**, [XT] **xtmixed postestimation**
Pintilie, M., [ST] **stcrreg**, [ST] **stcrreg postestimation**
Pisati, M., [TS] **time series**
Pischke, J.-S., [R] **ivregress**, [R] **ivregress
    postestimation**, [R] **qreg**, [R] **regress**,
    [U] **20.20 References**
Pitarakis, J.-Y., [TS] **vecrank**
Pitblado, J. S., [M-5] **deriv( )**, [M-5] **moptimize( )**,
    [P] **_robust**, [P] **intro**, [R] **frontier**, [R] **lpoly**,
    [R] **maximize**, [R] **ml**, [ST] **sts**, [SVY] **ml for
    svy**, [SVY] **survey**, [XT] **xtfrontier**
Plackett, R. L., [R] **ameans**, [R] **regress**, [R] **rologit**,
    [R] **summarize**, [R] **ttest**
Playfair, W., [G] **graph bar**, [G] **graph pie**
Plosser, C. I., [TS] **vecrank**
Plummer, D., [ST] **epitab**
Plummer Jr., W. D., [R] **sunflower**
Pocock, S. J., [R] **sampsi**
Poi, B. P., [R] **bootstrap**, [R] **bstat**, [R] **frontier**,
    [R] **ivregress**, [R] **ivregress postestimation**,
    [R] **nl**, [R] **nlsur**, [R] **reg3**, [XT] **xtrc**
Poirier, D. J., [R] **biprobit**
Poisson, S. D., [R] **poisson**

Ponce de Leon, A., [R] **roc**
Porter, T. M., [R] **correlate**
Posten, H. O., [D] **functions**
Powell, M. J. D., [M-5] **optimize( )**
Powers, D. A., [R] **logit**, [R] **probit**
Prais, S. J., [TS] **prais**
Preece, D. A., [R] **ttest**
Pregibon, D., [R] **glm**, [R] **linktest**, [R] **logistic**,
    [R] **logistic postestimation**, [R] **logit**, [R] **logit
    postestimation**
Prentice, R. L., [ST] **discrete**, [ST] **ltable**, [ST] **stcox**,
    [ST] **stcox PH-assumption tests**, [ST] **stcox
    postestimation**, [ST] **streg**, [ST] **sts**, [ST] **sts
    test**, [ST] **stset**, [XT] **xtgee**
Press, W. H., [D] **functions**, [P] **matrix symeigen**,
    [R] **dydx**, **vwls**, [TS] **arch**, [TS] **arima**
Prosser, R., [XT] **xtmixed**
Pryor, D. B., [ST] **stcox postestimation**
Punj, G. N., [R] **rologit**
Putter, H., [ST] **stcrreg**, [ST] **stcrreg postestimation**

Q

Qaqish, B., [XT] **xtgee**
Quesenberry, C. P., [MV] **discrim knn**

R

Rabe-Hesketh, S., [MV] **pca**, [MV] **screeplot**,
    [R] **anova**, [R] **gllamm**, [R] **glm**, [R] **heckprob**,
    [R] **ivprobit**, [R] **ivtobit**, [R] **logistic**,
    [R] **logit**, [R] **nbreg**, [R] **ologit**, [R] **oprobit**,
    [R] **poisson**, [R] **probit**, [XT] **xtcloglog**,
    [XT] **xtgee**, [XT] **xtintreg**, [XT] **xtlogit**,
    [XT] **xtmelogit**, [XT] **xtmelogit postestimation**,
    [XT] **xtmepoisson**, [XT] **xtmepoisson
    postestimation**, [XT] **xtmixed**, [XT] **xtpoisson**,
    [XT] **xtprobit**, [XT] **xtreg**, [XT] **xttobit**
Radmacher, R. D., [ST] **stpower**
Raftery, A., [R] **BIC note**, [R] **estat**, [R] **glm**
Raghunathan, T. E., [MI] **intro substantive**, [MI] **mi
    estimate**, [MI] **mi estimate postestimation**,
    [MI] **mi estimate using**, [MI] **mi impute**,
    [MI] **mi impute logit**, [MI] **mi impute mlogit**,
    [MI] **mi impute monotone**, [MI] **mi impute
    ologit**
Ramalheira, C., [R] **ameans**, [ST] **ltable**
Ramsey, J. B., [R] **regress postestimation**
Rao, C. R., [MV] **factor**, [MV] **hotelling**,
    [MV] **manova**, [XT] **xtmixed**
Rao, D. S. P., [XT] **xtfrontier**
Rao, J. N. K., [SVY] **direct standardization**,
    [SVY] **poststratification**, [SVY] **svy: tabulate
    twoway**, [SVY] **variance estimation**
Rao, T. R., [MV] *measure_option*
Raphson, J., [M-5] **optimize( )**
Rasbash, J., [XT] **xtmelogit**, [XT] **xtmixed**
Ratcliffe, S. J., [XT] **xtgee**
Ratkowsky, D. A., [R] **nl**, [R] **pk**, [R] **pkcross**

Raudenbush, S. W., [XT] **xtmelogit**, [XT] **xtmepoisson**,
    [XT] **xtmixed**
Redelmeier, D. A., [R] **brier**
Reichenheim, M. E., [R] **kappa**, [R] **roc**
Reid, C., [M-5] **Hilbert( )**, [R] **ci**
Reid, N., [ST] **stcox**
Reilly, M., [R] **logistic**, [ST] **epitab**
Reinfurt, K. H., [MV] **mvtest correlations**
Reinsch, C., [P] **matrix symeigen**
Reinsch, C. H., [M-5] **spline3( )**
Reinsel, G. C., [TS] **arima**, [TS] **corrgram**,
    [TS] **cumsp**, [TS] **dfuller**, [TS] **pergram**,
    [TS] **pperron**, [TS] **vec intro**, [TS] **xcorr**
Reiter, J. P., [MI] **intro**, [MI] **intro substantive**,
    [MI] **mi estimate**, [MI] **mi estimate
    postestimation**, [MI] **mi estimate using**
Relles, D. A., [R] **rreg**
Rencher, A. C., [MV] **biplot**, [MV] **ca**, [MV] **candisc**,
    [MV] **canon**, [MV] **canon postestimation**,
    [MV] **cluster**, [MV] **discrim**, [MV] **discrim
    estat**, [MV] **discrim knn**, [MV] **discrim
    lda**, [MV] **discrim lda postestimation**,
    [MV] **discrim logistic**, [MV] **discrim qda**,
    [MV] **discrim qda postestimation**, [MV] **factor**,
    [MV] **manova**, [MV] **mca**, [MV] **mvtest**,
    [MV] **mvtest correlations**, [MV] **mvtest
    covariances**, [MV] **mvtest means**, [MV] **mvtest
    normality**, [MV] **pca**, [MV] **screeplot**, [R] **anova
    postestimation**
Research Triangle Institute, [SVY] **svy: tabulate
    twoway**
Revankar, N. S., [R] **frontier**, [XT] **xtfrontier**
Richards, D. S. P., [MV] **mvtest means**
Richardson, W., [R] **ttest**
Richter, J. R., [ST] **stpower**
Riffenburgh, R. H., [R] **ksmirnov**, [R] **kwallis**
Riley, A. R., [D] **filefilter**, [D] **list**, [R] **net search**
Rivers, D., [R] **ivprobit**
Roberson, P. K., [R] **logistic postestimation**
Robins, J. M., [ST] **epitab**
Robins, R. P., [TS] **arch**
Robinson, A., [M-5] **Toeplitz( )**
Robyn, D. L., [G] **graph bar**, [G] **graph pie**,
    [G] **graph twoway histogram**, [R] **cumul**
Rodgers, J. L., [R] **correlate**
Rodríguez, G., [R] **nbreg**, [R] **poisson**, [R] **ztnb**,
    [XT] **xtmelogit**
Rogers, D. J., [MV] *measure_option*
Rogers, W. H., [D] **egen**, [P] **_robust**, [R] **brier**,
    [R] **glm**, [R] **heckman**, [R] **lincom**, [R] **mlogit**,
    [R] **mprobit**, [R] **mprobit postestimation**,
    [R] **nbreg**, [R] **poisson**, [R] **predictnl**, [R] **qreg**,
    [R] **regress**, [R] **rreg**, [R] **sktest**, [R] **slogit**,
    [R] **suest**, [ST] **stcox PH-assumption tests**,
    [ST] **stcox postestimation**, [U] **20.20 References**
Rogoff, K., [XT] **xtunitroot**
Rohlf, F. J., [MV] **cluster**, [MV] *measure_option*
Rombouts, J. V. K., [TS] **dvech**
Romney, A. K., [MV] **ca**

Ronchetti, E. M., [D] **egen**

Ronning, G., [R] **clogit**

Roodman, D., [D] **collapse**, [XT] **xtdpd**, [XT] **xtdpdsys**

Room, T., [TS] **arima**

Rosati, R. A., [ST] **stcox postestimation**

Rose, D. W., [MV] **discrim knn**

Rose, J. M., [R] **nlogit**

Rosen, H. S., [XT] **xtabond**, [XT] **xtdpd**, [XT] **xtdpdsys**

Rosner, B., [R] **sampsi**

Ross, G. J. S., [R] **nl**

Rossi, S. S., [ST] **stcrreg**

Rothenberg, T. J., [TS] **dfgls**, [TS] **Glossary**, [TS] **sspace**, [TS] **var svar**, [TS] **vec**

Rothkopf, E. Z., [MV] **mdslong**

Rothman, K. J., [R] **ci**, [R] **dstdize**, [R] **glogit**, [R] **poisson**, [ST] **epitab**

Rothstein, H. R., [R] **meta**

Rothwell, S. T., [SVY] **survey**, [SVY] **svy estimation**

Rousseeuw, P. J., [D] **egen**, [MV] **cluster**, [MV] **clustermat**, [MV] **matrix dissimilarity**, [MV] *measure_option*, [P] **matrix dissimilarity**, [R] **qreg**, [R] **regress postestimation**, [R] **rreg**

Rovine, M. J., [R] **correlate**

Roy, S. N., [MV] **canon**, [MV] **manova**

Royall, R. M., [P] **_robust**, [U] **20.20 References**

Royston, P., [D] **list**, [D] **sort**, [G] **graph twoway lowess**, [G] **graph twoway scatter**, [MI] **intro**, [MI] **intro substantive**, [MI] **mi estimate**, [MI] **mi export**, [MI] **mi export ice**, [MI] **mi import**, [MI] **mi import ice**, [MI] **mi impute**, [MI] **mi impute monotone**, [R] **centile**, [R] **cusum**, [R] **diagnostic plots**, [R] **dotplot**, [R] **dydx**, [R] **estat**, [R] **fracpoly**, [R] **fracpoly postestimation**, [R] **glm**, [R] **kdensity**, [R] **lnskew0**, [R] **lowess**, [R] **mfp**, [R] **ml**, [R] **nl**, [R] **regress**, [R] **sampsi**, [R] **sktest**, [R] **smooth**, [R] **swilk**, [ST] **epitab**, [ST] **stcox**, [ST] **stcox PH-assumption tests**, [ST] **stpower**, [ST] **stpower cox**, [ST] **streg**

Rubin, D. B., [MI] **intro substantive**, [MI] **mi estimate**, [MI] **mi estimate postestimation**, [MI] **mi estimate using**, [MI] **mi impute**, [MI] **mi impute logit**, [MI] **mi impute monotone**, [MI] **mi impute mvn**, [MI] **mi impute pmm**, [MI] **mi impute regress**, [XT] **xtmixed**

Rubin, H., [R] **ivregress postestimation**

Rubinfeld, D. L., [R] **biprobit**, [R] **heckprob**

Rubinstein, L. V., [ST] **stpower**, [ST] **stpower exponential**, [U] **1.4 References**

Rudebusch, G. D., [R] **ivregress postestimation**

Runkle, D. E., [TS] **arch**

Ruppert, D., [R] **boxcox**, [R] **rreg**, [XT] **xtmixed**

Rush, M., [D] **egen**

Russell, P. F., [MV] *measure_option*

Rutherford, E., [R] **poisson**

Ruud, P. A., [R] **gmm**, [R] **rologit**, [R] **suest**

Ryan, P., [D] **egen**, [D] **pctile**, [U] **11.7 References**

Ryan, T. P., [R] **qc**

S

Saikkonen, P., [TS] **vec intro**, [TS] **vecrank**

Sakamoto, Y., [R] **BIC note**

Salgado-Ugarte, I. H., [R] **kdensity**, [R] **lowess**, [R] **smooth**

Salim, A., [R] **logistic**, [ST] **epitab**

Salvador, M., [TS] **vecrank**

Samaniego, F. J., [TS] **varwle**

Sammon Jr., J. W., [MV] **Glossary**, [MV] **mds**, [MV] **mdslong**, [MV] **mdsmat**

Sampson, A. R., [MV] **hotelling**

Sanders, F., [R] **brier**

Santner, T. J., [ST] **stpower**, [ST] **stpower exponential**, [U] **1.4 References**

Santos Silva, J. M. C., [R] **gmm**

Sarafidis, V., [XT] **xtreg**

Sargan, J. D., [R] **ivregress postestimation**, [TS] **prais**

Sargent, T. J., [TS] **dfactor**

Särndal, C.-E., [SVY] **variance estimation**

Sasieni, P., [D] **list**, [D] **memory**, [R] **dotplot**, [R] **lowess**, [R] **nptrend**, [R] **smooth**, [ST] **stcox**

Satterthwaite, F. E., [R] **ttest**, [SVY] **variance estimation**

Sauerbrei, W., [R] **estat**, [R] **fracpoly**, [R] **mfp**

Savage, I. R., [ST] **sts test**

Savin, N. E., [R] **regress postestimation time series**

Saw, S. L. C., [R] **qc**

Sawa, T., [R] **estat**

Saxl, I., [R] **correlate**

Schaalje, G. B., [R] **anova postestimation**

Schabenberger, O., [XT] **xtmelogit**

Schafer, J. L., [MI] **intro substantive**, [MI] **mi estimate**, [MI] **mi impute**, [MI] **mi impute monotone**, [MI] **mi impute mvn**

Schaffer, C. M., [MV] **cluster**

Schaffer, M. E., [R] **ivregress**, [R] **ivregress postestimation**

Schank, T., [XT] **xtmelogit**, [XT] **xtmepoisson**, [XT] **xtmixed**, [XT] **xtreg**

Scheaffer, R. L., [SVY] **survey**

Scheffé, H., [R] **anova**, [R] **oneway**

Schenker, N., [MI] **intro substantive**, [MI] **mi impute**, [MI] **mi impute pmm**, [MI] **mi impute regress**

Schlesselman, J. J., [R] **boxcox**, [ST] **epitab**

Schlossmacher, E. J., [R] **qreg**

Schmeiser, B. W., [D] **functions**

Schmidt, C. H., [R] **brier**

Schmidt, P., [R] **frontier**, [R] **regress postestimation**, [XT] **xtfrontier**, [XT] **xtunitroot**

Schmidt, T. J., [D] **egen**

Schneider, H., [R] **sdtest**

Schneider, W., [TS] **sspace**

Schnell, D., [P] **_robust**, [R] **regress**, [SVY] **svy: tabulate twoway**

Schoenfeld, D. A., [ST] **stcox**, [ST] **stcox postestimation**, [ST] **stpower**, [ST] **stpower cox**, [ST] **stpower exponential**, [ST] **stpower logrank**, [ST] **streg**, [U] **1.4 References**

Schonlau, M., [R] **glm**, [R] **logistic**, [R] **logit**,
[R] **poisson**, [R] **regress**

Schuirmann, D. J., [R] **pkequiv**

Schumacher, M., [ST] **stcrreg**

Schumm, L. P., [D] **sort**

Schwarz, G., [MV] **factor postestimation**, [R] **BIC
note**, [R] **estat**

Schwert, G. W., [TS] **dfgls**

Scott, A. J., [SVY] **estat**, [SVY] **svy: tabulate twoway**

Scott, C., [SVY] **estat**, [SVY] **subpopulation
estimation**, [SVY] **svy estimation**

Scott, D. W., [R] **kdensity**

Scott, G. B., [R] **exlogistic**

Scotto, M. G., [R] **diagnostic plots**, [ST] **streg**

Searle, S. R., [R] **margins**, [XT] **xtmelogit**,
[XT] **xtmepoisson**, [XT] **xtmixed**

Sears, R. R., [ST] **epitab**

Seber, G. A. F., [MV] **biplot**, [MV] **manova**,
[MV] **mvtest**, [MV] **mvtest means**, [MV] **mvtest
normality**

Seed, P. T., [R] **ci**, [R] **correlate**, [R] **logistic
postestimation**, [R] **roc**, [R] **sampsi**, [R] **sdtest**,
[R] **spearman**

Seidler, J., [R] **correlate**

Self, S. G., [XT] **xtmelogit**, [XT] **xtmepoisson**,
[XT] **xtmixed**

Selvin, S., [R] **poisson**, [ST] **ltable**, [ST] **stcox**

Sempos, C. T., [R] **dstdize**, [ST] **epitab**, [ST] **ltable**,
[ST] **stcox**

Semykina, A., [R] **inequality**, [R] **qreg**

Seneta, E., [R] **correlate**, [U] **1.4 References**

Senn, S. J., [R] **glm**, [R] **ttest**

Serachitopol, D. M., [ST] **sts graph**

Serfling, R. J., [TS] **irf create**

Shah, B. V., [SVY] **direct standardization**,
[SVY] **poststratification**, [SVY] **variance
estimation**

Shanno, D. F., [M-5] **optimize( )**

Shao, J., [ST] **stpower**, [ST] **stpower exponential**,
[SVY] **survey**, [SVY] **svy jackknife**,
[SVY] **variance estimation**

Shapiro, S., [ST] **epitab**

Shapiro, S. S., [ST] **swilk**

Shaposhnikova, T., [D] **functions**

Shavelson, R. J., [R] **alpha**

Shea, J., [R] **ivregress postestimation**

Sheather, S. J., [R] **lpoly**

Sheldon, T. A., [R] **meta**

Shewhart, W. A., [R] **qc**

Shiboski, S. C., [R] **logistic**, [ST] **stcox**

Shiller, R. J., [R] **tobit**

Shimizu, M., [R] **kdensity**, [R] **lowess**

Shin, Y., [XT] **xtunitroot**

Shrout, P. E., [R] **kappa**

Shults, J., [XT] **xtgee**

Shumway, R. H., [TS] **arima**

Sibson, R., [MV] **cluster**

Šidák, Z., [R] **oneway**

Silvennoinen, A., [TS] **dvech**

Silverman, B. W., [R] **kdensity**

Silvey, S. D., [R] **ologit**, [R] **oprobit**

Simon, R., [ST] **stpower**

Simonoff, J. S., [R] **kdensity**, [R] **ztnb**, [R] **ztp**

Simor, I. S., [R] **kappa**

Simpson, T., [M-5] **optimize( )**

Sims, C. A., [TS] **dfactor**, [TS] **irf create**, [TS] **var
svar**, [TS] **vec**, [TS] **vec intro**, [TS] **vecrank**

Singleton, K. J., [R] **gmm**

Skinner, C. J., [SVY] **estat**, [SVY] **survey**, [SVY] **svy
estimation**, [SVY] **variance estimation**

Skovlund, E., [ST] **stpower**, [ST] **stpower cox**

Skrondal, A., [R] **gllamm**, [R] **glm**, [XT] **xtcloglog**,
[XT] **xtgee**, [XT] **xtintreg**, [XT] **xtlogit**,
[XT] **xtmelogit**, [XT] **xtmelogit postestimation**,
[XT] **xtmepoisson**, [XT] **xtmepoisson
postestimation**, [XT] **xtmixed**, [XT] **xtpoisson**,
[XT] **xtprobit**, [XT] **xttobit**

Slaymaker, E., [P] **file**

Slone, D., [ST] **epitab**

Smans, M., [XT] **xtmepoisson**

Smeeton, N. C., [R] **ranksum**, [R] **signrank**

Smirnov, N. V., [R] **ksmirnov**

Smith, A. F. M., [XT] **xtcloglog**, [XT] **xtintreg**,
[XT] **xtlogit**, [XT] **xtmelogit**, [XT] **xtmepoisson**,
[XT] **xtpoisson**, [XT] **xtprobit**, [XT] **xttobit**

Smith, B. T., [P] **matrix symeigen**

Smith, C. A. B., [MV] **discrim estat**, [MV] **discrim
qda**

Smith, H., [MV] **manova**, [R] **eivreg**, [R] **oneway**,
[R] **regress**, [R] **stepwise**

Smith, J. M., [R] **fracpoly**

Smith, R. J., [R] **ivprobit**

Smith, R. L., [ST] **streg**

Smith, T. M. F., [SVY] **survey**

Smith-Vikos, T., [MV] **discrim knn**

Smullyan, R., [MV] **mds**

Smythe, B., [ST] **sts**

Sneath, P. H. A., [MV] *measure_option*

Snedecor, G. W., [R] **ameans**, [R] **anova**, [R] **correlate**,
[R] **oneway**, [R] **ranksum**, [R] **signrank**

Snell, E. J., [R] **exlogistic**, [R] **expoisson**, [ST] **stcox**,
[ST] **stcox PH-assumption tests**, [ST] **streg
postestimation**

Sokal, R. R., [MV] *measure_option*

Solenberger, P., [MI] **intro substantive**, [MI] **mi
impute**, [MI] **mi impute logit**, [MI] **mi impute
mlogit**, [MI] **mi impute monotone**, [MI] **mi
impute ologit**

Song, F., [R] **meta**

Song, S. H., [XT] **xtmixed**

Soon, T. W., [R] **qc**

Sörbom, D., [MV] **factor postestimation**

Sorensen, D., [M-1] **LAPACK**, [M-5] **lapack( )**,
[P] **matrix eigenvalues**

Sørensen, T., [MV] *measure_option*

Sosa-Escudero, W., [XT] **xtreg**, [XT] **xtreg postestimation**, [XT] **xtregar**
Sotoca, S., [TS] **sspace**
Spanier, J., [D] **functions**
Späth, H., [MV] **cluster**
Spearman, C., [MV] **factor**, [R] **spearman**
Speed, F. M., [R] **margins**
Speed, T., [R] **diagnostic plots**
Spence, I., [G] **graph pie**
Sperling, R., [TS] **arch**, [TS] **arima**, [TS] **dfgls**, [TS] **wntestq**
Spiegelhalter, D. J., [R] **brier**
Spieldman, R. S., [R] **symmetry**
Spitzer, J. J., [R] **boxcox**
Sprent, P., [R] **ranksum**, [R] **signrank**
Sribney, W. M., [M-5] **deriv( )**, [M-5] **moptimize( )**, [P] **_robust**, [P] **intro**, [P] **matrix mkmat**, [R] **frontier**, [R] **maximize**, [R] **ml**, [R] **orthog**, [R] **ranksum**, [R] **signrank**, [R] **stepwise**, [R] **test**, [SVY] **estat**, [SVY] **ml for svy**, [SVY] **survey**, [SVY] **svy postestimation**, [SVY] **svy: tabulate twoway**, [SVY] **svydescribe**, [XT] **xtfrontier**
Staelin, R., [R] **rologit**
Stahel, W. A., [D] **egen**
Staiger, D., [R] **ivregress postestimation**
Starmer, C. F., [R] **vwls**
Startz, R., [R] **ivregress postestimation**
Stegun, I. A., [D] **functions**, [R] **orthog**, [XT] **xtmelogit**, [XT] **xtmepoisson**
Steichen, T. J., [D] **duplicates**, [R] **kappa**, [R] **kdensity**, [R] **sunflower**
Steiger, W., [R] **qreg**
Stein, C., [R] **bootstrap**
Stephenson, D. B., [MV] **pca**, [R] **brier**
Stepniewska, K. A., [R] **nptrend**
Stern, H. S., [MI] **intro substantive**, [MI] **mi impute mvn**, [MI] **mi impute regress**
Sterne, J. A. C., [MI] **intro**, [R] **dstdize**, [R] **meta**, [R] **summarize**, [ST] **stcox**
Stevens, E. H., [MV] **mvtest**
Stevenson, R. E., [R] **frontier**
Stewart, G. W., [P] **matrix svd**
Stewart, J., [ST] **ltable**
Stewart, M. B., [R] **intreg**, [R] **oprobit**, [R] **tobit**, [XT] **xtprobit**
Stigler, S. M., [R] **ameans**, [R] **ci**, [R] **correlate**, [R] **kwallis**, [R] **qreg**, [R] **regress**
Stillman, S., [R] **ivregress**, [R] **ivregress postestimation**
Stine, R., [R] **bootstrap**
Stock, J. H., [R] **ivregress**, [R] **ivregress postestimation**, [TS] **arch**, [TS] **dfactor**, [TS] **dfgls**, [TS] **Glossary**, [TS] **irf create**, [TS] **rolling**, [TS] **sspace**, [TS] **time series**, [TS] **var**, [TS] **var intro**, [TS] **var svar**, [TS] **vec**, [TS] **vec intro**, [TS] **vecrank**, [XT] **intro**, [XT] **xthtaylor**, [XT] **xtreg**, [U] **1.4 References**
Stoll, B. J., [ST] **epitab**

Stolley, P. D., [ST] **epitab**
Storer, B. E., [ST] **stcrreg**
Stork, D. G., [MV] **cluster**, [MV] **cluster stop**
Stoto, M. A., [R] **lv**
Støvring, H., [M-2] **pointers**
Stram, D. O., [XT] **xtmixed**
Street, J. O., [R] **rreg**
Stroup, W. W., [XT] **xtmelogit**
Stryhn, H., [R] **regress**, [ST] **epitab**
Stuart, A., [R] **centile**, [R] **mean**, [R] **proportion**, [R] **qreg**, [R] **ratio**, [R] **summarize**, [R] **symmetry**, [R] **tobit**, [R] **total**, [SVY] **survey**
Stuetzle, W., [R] **sunflower**
Sullivan, G., [P] **_robust**, [R] **regress**, [SVY] **svy: tabulate twoway**
Summers, R., [XT] **xtunitroot**
Sun, W., [MI] **intro substantive**
Sutton, A. J., [R] **meta**
Svennerholm, A. M., [ST] **epitab**
Swagel, P., [U] **21.6 Reference**
Swamy, P. A. V. B., [XT] **xtivreg**, [XT] **xtrc**, [XT] **xtreg**
Swed, F. S., [R] **runtest**
Sweeting, T. J., [ST] **streg**
Swensson, B., [SVY] **variance estimation**
Swets, J. A., [R] **logistic postestimation**
Szroeter, J., [R] **regress postestimation**

T

Tabachnick, B. G., [MV] **discrim**, [MV] **discrim lda**
Taka, M. T., [R] **pkcross**
Tamhane, A. C., [R] **oneway**, [R] **sampsi**
Tan, S. B., [ST] **stpower**, [ST] **stpower logrank**
Tan, S. H., [ST] **stpower**, [ST] **stpower logrank**
Tan, W. Y., [P] **_robust**, [U] **20.20 References**
Tanimoto, T. T., [MV] *measure_option*
Taniuchi, T., [R] **kdensity**
Tanner, M. A., [MI] **intro substantive**, [MI] **mi impute mvn**
Tanner Jr., W. P., [R] **logistic postestimation**
Tanur, J. M., [R] **kwallis**
Tapia, R. A., [R] **kdensity**
Tarlov, A. R., [MV] **factor**, [MV] **factor postestimation**, [R] **alpha**, [R] **lincom**, [R] **mlogit**, [R] **mprobit**, [R] **mprobit postestimation**, [R] **predictnl**, [R] **slogit**
Tarone, R. E., [ST] **epitab**, [ST] **sts test**, [U] **1.4 References**
Taub, A. J., [XT] **xtreg**
Taylor, C., [R] **gllamm**, [R] **glm**, [XT] **xtgee**, [XT] **xtreg**
Taylor, J. M. G., [MI] **intro substantive**, [MI] **mi impute**, [MI] **mi impute pmm**, [MI] **mi impute regress**
Taylor, W. E., [XT] **xthtaylor**
ten Berge, J. M. F., [MV] **procrustes**

ter Bogt, T., [MV] **mvtest**

Teräsvirta, T., [TS] **dvech**

Teukolsky, S. A., [D] **functions**, [P] **matrix symeigen**, [R] **dydx**, [R] **vwls**, [TS] **arch**, [TS] **arima**

Thall, P. F., [XT] **xtmepoisson**

Theil, H., [R] **ivregress**, [R] **reg3**, [TS] **prais**

Therneau, T. M., [ST] **stcox**, [ST] **stcox PH-assumption tests**, [ST] **stcox postestimation**, [ST] **stcrreg**

Thiele, T. N., [R] **summarize**

Thomas, D. C., [ST] **sttocc**

Thomas, D. G., [ST] **epitab**

Thomas, D. R., [SVY] **svy: tabulate twoway**

Thompson, B., [MV] **canon postestimation**

Thompson, J. C., [R] **diagnostic plots**

Thompson, J. R., [R] **kdensity**

Thompson, S. K., [SVY] **survey**

Thompson Jr., W. A., [XT] **xtmixed**

Thomson, G. H., [MV] **factor postestimation**, [MV] **Glossary**

Thorndike, F., [R] **poisson**

Thurstone, L. L., [MV] **rotate**, [R] **rologit**

Tibshirani, R. J., [MV] **discrim knn**, [R] **bootstrap**, [R] **qreg**

Tidmarsh, C. E., [R] **fracpoly**

Tierney, L., [XT] **xtmelogit**, [XT] **xtmepoisson**

Tilford, J. M., [R] **logistic postestimation**

Tilling, K., [ST] **stcox**

Timm, N. H., [MV] **manova**

Tippett, L. H. C., [ST] **streg**

Tobías, A., [R] **alpha**, [R] **logistic postestimation**, [R] **lrtest**, [R] **poisson**, [R] **roc**, [R] **sdtest**, [ST] **streg**

Tobin, J., [R] **tobit**

Toeplitz, O., [M-5] **Toeplitz( )**

Toman, R. J., [R] **stepwise**

Toplis, P. J., [R] **binreg**

Torgerson, W. S., [MV] **mds**, [MV] **mdslong**, [MV] **mdsmat**

Tosetto, A., [R] **logistic**, [R] **logit**

Toulopoulou, T., [XT] **xtmelogit**

Train, K. E., [R] **asmprobit**

Trapido, E., [R] **exlogistic**

Treiman, D. J., [R] **eivreg**, [R] **mlogit**

Trewn, J., [MV] **mds**

Trichopoulos, D., [ST] **epitab**

Trivedi, P. K., [R] **asclogit**, [R] **asmprobit**, [R] **bootstrap**, [R] **gmm**, [R] **heckman**, [R] **intreg**, [R] **ivregress**, [R] **ivregress postestimation**, [R] **logit**, [R] **mprobit**, [R] **nbreg**, [R] **ologit**, [R] **oprobit**, [R] **poisson**, [R] **probit**, [R] **qreg**, [R] **regress**, [R] **regress postestimation**, [R] **simulate**, [R] **sureg**, [R] **tobit**, [R] **ztnb**, [R] **ztp**, [XT] **xt**, [XT] **xtmixed**, [XT] **xtnbreg**, [XT] **xtpoisson**

Tsay, R. S., [TS] **varsoc**, [TS] **vec intro**

Tsiatis, A., [ST] **stcrreg**

Tsiatis, A. A., [R] **exlogistic**

Tu, D., [SVY] **survey**, [SVY] **svy jackknife**, [SVY] **variance estimation**

Tufte, E. R., [G] **graph bar**, [G] **graph pie**, [R] **stem**

Tukey, J. W., [D] **egen**, [G] **graph box**, [G] **graph matrix**, [P] **if**, [R] **jackknife**, [R] **ladder**, [R] **linktest**, [R] **lv**, [R] **regress**, [R] **regress postestimation**, [R] **rreg**, [R] **smooth**, [R] **spikeplot**, [R] **stem**, [SVY] **svy jackknife**

Tukey, P. A., [G] **by_option**, [G] **graph box**, [G] **graph matrix**, [R] **diagnostic plots**, [R] **lowess**, [U] **1.4 References**

Twisk, J. W. R., [XT] **xtgee**, [XT] **xtlogit**, [XT] **xtreg**

Tyler, D. E., [MV] **pca**

Tyler, J. H., [R] **regress**

Tzavalis, E., [XT] **xtunitroot**

U

Uebersax, J. S., [R] **tetrachoric**

Uhlendorff, A., [R] **asmprobit**, [R] **mlogit**, [R] **mprobit**

University Group Diabetes Program, [R] **glogit**, [ST] **epitab**

Upton, G., [U] **1.4 References**

Upward, R., [XT] **xtmelogit**, [XT] **xtmepoisson**, [XT] **xtmixed**, [XT] **xtreg**

Ureta, M., [XT] **xtreg**

Uthoff, V. A., [ST] **stpower cox**

Utts, J. M., [R] **ci**

V

Vach, W., [ST] **stcrreg**

Væth, M., [ST] **stpower**, [ST] **stpower cox**

Vail, S. C., [XT] **xtmepoisson**

Valliant, R., [SVY] **survey**

Valman, H. B., [R] **fracpoly**

Valsecchi, M. G., [ST] **stcrreg**, [ST] **stpower**, [ST] **stpower logrank**, [ST] **sts test**

van Belle, G., [MV] **factor**, [MV] **pca**, [R] **anova**, [R] **dstdize**, [R] **oneway**, [ST] **epitab**

van Buuren, S., [MI] **intro substantive**, [MI] **mi impute**, [MI] **mi impute logit**, [MI] **mi impute mlogit**, [MI] **mi impute monotone**, [MI] **mi impute ologit**

Van de Ven, W. P. M. M., [R] **biprobit**, [R] **heckprob**

van den Broeck, J., [R] **frontier**, [XT] **xtfrontier**

van der Ende, J., [MV] **mvtest**

Van der Heijden, P. G. M., [MV] **ca postestimation**

Van der Merwe, C. A., [MV] **mvtest**, [MV] **mvtest means**

van Doorslaer, E., [SVY] **svy estimation**, [SVY] **svyset**

van Dorsselaer, S., [MV] **mvtest**

Van Hoewyk, J., [MI] **intro substantive**, [MI] **mi impute**, [MI] **mi impute logit**, [MI] **mi impute mlogit**, [MI] **mi impute monotone**, [MI] **mi impute ologit**

Van Kerm, P., [MV] **ca**, [P] **postfile**, [R] **inequality**, [R] **kdensity**

Van Loan, C. F., [R] **orthog**, [R] **tetrachoric**

Van Pragg, B. M. S., [R] **biprobit**, [R] **heckprob**

Vandermonde, A.-T., [M-5] **Vandermonde( )**

Varadharajan-Krishnakumar, J., [XT] **xtivreg**

Velleman, P. F., [R] **regress postestimation**, [R] **smooth**

Verbeke, G., [XT] **xtmixed**

Verdurmen, J., [MV] **mvtest**

Vetterling, W. T., [D] **functions**, [P] **matrix symeigen**, [R] **dydx**, [R] **vwls**, [TS] **arch**, [TS] **arima**

Vidmar, S., [R] **ameans**, [ST] **epitab**

Vittinghoff, E., [R] **logistic**, [ST] **stcox**

Vollebergh, W. A. M., [MV] **mvtest**

von Bortkewitsch, L., [R] **poisson**

von Eye, A., [R] **correlate**

Von Storch, H., [R] **brier**

Vondráček, J., [R] **correlate**

Vuong, Q. H., [R] **ivprobit**, [R] **zinb**, [R] **zip**

W

Wacholder, S., [R] **binreg**

Wagner, H. M., [R] **qreg**

Wagner, M., [XT] **xtunitroot**

Wagner, T., [MV] **mvtest**

Wagstaff, A., [SVY] **svy estimation**, [SVY] **svyset**

Wainer, H., [G] **graph pie**

Wald, A., [TS] **varwle**

Walker, A. J., [D] **functions**, [M-5] **runiform( )**

Walker, A. M., [ST] **epitab**

Walker, S., [ST] **sts test**

Wallgren, A., [G] **graph intro**

Wallgren, B., [G] **graph intro**

Wallis, W. A., [R] **kwallis**

Walters, S. J., [R] **ci**, [R] **kappa**, [R] **tabulate twoway**

Wand, M. P., [R] **kdensity**, [XT] **xtmixed**

Wang, D., [D] **duplicates**, [R] **ci**, [R] **dstdize**, [R] **prtest**

Wang, H., [ST] **stpower**, [ST] **stpower exponential**

Wang, J.-L., [ST] **sts graph**

Wang, J. W., [ST] **streg**

Wang, Y., [R] **asmprobit**

Wang, Z., [R] **logistic postestimation**, [R] **lrtest**, [R] **stepwise**, [ST] **epitab**

Ward Jr., J. H., [MV] **cluster**

Ware, J. H., [ST] **sts test**, [XT] **xtmelogit**, [XT] **xtmepoisson**, [XT] **xtmixed**

Ware Jr., J. E., [MV] **factor**, [MV] **factor postestimation**, [R] **alpha**, [R] **lincom**, [R] **mlogit**, [R] **mprobit**, [R] **mprobit postestimation**, [R] **predictnl**, [R] **slogit**

Warren, K., [ST] **epitab**

Waterson, E. J., [R] **binreg**

Watson, G. S., [R] **lpoly**, [R] **regress postestimation time series**, [TS] **Glossary**, [TS] **prais**

Watson, M. W., [R] **ivregress**, [TS] **arch**, [TS] **dfactor**, [TS] **dfgls**, [TS] **irf create**, [TS] **rolling**, [TS] **sspace**, [TS] **time series**, [TS] **var**, [TS] **var intro**, [TS] **var svar**, [TS] **vec**,

Watson, M. W., *continued* [TS] **vec intro**, [TS] **vecrank**, [XT] **intro**, [XT] **xtreg**, [U] **1.4 References**

Webster, A. D., [R] **fracpoly**

Wedderburn, R. W. M., [R] **glm**, [XT] **xtgee**

Weesie, J., [D] **generate**, [D] **joinby**, [D] **label**, [D] **label language**, [D] **labelbook**, [D] **list**, [D] **merge**, [D] **mvencode**, [D] **order**, [D] **recode**, [D] **rename**, [D] **reshape**, [D] **sample**, [MV] **ca postestimation**, [MV] **pca**, [P] **matrix define**, [R] **alpha**, [R] **constraint**, [R] **hausman**, [R] **ladder**, [R] **logistic postestimation**, [R] **reg3**, [R] **regress**, [R] **regress postestimation**, [R] **rologit**, [R] **simulate**, [R] **suest**, [R] **sureg**, [R] **tabstat**, [R] **tabulate twoway**, [R] **test**, [R] **tetrachoric**, [ST] **stsplit**, [U] **20.20 References**

Wei, L. J., [P] **_robust**, [ST] **stcox**, [ST] **stcrreg**, [SVY] **svy estimation**, [U] **20.20 References**

Weibull, W., [ST] **streg**

Weisberg, H. F., [R] **summarize**

Weisberg, S., [R] **boxcox**, [R] **regress**, [R] **regress postestimation**

Weiss, J., [MV] **mdsmat**

Welch, B. L., [R] **ttest**

Welch, K. B., [XT] **xtmixed**

Weller, S. C., [MV] **ca**

Wellington, J. F., [R] **qreg**

Wells, K. B., [R] **lincom**, [R] **mlogit**, [R] **mprobit**, [R] **mprobit postestimation**, [R] **predictnl**, [R] **slogit**

Welsch, R. E., [R] **estat**, [R] **regress postestimation**, [U] **18.13 References**

Wernow, J. B., [D] **destring**

West, B. T., [SVY] **subpopulation estimation**, [XT] **xtmixed**

West, K. D., [R] **glm**, [R] **gmm**, [R] **ivregress**, [TS] **newey**, [TS] **pperron**, [XT] **xtunitroot**

West, S., [ST] **epitab**

West, S. G., [R] **pcorr**

Westfall, R. S., [M-5] **optimize( )**

Westlake, W. J., [R] **pkequiv**

White, H., [P] **_robust**, [R] **regress**, [R] **regress postestimation**, [R] **suest**, [TS] **newey**, [TS] **prais**, [XT] **xtivreg**, [U] **20.20 References**

White, I. R., [MI] **intro**, [MI] **intro substantive**, [MI] **mi estimate**, [ST] **sts test**

White, K. J., [R] **boxcox**, [R] **regress postestimation time series**

White, P. O., [MV] **Glossary**, [MV] **rotate**, [MV] **rotatemat**

Whitehead, A., [XT] **xtunitroot**

Whitehouse, E., [R] **inequality**

Whitney, D. R., [R] **kwallis**, [R] **ranksum**

Whittaker, J., [D] **functions**, [MV] **ca**, [MV] **factor**, [MV] **mca**, [MV] **pca**

Wichern, D. W., [MV] **canon**, [MV] **discrim**, [MV] **discrim estat**, [MV] **discrim lda**, [MV] **discrim lda postestimation**, [MV] **mvtest**,

Wichern, D. W., *continued*
  [MV] **mvtest correlations**, [MV] **mvtest covariances**, [MV] **mvtest means**
Wichura, M. J., [D] **functions**
Wiesner, R. H., [ST] **stcrreg**
Wiggins, V. L., [R] **regress postestimation**, [R] **regress postestimation time series**, [TS] **arch**, [TS] **arima**, [TS] **sspace**
Wilcox, D. W., [R] **ivregress postestimation**
Wilcox, R. R., [D] **egen**
Wilcoxon, F., [R] **kwallis**, [R] **ranksum**, [R] **signrank**, [ST] **sts test**
Wilde, J., [R] **gmm**
Wilk, M. B., [R] **cumul**, [R] **diagnostic plots**, [R] **swilk**
Wilkinson, J. H., [P] **matrix symeigen**
Wilkinson, L., [ST] **sts**
Wilks, D. S., [R] **brier**
Wilks, S. S., [MV] **canon**, [MV] **hotelling**, [MV] **manova**
Williams, B., [SVY] **survey**
Williams, B. K., [MV] **discrim lda**
Williams, R., [R] **ologit**, [R] **pcorr**, [R] **stepwise**
Williams, W. T., [MV] **cluster**
Wilson, E. B., [MV] **mvtest normality**, [R] **ci**
Wilson, M., [MV] **rotate**
Wilson, S. R., [R] **bootstrap**
Windmeijer, F., [R] **gmm**, [XT] **xtabond**, [XT] **xtdpd**, [XT] **xtdpdsys**
Winer, B. J., [R] **anova**, [R] **loneway**, [R] **oneway**
Winsten, C. B., [TS] **prais**
Winter, N. J. G., [G] **graph twoway scatter**, [P] **levelsof**, [SVY] **survey**
Winters, P. R., [TS] **tssmooth**, [TS] **tssmooth dexponential**, [TS] **tssmooth exponential**, [TS] **tssmooth hwinters**, [TS] **tssmooth shwinters**
Wish, M., [MV] **mds**, [MV] **mdslong**, [MV] **mdsmat**
Wittes, J., [ST] **stpower**
Wolfe, F., [R] **correlate**, [R] **spearman**
Wolfe, R., [R] **ologit**, [R] **oprobit**, [R] **tabulate twoway**
Wolfinger, R. D., [XT] **xtmelogit**
Wolfowitz, J., [TS] **varwle**
Wolfram, S., [ST] **streg**, [XT] **xtmelogit postestimation**
Wolfson, C., [R] **kappa**
Wolk, A., [ST] **epitab**
Wolpin, K. I., [R] **asmprobit**
Wolter, K. M., [SVY] **survey**, [SVY] **variance estimation**
Wong, W. H., [MI] **intro substantive**, [MI] **mi impute mvn**
Wood, F. S., [R] **diagnostic plots**
Woodard, D. E., [MV] **manova**
Wooldridge, J. M., [R] **gmm**, [R] **intreg**, [R] **ivprobit**, [R] **ivregress**, [R] **ivregress postestimation**, [R] **ivtobit**, [R] **margins**, [R] **regress**, [R] **regress postestimation time series**, [R] **tobit**, [TS] **arch**, [TS] **dvech**, [TS] **prais**, [XT] **xt**, [XT] **xtivreg**, [XT] **xtreg**

Woolf, B., [ST] **epitab**
Working, H., [R] **roc**, [R] **rocfit**
Wretman, J., [SVY] **variance estimation**
Wright, J. H., [R] **ivregress**, [R] **ivregress postestimation**, [XT] **xthtaylor**
Wright, J. T., [R] **binreg**
Wright, P. G., [R] **ivregress**
Wu, C. F. J., [R] **qreg**
Wu, D.-M., [R] **ivregress postestimation**
Wu, P. X., [XT] **xtregar**
Wu, S., [XT] **xtunitroot**

X

Xie, Y., [R] **logit**, [R] **probit**
Xu, J., [R] **cloglog**, [R] **logit**, [R] **logistic**, [R] **mlogit**, [R] **ologit**, [R] **oprobit**, [R] **probit**

Y

Yang, K., [MV] **mds**
Yar, M., [TS] **tssmooth**, [TS] **tssmooth dexponential**, [TS] **tssmooth exponential**, [TS] **tssmooth hwinters**, [TS] **tssmooth shwinters**
Yates, F., [P] **levelsof**
Yates, J. F., [R] **brier**
Yee, T. W., [R] **slogit**
Yellott Jr., J. I., [R] **rologit**
Yen, S., [ST] **epitab**
Yen, W. M., [R] **alpha**
Yogo, M., [R] **ivregress**, [R] **ivregress postestimation**, [XT] **xthtaylor**
Young, F. W., [MV] **mds**, [MV] **mdslong**, [MV] **mdsmat**
Young, G., [MV] **mds**, [MV] **mdslong**, [MV] **mdsmat**
Ypma, T. J., [M-5] **optimize( )**
Yu, J., [MV] **mvtest**, [MV] **mvtest means**
Yule, G. U., [MV] *measure_option*

Z

Zabell, S., [R] **kwallis**
Zakoian, J. M., [TS] **arch**
Zappasodi, P., [MV] **manova**
Zavoina, W., [R] **ologit**
Zeger, S. L., [XT] **xtcloglog**, [XT] **xtgee**, [XT] **xtlogit**, [XT] **xtmixed**, [XT] **xtnbreg**, [XT] **xtpoisson**, [XT] **xtprobit**
Zelen, M., [R] **ttest**
Zellner, A., [R] **frontier**, [R] **nlsur**, [R] **reg3**, [R] **sureg**, [TS] **prais**, [XT] **xtfrontier**
Zelterman, D., [R] **tabulate twoway**
Zhao, L. P., [XT] **xtgee**
Zimmerman, F., [R] **regress**
Zirkler, B., [MV] **mvtest**, [MV] **mvtest normality**
Zubin, J., [MV] *measure_option*

Zubkoff, M., [MV] **factor**, [MV] **factor postestimation**,
    [R] **alpha**, [R] **lincom**, [R] **mlogit**, [R] **mprobit**,
    [R] **mprobit postestimation**, [R] **predictnl**,
    [R] **slogit**
Zwiers, F. W., [R] **brier**

Subject index

! (not), *see* logical operators

!= (not equal), *see* relational operators

& (and), *see* logical operators

*, clear subcommand, [D] **clear**

* abbreviation character, *see* abbreviations

* comment indicator, [P] **comments**

− abbreviation character, *see* abbreviations

−> operator, [M-2] **struct**

., class, [P] **class**

/* */ comment delimiter, [M-2] **comments**,
    [P] **comments**

// comment indicator, [M-2] **comments**, [P] **comments**

/// comment indicator, [P] **comments**

; delimiter, [P] **#delimit**

< (less than), *see* relational operators

<= (less than or equal), *see* relational operators

== (equality), *see* relational operators

> (greater than), *see* relational operators

>= (greater than or equal), *see* relational operators

? abbreviation characters, *see* abbreviations

| (or), *see* logical operators

~ (not), *see* logical operators

~ abbreviation character, *see* abbreviations

~= (not equal), *see* relational operators

100% sample, [SVY] **Glossary**

A

.a, .b, . . . , .z, *see* missing values

Aalen–Nelson cumulative hazard, *see* Nelson–Aalen
    cumulative hazard

abbrev( ) function, [D] **functions**, [M-5] **abbrev( )**

abbreviations, [U] **11.2 Abbreviation rules**;
    [U] **11.1.1 varlist**, [U] **11.4 varlists**
    unabbreviating command names, [P] **unabcmd**
    unabbreviating variable list, [P] **unab**; [P] **syntax**

abond, estat subcommand, [XT] **xtabond
    postestimation**, [XT] **xtdpd postestimation**,
    [XT] **xtdpdsys postestimation**

aborting command execution, [U] **9 The Break key**,
    [U] **10 Keyboard use**

about command, [R] **about**

abs( ) function, [D] **functions**, [M-5] **abs( )**

absolute value dissimilarity measure,
    [MV] *measure_option*

absolute value function, *see* abs( ) function

absorption in regression, [R] **areg**

ac command, [TS] **corrgram**

accelerated failure-time model, [ST] **Glossary**,
    [ST] **streg**

Access, Microsoft, reading data from, [D] **odbc**,
    [U] **21.4 Transfer programs**

accrual period, [ST] **Glossary**, [ST] **stpower
    exponential**, [ST] **stpower logrank**

accum, matrix subcommand, [P] **matrix accum**

acos( ) function, [D] **functions**, [M-5] **sin( )**

acosh( ) function, [D] **functions**, [M-5] **sin( )**

acprplot command, [R] **regress postestimation**

actuarial tables, *see* life tables

add, irf subcommand, [TS] **irf add**

add, mi subcommand, [MI] **mi add**

add, return subcommand, [P] **return**

added lines, $y=x$, [G] **graph twoway function**

addedlinestyle, [G] *addedlinestyle*

added-variable plots, [G] **graph other**, [R] **regress
    postestimation**

adding
    fits, *see* fits, adding
    lines, *see* lines, adding
    text, *see* text, adding

addition, [M-2] **op_arith**, [M-2] **op_colon**

addition across
    observations, [D] **egen**
    variables, [D] **egen**

addition operator, *see* arithmetic operators

addplot( ) option, [G] *addplot_option*

adjoint matrix, [M-2] **op_transpose**, [M-5] **conj( )**

adjugate matrix, [M-2] **op_transpose**, [M-5] **conj( )**

adjusted Kaplan–Meier survivor function, [ST] **sts**

adjusted margins, [R] **margins**

adjusted means, [R] **margins**

adjusted partial residual plot, [R] **regress
    postestimation**

administrative censoring, [ST] **Glossary**

ado,
    clear subcommand, [D] **clear**
    update subcommand, [R] **update**
    view subcommand, [R] **view**

ado
    command, [R] **net**
    describe command, [R] **net**
    dir command, [R] **net**
    uninstall command, [R] **net**

.ado filename suffix, [U] **11.6 File-naming conventions**

ado_d, view subcommand, [R] **view**

ado-files, [M-1] **ado**, [P] **sysdir**, [P] **version**,
    [U] **3.5 The Stata Journal**, [U] **17 Ado-files**,
    [U] **18.11 Ado-files**
    adding comments to, [P] **comments**
    debugging, [P] **trace**
    downloading, *see* files, downloading
    editing, [R] **doedit**
    installing, [R] **net**, [R] **sj**, [R] **ssc**, [U] **17.6 How do
    I install an addition?**
    location of, [R] **which**, [U] **17.5 Where does Stata
    look for ado-files?**
    long lines, [P] **#delimit**, [U] **18.11.2 Comments and
    long lines in ado-files**
    official, [R] **update**, [U] **28 Using the Internet to
    keep up to date**

ado-path, [M-5] **adosubdir( )**

adopath command, [P] **sysdir**, [U] **17.5 Where does
    Stata look for ado-files?**

adosize, [U] **18.11 Ado-files**

adosize, set subcommand, [R] **set**

adosize set command, [P] **sysdir**

adosubdir( ) function, [M-5] **adosubdir( )**

adosubdir macro extended function, [P] **macro**

adoupdate command, [R] **adoupdate**

agglomerative hierarchical clustering methods,
    [MV] **Glossary**

aggregate

    functions, [D] **egen**

    statistics, dataset of, [D] **collapse**

agreement, interrater, [R] **kappa**

AIC, [R] **BIC note**, [R] **estat**, [R] **estimates stats**,
    [R] **glm**, [ST] **streg**, [TS] **varsoc**

Akaike information criterion, *see* AIC

algebraic expressions, functions, and operators,
    [P] **matrix define**, [U] **13 Functions and**
    expressions, [U] **13.3 Functions**

alignment of text, [G] *textbox_options*

alignmentstyle, [G] *alignmentstyle*

all( ) function, [M-5] **all( )**

all,

    clear subcommand, [D] **clear**   update
    subcommand, [R] **update**

_all, [U] **11.1.1 varlist**

all macro extended function, [P] **macro**

allof( ) function, [M-5] **all( )**

alpha coefficient, Cronbach's, [R] **alpha**

alpha command, [R] **alpha**

alphabetizing

    observations, [D] **sort**; [D] **gsort**

    variable names, [D] **order**

    variables, [D] **sort**

alphanumeric variables, *see* string variables, parsing

alternative-specific multinomial probit regression,
    [R] **asmprobit**

alternatives, estat subcommand, [R] **asclogit**
    postestimation, [R] **asmprobit postestimation**,
    [R] **asroprobit postestimation**

ameans command, [R] **ameans**

analysis of covariance, *see* ANCOVA

analysis of variance, *see* ANOVA

analysis step, [MI] **intro substantive**, [MI] **mi estimate**,
    also see estimation

analysis time, [ST] **Glossary**

analysis-of-variance test of normality, [R] **swilk**

analytic weights, [U] **11.1.6 weight**,
    [U] **20.18.2 Analytic weights**

ANCOVA, [R] **anova**

and operator, [U] **13.2.4 Logical operators**

Anderberg coefficient similarity measure,
    [MV] *measure_option*

angle of text, [G] *anglestyle*

anglestyle, [G] *anglestyle*

angular similarity measure, [MV] *measure_option*

ANOVA, [R] **anova**, [R] **loneway**, [R] **oneway**

    Kruskal–Wallis, [R] **kwallis**

    repeated measures, [R] **anova**

anova, estat subcommand, [MV] **discrim lda**
    postestimation

anova command, [R] **anova**, [R] **anova postestimation**,
    also see postestimation command

anti, estat subcommand, [MV] **factor**
    postestimation, [MV] **pca postestimation**

anti-image

    correlation matrix, [MV] **factor postestimation**,
        [MV] **Glossary**, [MV] **pca postestimation**

    covariance matrix, [MV] **factor postestimation**,
        [MV] **Glossary**, [MV] **pca postestimation**

any( ) function, [M-5] **all( )**

anycount( ), egen function, [D] **egen**

anymatch( ), egen function, [D] **egen**

anyof( ) function, [M-5] **all( )**

anyvalue( ), egen function, [D] **egen**

A-PARCH, [TS] **arch**

append, mi subcommand, [MI] **mi append**

append command, [D] **append**, [U] **22 Combining**
    datasets

_append variable, [D] **append**

appending

    data, [D] **append**, [MI] **mi append**,
        [U] **22 Combining datasets**

    files, [D] **copy**

    rows and columns to matrix, [P] **matrix define**

apply recording, [G] **graph play**

approximating Euclidean distances, [MV] **mds**
    postestimation

AR, [TS] **dfactor**, [TS] **sspace**

arbitrary pattern of missing values, [MI] **mi impute**
    mvn, *also see* pattern of missingness

arccosine, arcsine, and arctangent functions,
    [D] **functions**

ARCH

    effects, estimation, [TS] **arch**

    effects, testing for, [R] **regress postestimation time**
        series

    model, [TS] **dvech**, [TS] **Glossary**

    postestimation, [TS] **arch postestimation**

    regression, [TS] **arch**

arch command, [TS] **arch**, [TS] **arch postestimation**

archlm, estat subcommand, [R] **regress**
    postestimation time series

area, graph twoway subcommand, [G] **graph twoway**
    area

areas, [G] *colorstyle*, *also see* fill areas and colors

areastyle, [G] *areastyle*

areg command, [R] **areg**, [R] **areg postestimation**,
    also see postestimation command

Arellano–Bond

    estimator, [XT] **Glossary**, [XT] **xtabond**

    postestimation, [XT] **xtabond postestimation**,
        [XT] **xtdpd postestimation**, [XT] **xtdpdsys**
        postestimation

Arellano–Bover estimator, [XT] **xtdpd**, [XT] **xtdpdsys**

arg( ) function, [M-5] **sin( )**

args( ) function, [M-5] **args( )**

args command, [P] **syntax**
arguments,
    program, [M-2] **declarations**, [M-6] **Glossary**
    values returned in, [M-1] **returnedargs**
    varying number, [M-2] **optargs**, [M-5] **args( )**
ARIMA
    postestimation, [TS] **arima postestimation**
    regression, [TS] **arima**
arima command, [TS] **arima**, [TS] **arima**
    postestimation
arithmetic operators, [M-2] **op_arith**, [P] **matrix define**,
    [U] **13.2.1 Arithmetic operators**
ARMA, [TS] **arch**, [TS] **arima**, [TS] **dfactor**,
    [TS] **Glossary**, [TS] **sspace**
ARMAX, [TS] **dfactor**, [TS] **sspace**
ARMAX model, [TS] **arima**, [TS] **Glossary**
arrays, class, [P] **class**
.Arrdropall built-in class modifier, [P] **class**
.Arrdropel built-in class modifier, [P] **class**
.arrindexof built-in class function, [P] **class**
.arrnels built-in class function, [P] **class**
arrows, [G] **graph twoway pcarrow**
.Arrpop built-in class modifier, [P] **class**
.Arrpush built-in class modifier, [P] **class**
as error, display directive, [P] **display**
as input, display directive, [P] **display**
as result, display directive, [P] **display**
as text, display directive, [P] **display**
as txt, display directive, [P] **display**
asarray( ) function, [M-5] **asarray( )**
asarray_contains( ) function, [M-5] **asarray( )**
asarray_contents( ) function, [M-5] **asarray( )**
asarray_create( ) function, [M-5] **asarray( )**
asarray_elements( ) function, [M-5] **asarray( )**
asarray_first( ) function, [M-5] **asarray( )**
asarray_key( ) function, [M-5] **asarray( )**
asarray_keys( ) function, [M-5] **asarray( )**
asarray_next( ) function, [M-5] **asarray( )**
asarray_notfound( ) function, [M-5] **asarray( )**
asarray_remove( ) function, [M-5] **asarray( )**
ascategory( ) option, [G] **graph bar**, [G] **graph box**,
    [G] **graph dot**
ascii( ) function, [M-5] **ascii( )**
ASCII,
    codes, [M-5] **ascii( )**
    reading data in, [D] **infile**, [D] **infile (fixed format)**,
        [D] **infile (free format)**, [D] **infix (fixed format)**,
        [D] **insheet**
    saving data in, [D] **outfile**, [D] **outsheet**
    text files, writing and reading, [P] **file**
asclogit command, [R] **asclogit**, [R] **asclogit**
    postestimation, *also see* postestimation command
asin( ) function, [D] **functions**, [M-5] **sin( )**
asinh( ) function, [D] **functions**, [M-5] **sin( )**
_asis, display directive, [P] **display**
asis print color mapping, [G] **set printcolor**

asmprobit command, [R] **asmprobit**, [R] **asmprobit**
    postestimation, *also see* postestimation command
aspect ratio, [G] *aspect_option*
    changing, [G] **graph display**
    controlling, [G] **graph combine**
asroprobit command, [R] **asroprobit**, [R] **asroprobit**
    postestimation, *also see* postestimation command
assert( ) function, [M-5] **assert( )**
assert command, [D] **assert**
asserteq( ) function, [M-5] **assert( )**
assignment, class, [P] **class**
assignment operator, [M-2] **op_assignment**
association tests, *see* tests, association
association, measures of, [R] **tabulate twoway**
associative arrays, [M-5] **asarray( )**
asymmetry, [R] **lnskew0**, [R] **lv**, [R] **sktest**,
    [R] **summarize**
asyvars option, [G] **graph bar**, [G] **graph box**,
    [G] **graph dot**
at risk, [ST] **Glossary**
atan( ) function, [D] **functions**, [M-5] **sin( )**
atan2( ) function, [D] **functions**, [M-5] **sin( )**
atanh( ) function, [D] **functions**, [M-5] **sin( )**
at-risk table, [ST] **sts graph**
attributable fraction, [ST] **epitab**, [ST] **Glossary**
attributable proportion, [ST] **epitab**
attribute tables, [R] **table**, [R] **tabulate, summarize()**,
    [R] **tabulate twoway**
AUC, [R] **pk**
augmented
    component-plus-residual plot, [R] **regress**
        postestimation
    partial residual plot, [R] **regress postestimation**
Author Support Program, [U] **3.8.2 For authors**
autocode( ) function, [D] **functions**,
    [U] **25.1.2 Converting continuous variables to**
    categorical variables
autocorrelation, [R] **regress postestimation time**
    series, [TS] **arch**, [TS] **arima**, [TS] **corrgram**,
    [TS] **dfactor**, [TS] **Glossary**, [TS] **newey**,
    [TS] **prais**, [TS] **sspace**, [TS] **var**, [TS] **varlmar**,
    [XT] **xtabond**, [XT] **xtgee**, [XT] **xtgls**,
    [XT] **xtpcse**, [XT] **xtregar**
auto.dta, [U] **1.2.2 Example datasets**
automatic print color mapping, [G] **set printcolor**
Automation, [P] **automation**
autoregressive
    conditional heteroskedasticity, [TS] **arch**, [TS] **dvech**
        testing for, [R] **regress postestimation time**
        series
    integrated moving average, [TS] **arch**, [TS] **arima**
    model, [TS] **dfactor**, [TS] **sspace**
    moving average, [TS] **arch**, [TS] **arima**,
        [TS] **dfactor**, [TS] **sspace**
    process, [TS] **Glossary**, [XT] **Glossary**,
        [XT] **xtabond**, [XT] **xtdpd**, [XT] **xtdpdsys**
autotabgraphs, set subcommand, [R] **set**
available area, [G] *region_options*

available-case analysis, [MI] **intro substantive**
average marginal effects, [R] **margins**
average partial effects (APEs), [R] **margins**
average predictions, [R] **margins**
average RVI, [MI] **Glossary**, [MI] **mi estimate**
averagelinkage, cluster subcommand,
    [MV] **cluster linkage**
average-linkage clustering, [MV] **cluster**, [MV] **cluster
    linkage**, [MV] **clustermat**, [MV] **Glossary**
averages, *see* means
avplot and avplots commands, [R] **regress
    postestimation**
[aweight=*exp*] modifier, [U] **11.1.6 weight**,
    [U] **20.18.2 Analytic weights**
axes
    multiple scales, [G] *axis_choice_options*
    setting offset between and plot region,
        [G] *region_options*
    suppressing, [G] *axis_scale_options*
axis
    labeling, [G] *axis_label_options*, [G] *axis_options*
    line, look of, [G] *axis_scale_options*,
        [G] *cat_axis_label_options*,
        [G] *cat_axis_line_options*
    log, [G] *axis_scale_options*
    overall look, [G] *axisstyle*
    range, [G] *axis_scale_options*
    reversed, [G] *axis_scale_options*
    scale, [G] *axis_options*, [G] *axis_scale_options*
    selection of, [G] *axis_choice_options*
    suppressing, [G] *axis_scale_options*
    ticking, [G] *axis_label_options*
    titling, [G] *axis_options*, [G] *axis_title_options*
        suppressing, [G] *axis_title_options*
axisstyle, [G] *axisstyle*

B

b() function, [D] **functions**
_b[], [U] **13.5 Accessing coefficients and standard
    errors**
b1title() option, [G] *title_options*
b2title() option, [G] *title_options*
background color, [G] **schemes intro**
    setting, [G] *region_options*
balanced data, [XT] **Glossary**
balanced repeated replication, [SVY] *brr_options*,
    [SVY] **Glossary**, [SVY] **svy brr**,
    [SVY] **variance estimation**
bar
    graph subcommand, [G] **graph bar**
    graph twoway subcommand, [G] **graph twoway
        bar**
bar charts, [G] **graph bar**
barbsize option, [G] **graph twoway pcarrow**
barlook options, [G] *barlook_options*

bars
    labeling, [G] *blabel_option*
    look of, [G] *barlook_options*
Bartlett scoring, [MV] **factor postestimation**
Bartlett's
    bands, [TS] **corrgram**
    periodogram test, [TS] **wntestb**
    test for equal variances, [R] **oneway**
base conversion, [M-5] **inbase( )**
BASE directory, [P] **sysdir**, [U] **17.5 Where does Stata
    look for ado-files?**
base level, [U] **11.4.3 Factor variables**
base plottypes, [G] *advanced_options*
baseline, [ST] **Glossary**
baseline dataset, [ST] **stbase**
baseline hazard and survivor functions, [ST] **stcox**,
    [ST] **stcox PH-assumption tests**, [ST] **stcrreg**
baseline suboption, [G] *alignmentstyle*
basis, orthonormal, [P] **matrix svd**
Battese–Coelli parameterization, [XT] **xtfrontier**
Bayes' theorem, [MV] **Glossary**
Bayesian concepts, [MI] **intro substantive**
Bayesian information criterion, *see* BIC
bcskew0 command, [R] **lnskew0**
Bentler rotation, [MV] **rotate**, [MV] **rotatemat**
Bentler's invariant pattern simplicity rotation,
    [MV] **Glossary**
Berndt–Hall–Hall–Hausman algorithm,
    [M-5] **moptimize( )**, [M-5] **optimize( )**, [R] **ml**
beta
    coefficients, [R] **gmm**, [R] **ivregress**, [R] **nlsur**,
        [R] **regress**, [R] **regress postestimation**
    density,
        central, [D] **functions**
        noncentral, [D] **functions**
    distribution,
        cumulative, [D] **functions**
        cumulative noncentral, [D] **functions**
    function, [M-5] **normal( )**
        complement to incomplete, [D] **functions**
        incomplete, [D] **functions**
        inverse cumulative, [D] **functions**
        inverse cumulative noncentral, [D] **functions**
        inverse reverse cumulative, [D] **functions**
        reverse cumulative, [D] **functions**
betaden() function, [D] **functions**, [M-5] **normal( )**
between estimators, [XT] **Glossary**, [XT] **xtivreg**,
    [XT] **xtreg**
between-cell means and variances, [XT] **xtdescribe**,
    [XT] **xtsum**
between-imputation variability, [MI] **mi estimate**
between matrix, [MV] **Glossary**
BFGS algorithm, *see* Broyden–Fletcher–Goldfarb–
    Shanno algorithm
bgodfrey, estat subcommand, [R] **regress
    postestimation time series**
BHHH algorithm, *see* Berndt–Hall–Hall–Hausman
    algorithm

bias corrected and accelerated, [R] **bootstrap postestimation**

BIC, [R] **BIC note**, [R] **estat**, [R] **estimates stats**, [R] **glm**, [ST] **streg**,

bin() option, [G] **graph twoway histogram**

binary files, writing and reading, [P] **file**

binary I/O, [M-5] **bufio( )**

binary outcome models, *see* dichotomous outcome models

binary variable imputation, *see* imputation, binary

binomial
    distribution,
        confidence intervals, [R] **ci**
            cumulative, [D] **functions**
            inverse cumulative, [D] **functions**
            inverse reverse cumulative, [D] **functions**
            reverse cumulative, [D] **functions**
        family regression, [R] **binreg**
        probability mass function, [D] **functions**
        probability test, [R] **bitest**

binomial() function, [D] **functions**, [M-5] **normal( )**

binomialp() function, [D] **functions**, [M-5] **normal( )**

binomialtail() function, [D] **functions**, [M-5] **normal( )**

binormal() function, [D] **functions**, [M-5] **normal( )**

binreg command, [R] **binreg postestimation**, [R] **binreg**, *also see* postestimation command

bioequivalence tests, [R] **pk**, [R] **pkequiv**

biopharmaceutical data, *see* pk (pharmacokinetic data)

biplot, [MV] **Glossary**

biplot command, [MV] **biplot**

biplots, [MV] **biplot**, [MV] **ca postestimation**

biprobit command, [R] **biprobit**, [R] **biprobit postestimation**, *also see* postestimation command

biquartimax rotation, [MV] **Glossary**, [MV] **rotate**, [MV] **rotatemat**

biquartimin rotation, [MV] **Glossary**, [MV] **rotate**, [MV] **rotatemat**

bitest and bitesti commands, [R] **bitest**

bitmap, [G] *png_options*, [G] *tif_options*

bivariate normal function, [D] **functions**

bivariate probit regression, [R] **biprobit**, [SVY] **svy estimation**

biweight regression estimates, [R] **rreg**

biyearly() function, [U] **25 Working with categorical data and factor variables**

blanks, removing from strings, [D] **functions**

block diagonal covariance, [MV] **mvtest covariances**

block diagonal matrix, [M-5] **blockdiag( )**

block exogeneity, [TS] **vargranger**

blockdiag() function, [M-5] **blockdiag( )**

blogit command, [R] **glogit**, [R] **glogit postestimation**, *also see* postestimation command

Blundell–Bond estimator, [XT] **xtdpd**, [XT] **xtdpdsys**

BLUPs, [XT] **Glossary**

bold font, [G] *text*

Bonferroni
    adjustment, [R] **correlate**, [R] **spearman**, [R] **test**, [R] **testnl**
    multiple comparison test, [R] **anova postestimation**, [R] **oneway**, [R] **regress postestimation**

bootstrap
    sampling and estimation, [P] **postfile**, [R] **bootstrap**, [R] **bsample**, [R] **bstat**, [R] **qreg**, [R] **simulate**
    standard errors, [R] *vce_option*, [XT] *vce_options*

bootstrap, estat subcommand, [R] **bootstrap postestimation**

bootstrap prefix command, [R] **bootstrap**, [R] **bootstrap postestimation**, *also see* postestimation command

border around plot region, suppressing, [G] *region_options*

borders
    misplacement of, [G] *added_text_options*
    suppressing, [G] *linestyle*
    suppressing around plot region, [G] *region_options*

Boston College archive, *see* SSC archive

bottom suboption, [G] *alignmentstyle*

boundary solution, [MV] **Glossary**

box, graph subcommand, [G] **graph box**

Box M test, [MV] **mvtest covariances**

box plots, [G] **graph box**

Box–Cox
    power transformations, [R] **lnskew0**
    regression, [R] **boxcox**

boxcox command, [R] **boxcox**, [R] **boxcox postestimation**, *also see* postestimation command

Box's conservative epsilon, [R] **anova**

bprobit command, [R] **glogit**, [R] **glogit postestimation**, *also see* postestimation command

break, [M-2] **break**

break command, [P] **break**

Break key, [U] **9 The Break key**, [U] **16.1.4 Error handling in do-files**
    interception, [P] **break**, [P] **capture**
    processing, [M-5] **setbreakintr( )**

breakkey() function, [M-5] **setbreakintr( )**

breakkeyreset() function, [M-5] **setbreakintr( )**

breitung, xtunitroot subcommand, [XT] **xtunitroot**

Breitung test, [XT] **xtunitroot**

Breusch–Godfrey test, [R] **regress postestimation time series**

Breusch–Pagan Lagrange multiplier test, [XT] **xtreg postestimation**

Breusch–Pagan test of independence, [R] **mvreg**, [R] **sureg**

brier command, [R] **brier**

Brier score decomposition, [R] **brier**

broad type, [M-6] **Glossary**

browse, view subcommand, [R] **view**

browse command, [D] **edit**

Broyden–Fletcher–Goldfarb–Shanno algorithm, [M-5] **moptimize( )**, [M-5] **optimize( )**, [R] **ml**

BRR, *see* balanced repeated replication

brr_options, [SVY] *brr_options*

bsample command, [R] **bsample**

bsqreg command, [R] **qreg**, [R] **qreg postestimation**, *also see* postestimation command

bstat command, [R] **bstat**

bstyle() option, [G] *barlook_options*

bufbfmtisnum() function, [M-5] **bufio( )**

bufbfmtlen() function, [M-5] **bufio( )**

bufbyteorder() function, [M-5] **bufio( )**

buffered I/O, [M-5] **bufio( )**

bufget() function, [M-5] **bufio( )**

bufio() function, [M-5] **bufio( )**

bufmissingvalue() function, [M-5] **bufio( )**

bufput() function, [M-5] **bufio( )**

building a graph, [G] **graph intro**

built-in, class, [P] **class**

built-in variables, [U] **11.3 Naming conventions**, [U] **13.4 System variables (_variables)**

bullet symbol, [G] *text*

_by() function, [P] **byable**

by(), use of legends with, [G] *by_option*, [G] *legend_option*

by() option, [G] *by_option*, [G] **graph bar**

by *varlist*: prefix, [D] **by**, [P] **byable**, [U] **11.5 by varlist: construct**, [U] **13.7 Explicit subscripting**, [U] **27.2 The by construct**

byable, [P] **byable**

by-graphs, look of, [G] *bystyle*

by-groups, [D] **by**, [D] **statsby**, [P] **byable**, [U] **11.5 by varlist: construct**

_byindex() function, [P] **byable**

_bylastcall() function, [P] **byable**

_byn1() function, [P] **byable**

_byn2() function, [P] **byable**

bysort *varlist*: prefix, [D] **by**

bystyle, [G] *bystyle*

byte, [D] **data types**, [I] **data types**

byte (storage type), [U] **12.2.2 Numeric storage types**

byteorder() function, [D] **functions**, [M-5] **byteorder( )**

C

C() function, [M-5] **C( )**

c() function, [D] **functions**, [M-5] **c( )**

c(adopath) c-class value, [P] **creturn**, [P] **sysdir**

c(adosize) c-class value, [P] **creturn**, [P] **sysdir**

c(ALPHA) c-class value, [P] **creturn**

c(alpha) c-class value, [P] **creturn**

c(autotabgraphs) c-class value, [P] **creturn**

c(born_date) c-class value, [P] **creturn**

c(byteorder) c-class value, [P] **creturn**

c(changed) c-class value, [P] **creturn**

c(checksum) c-class value, [D] **checksum**, [P] **creturn**

c(cmdlen) c-class value, [P] **creturn**

c(console) c-class value, [P] **creturn**

c(copycolor) c-class value, [P] **creturn**

c(current_date) c-class value, [P] **creturn**

c(current_time) c-class value, [P] **creturn**

c(dirsep) c-class value, [P] **creturn**

c(dockable) c-class value, [P] **creturn**

c(dockingguides) c-class value, [P] **creturn**

c(doublebuffer) c-class value, [P] **creturn**

c(dp) c-class value, [D] **format**, [P] **creturn**

c(eolchar) c-class value, [P] **creturn**

c(epsdouble) c-class value, [P] **creturn**

c(epsfloat) c-class value, [P] **creturn**

c(eqlen) c-class value, [P] **creturn**

c(fastscroll) c-class value, [P] **creturn**

c(filedate) c-class value, [P] **creturn**

c(filename) c-class value, [P] **creturn**

c(flavor) c-class value, [P] **creturn**

c(graphics) c-class value, [P] **creturn**

c(httpproxy) c-class value, [P] **creturn**

c(httpproxyauth) c-class value, [P] **creturn**

c(httpproxyhost) c-class value, [P] **creturn**

c(httpproxyport) c-class value, [P] **creturn**

c(httpproxypw) c-class value, [P] **creturn**

c(httpproxyuser) c-class value, [P] **creturn**

c(k) c-class value, [P] **creturn**

c(level) c-class value, [P] **creturn**

c(linegap) c-class value, [P] **creturn**

c(linesize) c-class value, [P] **creturn**

c(locksplitters) c-class value, [P] **creturn**

c(logtype) c-class value, [P] **creturn**

c(macgphengine) c-class value, [P] **creturn**

c(machine_type) c-class value, [P] **creturn**

c(macrolen) c-class value, [P] **creturn**

c(matacache) c-class value, [P] **creturn**

c(matafavor) c-class value, [P] **creturn**

c(matalibs) c-class value, [P] **creturn**

c(matalnum) c-class value, [P] **creturn**

c(matamofirst) c-class value, [P] **creturn**

c(mataoptimize) c-class value, [P] **creturn**

c(matastrict) c-class value, [P] **creturn**

c(matsize) c-class value, [P] **creturn**

c(maxbyte) c-class value, [P] **creturn**

c(max_cmdlen) c-class value, [P] **creturn**

c(maxdb) c-class value, [P] **creturn**

c(maxdouble) c-class value, [P] **creturn**

c(maxfloat) c-class value, [P] **creturn**

c(maxint) c-class value, [P] **creturn**

c(maxiter) c-class value, [P] **creturn**

c(max_k_current) c-class value, [P] **creturn**

c(max_k_theory) c-class value, [P] **creturn**

c(maxlong) c-class value, [P] **creturn**

c(max_macrolen) c-class value, [P] **creturn**

c(max_matsize) c-class value, [P] **creturn**

c(max_N_current) c-class value, [P] **creturn**

c(max_N_theory) c-class value, [P] **creturn**

c(maxstrvarlen) c-class value, [P] **creturn**

c(maxvar) c-class value, [D] **memory**, [P] **creturn**

c(max_width_current) c-class value, [P] **creturn**

c(max_width_theory) c-class value, [P] **creturn**

c(memory) c-class value, [D] **memory**, [P] **creturn**

c(minbyte) c-class value, [P] **creturn**
c(mindouble) c-class value, [P] **creturn**
c(minfloat) c-class value, [P] **creturn**
c(minint) c-class value, [P] **creturn**
c(minlong) c-class value, [P] **creturn**
c(min_matsize) c-class value, [P] **creturn**
c(mode) c-class value, [P] **creturn**
c(Mons) c-class value, [P] **creturn**
c(Months) c-class value, [P] **creturn**
c(more) c-class value, [P] **creturn**, [P] **more**
c(MP) c-class value, [P] **creturn**
c(N) c-class value, [P] **creturn**
c(namelen) c-class value, [P] **creturn**
c(noisily) c-class value, [P] **creturn**
c(notifyuser) c-class value, [P] **creturn**
c(odbcmgr) c-class value, [P] **creturn**
c(os) c-class value, [P] **creturn**
c(osdtl) c-class value, [P] **creturn**
c(pagesize) c-class value, [P] **creturn**
c(persistfv) c-class value, [P] **creturn**
c(persistvtopic) c-class value, [P] **creturn**
c(pi) c-class value, [P] **creturn**
c(pinnable) c-class value, [P] **creturn**
c(playsnd) c-class value, [P] **creturn**
c(printcolor) c-class value, [P] **creturn**
c(processors) c-class value, [P] **creturn**
c(processors_lic) c-class value, [P] **creturn**
c(processors_mach) c-class value, [P] **creturn**
c(processors_max) c-class value, [P] **creturn**
c(pwd) c-class value, [P] **creturn**
c(rc) c-class value, [P] **capture**, [P] **creturn**
c(reventries) c-class value, [P] **creturn**
c(revkeyboard) c-class value, [P] **creturn**
c(revwindow) c-class value, [P] **creturn**
c(rmsg) c-class value, [P] **creturn**, [P] **rmsg**
c(rmsg_time) c-class value, [P] **creturn**
c(scheme) c-class value, [P] **creturn**
c(scrollbufsize) c-class value, [P] **creturn**
c(SE) c-class value, [P] **creturn**
c(searchdefault) c-class value, [P] **creturn**
c(seed) c-class value, [P] **creturn**, [R] **set emptycells**,
    [R] **set seed**
c(smallestdouble) c-class value, [P] **creturn**
c(smoothfonts) c-class value, [P] **creturn**
c(smoothsize) c-class value, [P] **creturn**
c(stata_version) c-class value, [P] **creturn**
c(sysdir_base) c-class value, [P] **creturn**, [P] **sysdir**
c(sysdir_oldplace) c-class value, [P] **creturn**,
    [P] **sysdir**
c(sysdir_personal) c-class value, [P] **creturn**,
    [P] **sysdir**
c(sysdir_plus) c-class value, [P] **creturn**, [P] **sysdir**
c(sysdir_site) c-class value, [P] **creturn**, [P] **sysdir**
c(sysdir_stata) c-class value, [P] **creturn**,
    [P] **sysdir**
c(sysdir_updates) c-class value, [P] **creturn**,
    [P] **sysdir**

c(timeout1) c-class value, [P] **creturn**
c(timeout2) c-class value, [P] **creturn**
c(tmpdir) c-class value, [P] **creturn**
c(trace) c-class value, [P] **creturn**, [P] **trace**
c(tracedepth) c-class value, [P] **creturn**, [P] **trace**
c(traceexpand) c-class value, [P] **creturn**, [P] **trace**
c(tracehilite) c-class value, [P] **creturn**, [P] **trace**
c(traceindent) c-class value, [P] **creturn**, [P] **trace**
c(tracenumber) c-class value, [P] **creturn**, [P] **trace**
c(tracesep) c-class value, [P] **creturn**, [P] **trace**
c(type) c-class value, [D] **generate**, [P] **creturn**
c(update_interval) c-class value, [P] **creturn**
c(update_prompt) c-class value, [P] **creturn**
c(update_query) c-class value, [P] **creturn**
c(use_atsui_graph) c-class value, [P] **creturn**
c(use_qd_text) c-class value, [P] **creturn**
c(username) c-class value, [P] **creturn**
c(varabbrev) c-class value, [P] **creturn**
c(varkeyboard) c-class value, [P] **creturn**
c(varlabelpos) c-class value, [P] **creturn**
c(varwindow) c-class value, [P] **creturn**
c(version) c-class value, [P] **creturn**, [P] **version**
c(virtual) c-class value, [D] **memory**, [P] **creturn**
c(Wdays) c-class value, [P] **creturn**
c(Weekdays) c-class value, [P] **creturn**
c(width) c-class value, [P] **creturn**
C charts, [G] **graph other**
CA, [MV] **Glossary**
ca command, [MV] **ca**
cabiplot command, [MV] **ca postestimation**
calculator, [R] **display**
Caliński and Harabasz index stopping rules,
    [MV] **cluster stop**
_caller() pseudofunction, [D] **functions**
callersversion() function, [M-5] **callersversion( )**
camat command, [MV] **ca**
Canberra dissimilarity measure, [MV] *measure_option*
cancer data, [MI] **mi estimate**
candisc command, [MV] **candisc**
canon command, [MV] **canon**
canonical correlation analysis, [MV] **Glossary**
canonical correlations, [MV] **canon**, [MV] **canon
    postestimation**
canonical discriminant analysis, [MV] **candisc**,
    [MV] **Glossary**
canonical link, [XT] **Glossary**
canonical loadings, [MV] **Glossary**
canonical variate set, [MV] **Glossary**
canontest, estat subcommand, [MV] **discrim lda
    postestimation**
capped spikes, [G] *rcap_options*
caprojection command, [MV] **ca postestimation**
caption() option, [G] *title_options*
capture command, [P] **capture**
case–cohort data, [ST] **sttocc**

case–control
data, [R] **clogit**, [R] **logistic**, [R] **symmetry**,
[ST] **epitab**, [ST] **sttocc**
studies, [ST] **Glossary**
casement displays, [G] *by_option*
casewise deletion, [D] **egen**, [P] **mark**, *see* listwise
deletion
cat( ) function, [M-5] **cat( )**
cat command, [D] **type**
categorical axis, look of
labels, [G] *cat_axis_label_options*
line, [G] *cat_axis_line_options*
categorical data, [D] **egen**, [D] **recode**, [MV] **ca**,
[MV] **manova**, [MV] **mca**, [ST] **epitab**,
[SVY] **svy estimation**, [SVY] **svy: tabulate**
oneway, [SVY] **svy: tabulate twoway**
agreement, measures for, [R] **kappa**
categorical variable imputation, *see* imputation,
categorical
categorical variables, [U] **25.1.2 Converting continuous**
variables to categorical variables
cause-specific hazard, [ST] **Glossary**
cc command, [ST] **epitab**
cchart command, [R] **qc**
cci command, [ST] **epitab**
c-class command, [P] **creturn**
c-conformability, [M-2] **op_colon**, [M-6] **Glossary**
cd command, [D] **cd**
cd, net subcommand, [R] **net**
Cdhms( ) function, [D] **dates and times**, [D] **functions**,
[M-5] **date( )**
cdir, classutil subcommand, [P] **classutil**
ceil( ) function, [D] **functions**, [M-5] **trunc( )**
ceiling function, [D] **functions**
censored-normal regression, [SVY] **svy estimation**, *see*
interval regression
census, [SVY] **Glossary**
census data, [SVY] **direct standardization**,
[SVY] **survey**, [SVY] **variance estimation**
center suboption, [G] *justificationstyle*
centered data, [MV] **Glossary**
centile command, [R] **centile**
centiles, *see* percentiles, displaying, *see* percentiles
central tendency, measures of, [R] **summarize**;
[R] **ameans**, [R] **lv**, [R] **mean**
centroidlinkage, cluster subcommand,
[MV] **cluster linkage**
centroid-linkage clustering, [MV] **cluster**, [MV] **cluster**
linkage, [MV] **clustermat**, [MV] **Glossary**
certainty strata, [SVY] **estat**
certainty units, [SVY] **variance estimation**
certifying data, [D] **assert**, [D] **count**,
[D] **datasignature**, [D] **inspect**,
[P] **_datasignature**, [P] **signestimationsample**
certifying mi data are consistent, [MI] **mi update**
cf command, [D] **cf**
cgraph, irf subcommand, [TS] **irf cgraph**
changeeol command, [D] **changeeol**

changing
data, *see* editing data
directories, [D] **cd**
_char(#), display directive, [P] **display**
char
command, [U] **12.8 Characteristics**
define command, [P] **char**
list command, [P] **char**
macro extended function, [P] **macro**
rename command, [P] **char**
char( ) function, [M-5] **ascii( )**, [D] **functions**
character
data, *see* string variables
variables, [D] **infile (free format)**
character data, *see* string variables
characteristic roots, [M-5] **eigensystem( )**
characteristics, [P] **char**, [U] **12.8 Characteristics**,
[U] **18.3.6 Extended macro functions**,
[U] **18.3.13 Referring to characteristics**
charset, set subcommand, [P] **smcl**
chdir command, [D] **cd**
_chdir( ) function, [M-5] **chdir( )**
chdir( ) function, [M-5] **chdir( )**
check,
icd9 subcommand, [D] **icd9**
icd9p subcommand, [D] **icd9**
ml subcommand, [R] **ml**
checkestimationsample command,
[P] **signestimationsample**
checking data, [D] **assert**
checkpoint, [D] **snapshot**
checksum, set subcommand, [D] **checksum**, [R] **set**
checksum command, [D] **checksum**
checksums of data, [D] **checksum**, [D] **datasignature**,
[P] **_datasignature**, [P] **signestimationsample**
chelp command, [R] **help**
chi2( ) function, [D] **functions**, [M-5] **normal( )**
chi2tail( ) function, [D] **functions**, [M-5] **normal( )**
chi-squared
hypothesis test, [R] **sdtest**, [R] **test**, [R] **testnl**
probability plot, [R] **diagnostic plots**
test of independence, [R] **tabulate twoway**,
[SVY] **svy: tabulate twoway**
chi-squared, test of independence, [ST] **epitab**
chi-squared distribution,
cumulative, [D] **functions**
cumulative noncentral, [D] **functions**
inverse cumulative, [D] **functions**
inverse cumulative noncentral, [D] **functions**
inverse reverse cumulative, [D] **functions**
reverse cumulative, [D] **functions**
chi-squared probability and quantile plots, [G] **graph**
other
Chms( ) function, [D] **dates and times**, [D] **functions**,
[M-5] **date( )**
choice models, [R] **asclogit**, [R] **asmprobit**,
[R] **asroprobit**, [R] **clogit**, [R] **cloglog**,
[R] **exlogistic**, [R] **glogit**, [R] **heckprob**,

choice models, *continued*
>   [R] **hetprob**, [R] **ivprobit**, [R] **logistic**, [R] **logit**,
>   [R] **mlogit**, [R] **mprobit**, [R] **nlogit**, [R] **ologit**,
>   [R] **oprobit**, [R] **probit**, [R] **rologit**, [R] **scobit**,
>   [R] **slogit**

_cholesky() function, [M-5] **cholesky( )**

cholesky() function, [D] **functions**, [M-5] **cholesky( )**,
>   [P] **matrix define**

Cholesky decomposition, [M-5] **cholesky( )**, [P] **matrix
>   define**

Cholesky ordering, [TS] **Glossary**

_cholinv() function, [M-5] **cholinv( )**

cholinv() function, [M-5] **cholinv( )**

_cholsolve() function, [M-5] **cholsolve( )**

cholsolve() function, [M-5] **cholsolve( )**

chop() function, [D] **functions**

Chow test, [R] **anova**

ci and cii commands, [R] **ci**

CIF, *also see* cumulative incidence function

class
>   definition, [P] **class**
>   programming, [M-6] **Glossary**, [P] **class**
>   programming utilities, [P] **classutil**

class, [M-2] **class**

class exit command, [P] **class exit**

classes, [M-2] **class**

classfunctions, estat subcommand, [MV] **discrim
>   lda postestimation**

classical scaling, [MV] **Glossary**

classification, *see* discriminant analysis or cluster
>   analysis
>   function, [MV] **Glossary**
>   table, [MV] **Glossary**

classification, estat subcommand, [R] **logistic
>   postestimation**

.classmv built-in class function, [P] **class**

.classname built-in class function, [P] **class**

classtable, estat subcommand, [MV] **discrim
>   estat**, [MV] **discrim knn postestimation**,
>   [MV] **discrim lda postestimation**, [MV] **discrim
>   logistic postestimation**, [MV] **discrim qda
>   postestimation**

classutil
>   cdir command, [P] **classutil**
>   describe command, [P] **classutil**
>   dir command, [P] **classutil**
>   drop command, [P] **classutil**
>   which command, [P] **classutil**

classwide variable, [P] **class**

clean,
>   icd9 subcommand, [D] **icd9**
>   icd9p subcommand, [D] **icd9**

clear
>   * command, [D] **clear**
>   ado command, [D] **clear**
>   all command, [D] **clear**
>   command, [D] **clear**
>   mata command, [D] **clear**

clear, *continued*
>   matrix command, [D] **clear**
>   programs command, [D] **clear**
>   results command, [D] **clear**

clear,
>   datasignature subcommand, [D] **datasignature**
>   ereturn subcommand, [P] **ereturn**; [P] **return**
>   _estimates subcommand, [P] **_estimates**
>   estimates subcommand, [R] **estimates store**
>   ml subcommand, [R] **ml**
>   postutil subcommand, [P] **postfile**
>   return subcommand, [P] **return**
>   serset subcommand, [P] **serset**
>   sreturn subcommand, [P] **return**; [P] **program**
>   timer subcommand, [P] **timer**

clear, [M-3] **mata clear**

clear option, [U] **11.2 Abbreviation rules**

clearing estimation results, [R] **estimates store**;
>   [P] **ereturn**, [P] **_estimates**

clearing memory, [D] **clear**

clinical trial, [ST] **stpower**

clip() function, [D] **functions**

clist command, [D] **list**

Clock() function, [D] **dates and times**, [D] **functions**,
>   [M-5] **date( )**

clock() function, [D] **dates and times**, [D] **functions**,
>   [M-5] **date( )**

clock position, [G] *clockposstyle*

clock time, [TS] **tsset**

clogit command, [R] **clogit**, [R] **clogit
>   postestimation**, [R] **mlogit**, [R] **rologit**, *also see*
>   postestimation command

cloglog() function, [D] **functions**, [M-5] **logit( )**

cloglog command, [R] **cloglog**, [R] **cloglog
>   postestimation**, *also see* postestimation command

clonevar command, [D] **clonevar**

close, file subcommand, [P] **file**

clstyle() option, [G] *connect_options*

cluster, [SVY] **Glossary**, [SVY] **survey**, [SVY] **svy
>   estimation**, [SVY] **svyset**, [SVY] **variance
>   estimation**

cluster analysis, [MV] **cluster**, [MV] **cluster
>   dendrogram**, [MV] **cluster generate**,
>   [MV] **cluster kmeans and kmedians**,
>   [MV] **cluster linkage**, [MV] **cluster stop**,
>   [MV] **cluster utility**, [MV] **Glossary**
>   dendrograms, [MV] **cluster dendrogram**
>   dir, [MV] **cluster utility**
>   drop, [MV] **cluster utility**
>   hierarchical, [MV] **cluster**, [MV] **cluster linkage**,
>   [MV] **clustermat**
>   kmeans, [MV] **cluster kmeans and kmedians**
>   kmedians, [MV] **cluster kmeans and kmedians**
>   list, [MV] **cluster utility**
>   notes, [MV] **cluster notes**
>   programming, [MV] **cluster programming
>   subroutines**, [MV] **cluster programming
>   utilities**

cluster analysis, *continued*

rename, [MV] **cluster utility**

renamevar, [MV] **cluster utility**

stopping rules, [MV] **cluster**, [MV] **cluster stop**

trees, [MV] **cluster dendrogram**

use, [MV] **cluster utility**

cluster estimator of variance, [P] **_robust**,
[R] *vce_option*, [XT] *vce_options*

alternative-specific conditional logit model,
[R] **asclogit**

alternative-specific multinomial probit regression,
[R] **asmprobit**

alternative-specific rank-ordered probit regression,
[R] **asroprobit**

bivariate probit regression, [R] **biprobit**

complementary log-log regression, [R] **cloglog**

conditional logistic regression, [R] **clogit**

constrained linear regression, [R] **cnsreg**

Cox proportional hazards model, [ST] **stcox**,
[ST] **stcrreg**

estimate mean, [R] **mean**

estimate proportions, [R] **proportion**

estimate ratios, [R] **ratio**

estimate totals, [R] **total**

fit population-averaged panel-data models by using
GEE, [XT] **xtgee**

fixed- and random-effects linear models, [XT] **xtreg**

generalized linear models, [R] **glm**

generalized linear models for binomial family,
[R] **binreg**

generalized methods of moments, [R] **gmm**

heckman selection model, [R] **heckman**

heteroskedastic probit model, [R] **hetprob**

instrumental-variables regression, [R] **ivregress**

interval regression, [R] **intreg**

linear regression, [R] **regress**

linear regression with dummy-variable set, [R] **areg**

logistic regression, [R] **logistic**, [R] **logit**

logit and probit estimation for grouped data,
[R] **glogit**

maximum likelihood estimation, [R] **ml**

multinomial logistic regression, [R] **mlogit**

multinomial probit regression, [R] **mprobit**

negative binomial regression, [R] **nbreg**

nested logit regression, [R] **nlogit**

nonlinear least-squares estimation, [R] **nl**

nonlinear systems of equations, [R] **nlsur**

ordered logistic regression, [R] **ologit**

ordered probit regression, [R] **oprobit**

parametric survival models, [ST] **streg**

Poisson regression, [R] **poisson**

population-averaged cloglog models, [XT] **xtcloglog**

population-averaged logit models, [XT] **xtlogit**

population-averaged negative binomial models,
[XT] **xtnbreg**

population-averaged Poisson models, [XT] **xtpoisson**

population-averaged probit models, [XT] **xtprobit**

cluster estimator of variance, *continued*

Prais–Winsten and Cochrane–Orcutt regression,
[TS] **prais**

probit model with endogenous regressors,
[R] **ivprobit**

probit model with sample selection, [R] **heckprob**

probit regression, [R] **probit**

rank-ordered logistic regression, [R] **rologit**

skewed logistic regression, [R] **scobit**

stereotype logistic regression, [R] **slogit**

tobit model with endogenous regressors, [R] **ivtobit**

treatment-effects model, [R] **treatreg**

truncated regression, [R] **truncreg**

zero-inflated negative binomial regression, [R] **zinb**

zero-inflated Poisson regression, [R] **zip**

zero-truncated negative binomial regression, [R] **ztnb**

zero-truncated Poisson regression, [R] **ztp**

cluster sampling, [P] **_robust**; [R] **bootstrap**,
[R] **bsample**, [R] **gmm**, [R] **jackknife**

cluster tree, [MV] **Glossary**

clustering, [MV] **Glossary**

clustermat command, [MV] **clustermat**

clusters, duplicating, [D] **expandcl**

cmdlog command, [R] **log**, [U] **15 Saving and printing
output—log files**

Cmdyhms() function, [D] **dates and times**,
[D] **functions**, [M-5] **date( )**

cmissing() option, [G] *cline_options*,
[G] *connect_options*

cnsreg command, [R] **cnsreg**, [R] **cnsreg
postestimation**, *also see* postestimation command

Cochrane–Orcutt regression, [TS] **Glossary**, [TS] **prais**

code, timing, [P] **timer**

codebook command, [D] **codebook**

_coef[], [U] **13.5 Accessing coefficients and
standard errors**

coefficient

alpha, [R] **alpha**

of variation, [R] **tabstat**

coefficients (from estimation),

accessing, [P] **ereturn**, [P] **matrix get**, [R] **estimates
store**, [U] **13.5 Accessing coefficients and
standard errors**

estimated linear combinations of, *see* linear
combinations of estimators

testing equality of, [R] **test**, [R] **testnl**

Cofc() function, [D] **dates and times**, [D] **functions**,
[M-5] **date( )**

cofC() function, [D] **dates and times**, [D] **functions**,
[M-5] **date( )**

Cofd() function, [D] **dates and times**, [D] **functions**,
[M-5] **date( )**

cofd() function, [D] **dates and times**, [D] **functions**,
[M-5] **date( )**

cohort studies, [ST] **Glossary**

cointegration, [TS] **fcast compute**, [TS] **fcast graph**,
[TS] **Glossary**, [TS] **vec**, [TS] **vec intro**,

cointegration, *continued*
> [TS] **veclmar**, [TS] **vecnorm**, [TS] **vecrank**, [TS] **vecstable**

coleq, matrix subcommand, [P] **matrix rownames**

coleq macro extended function, [P] **macro**

colfullnames macro extended function, [P] **macro**

collapse command, [D] **collapse**

_collate() function, [M-5] **sort( )**

collect statistics, [D] **statsby**

collinear variables, removing, [P] **_rmcoll**

collinearity, handling by regress, [R] **regress**

colmax() function, [M-5] **minmax( )**

colmaxabs() function, [M-5] **minmax( )**

colmin() function, [M-5] **minmax( )**

colminmax() function, [M-5] **minmax( )**

colmissing() function, [M-5] **missing( )**

colnames, matrix subcommand, [P] **matrix rownames**

colnames macro extended function, [P] **macro**

colnonmissing() function, [M-5] **missing( )**

colnumb() matrix function, [D] **functions**, [P] **matrix define**

colon operators, [M-2] **op_colon**, [M-6] **Glossary**

color, [G] *colorstyle*, [G] **palette**
> background, [G] **schemes intro**
> dimming and brightening, [G] *colorstyle*, [G] **graph twoway histogram**, [G] **graph twoway kdensity**
> foreground, [G] **schemes intro**
> intensity adjustment, [G] *colorstyle*, [G] **graph twoway histogram**, [G] **graph twoway kdensity**
> of bars, [G] *barlook_options*
> of connecting lines, [G] *connect_options*
> of markers, [G] *marker_options*
> of pie slices, [G] **graph pie**
> of text, [G] *textbox_options*
> setting background and fill, [G] *region_options*

color() option, [G] **graph twoway histogram**, [G] **graph twoway kdensity**

color, palette subcommand, [G] **palette**

colors, specifying in programs, [P] **display**

colorstyle, [G] *colorstyle*

cols() function, [M-5] **rows( )**

colscalefactors() function, [M-5] **_equilrc( )**

colshape() function, [M-5] **rowshape( )**

colsof() matrix function, [D] **functions**, [P] **matrix define**

colsum() function, [M-5] **sum( )**

_column(#), display directive, [P] **display**

column-join operator, [M-2] **op_join**

column-major order, [M-6] **Glossary**

column of matrix, selecting, [M-5] **select( )**

columns of matrix,
> appending to, [P] **matrix define**
> names of, [P] **ereturn**, [P] **matrix define**, [P] **matrix rownames**, [U] **14.2 Row and column names**
> operators on, [P] **matrix define**

colvector, [M-2] **declarations**, [M-6] **Glossary**

comb() function, [D] **functions**, [M-5] **comb( )**

combination step, [MI] **intro substantive**, [MI] **mi estimate** [MI] **mi estimate postestimation**, [MI] **mi estimate using**,

combinatorial function, [M-5] **comb( )**

combinatorials, calculating, [D] **functions**

combine, graph subcommand, [G] **graph combine**

combining
> data, [MI] **mi add**, [MI] **mi append**, [MI] **mi merge**
> datasets, [D] **append**, [D] **cross**, [D] **joinby**, [D] **merge**, [U] **22 Combining datasets**
> graphs, [G] **graph combine**

command
> arguments, [P] **gettoken**, [P] **syntax**, [P] **tokenize**, [U] **18.4 Program arguments**
> line, launching dialog box from, [R] **db**
> parsing, [P] **gettoken**, [P] **syntax**, [P] **tokenize**, [U] **18.4 Program arguments**
> timings, [U] **8 Error messages and return codes**

commands,
> abbreviating, [U] **11.2 Abbreviation rules**
> aborting, [P] **continue**, [U] **9 The Break key**, [U] **10 Keyboard use**
> editing and repeating, [U] **10 Keyboard use**
> immediate, [U] **19 Immediate commands**
> repeating automatically, [D] **by**, [P] **byable**, [P] **continue**, [P] **foreach**, [P] **forvalues**, [P] **while**
> reviewing, [R] **#review**
> unabbreviating names of, [P] **unabcmd**

commas, reading data separated by, [D] **insheet**; [D] **infile (fixed format)**, [D] **infile (free format)**

comments, [M-2] **comments**
> adding to programs, [P] **comments**
> in programs, do-files, etc., [U] **16.1.2 Comments and blank lines in do-files**, [U] **18.11.2 Comments and long lines in ado-files**
> with data, [D] **notes**

common, estat subcommand, [MV] **factor postestimation**

common factors, [MV] **Glossary**

communalities, [MV] **factor**, [MV] **factor postestimation**, [MV] **Glossary**

commutation matrix, [M-5] **Kmatrix( )**

comparative scatterplot, [R] **dotplot**

compare, estat subcommand, [MV] **procrustes postestimation**

compare command, [D] **compare**

comparing two
> files, [D] **cf**
> variables, [D] **compare**

compassdirstyle, [G] *compassdirstyle*

compatibility of Stata programs across releases, [P] **version**

competing risks, [ST] **Glossary**, [ST] **stcrreg**

complementary log-log
> postestimation, [XT] **xtcloglog postestimation**
> regression, [R] **cloglog**, [R] **glm**, [SVY] **svy estimation**, [XT] **xtcloglog**, [XT] **xtgee**

complete degrees of freedom for coefficients,
[MI] **Glossary**, [MI] **mi estimate**
complete observations, [MI] **Glossary**
completed-data analysis, [MI] **Glossary**, [MI] **intro
substantive**, [MI] **mi estimate**
completelinkage, cluster subcommand,
[MV] **cluster linkage**
complete-linkage clustering, [MV] **cluster**,
[MV] **cluster linkage**, [MV] **clustermat**,
[MV] **Glossary**
completely determined outcomes, [R] **logit**
complex, [M-2] **declarations**, [M-6] **Glossary**
component
analysis, [MV] **factor**, [MV] **pca**, [MV] **rotate**,
[MV] **rotatemat**
loading plot, [MV] **scoreplot**
plot, [MV] **scoreplot**
scores, [MV] **Glossary**
component-plus-residual plot, [G] **graph other**,
[R] **regress postestimation**
compound double quotes, [P] **macro**
compound symmetric
correlation matrix, [MV] **mvtest correlations**
covariance matrix, [MV] **mvtest covariances**
compress command, [D] **compress**
compress files, [D] **zipfile**
compute, fcast subcommand, [TS] **fcast compute**
Comrey's tandem rotation, [MV] **Glossary**,
[MV] **rotate**, [MV] **rotatemat**
concat(), egen function, [D] **egen**
concatenating strings, [U] **13.2.2 String operators**
concordance, estat subcommand, [ST] **stcox
postestimation**
cond() function, [D] **functions**, [M-5] **cond( )**
condition number, [M-5] **cond( )**, [M-6] **Glossary**
condition statement, [P] **if**
conditional
(fixed-effects) logistic regression, [SVY] **svy
estimation**
fixed-effects model, [XT] **Glossary**
logistic regression, [R] **asclogit**, [R] **clogit**,
[R] **rologit**, [R] **slogit**, [XT] **xtlogit**
marginal effects, [R] **margins**
margins, [R] **margins**
operator, [M-2] **op_conditional**
variance, [TS] **arch**, [TS] **Glossary**
confidence intervals, [SVY] **variance estimation**,
[U] **20.7 Specifying the width of confidence
intervals**
for bootstrap statistics, [R] **bootstrap postestimation**
for linear combinations of coefficients, [R] **lincom**
for means, proportions, and counts, [R] **ci**,
[R] **mean**, [R] **proportion**, [R] **ttest**
for medians and percentiles, [R] **centile**
for nonlinear combinations of coefficients, [R] **nlcom**
for odds and risk ratios, [R] **lincom**, [R] **nlcom**,
[ST] **epitab**, [ST] **stci**
for ratios, [R] **ratio**

confidence intervals, *continued*
for tabulated proportions, [SVY] **svy: tabulate
twoway**
for totals, [R] **total**
linear combinations, [SVY] **svy postestimation**
set default, [R] **level**
config, estat subcommand, [MV] **mds
postestimation**
configuration, [MV] **Glossary**
plot, [MV] **Glossary**, [MV] **mds postestimation**
confirm
existence command, [P] **confirm**
file command, [P] **confirm**
format command, [P] **confirm**
matrix command, [P] **confirm**
names command, [P] **confirm**
number command, [P] **confirm**
scalar command, [P] **confirm**
variable command, [P] **confirm**
confirm, datasignature subcommand,
[D] **datasignature**
conformability, [M-2] **void**, [M-6] **Glossary**, *also
see* c-conformability, p-conformability, and r-
conformability
confounding, [ST] **Glossary**
confusion matrix, [MV] **Glossary**
_conj() function, [M-5] **conj( )**
conj() function, [M-5] **conj( )**
conjoint analysis, [R] **rologit**
conjugate, [M-5] **conj( )**, [M-6] **Glossary**
transpose, [M-2] **op_transpose**, [M-5] **conj( )**,
[M-6] **Glossary**
connect() option, [G] *cline_options*,
[G] *connect_options*, [G] *connectstyle*
connected, graph twoway subcommand, [G] **graph
twoway connected**
connectstyle, [G] *connectstyle*
conren, set subcommand, [R] **set**
console,
controlling scrolling of output, [P] **more**, [R] **more**
obtaining input from, [P] **display**
constrained estimation, [R] **constraint**, [R] **estimation
options**
alternative-specific multinomial probit regression,
[R] **asmprobit**
ARCH, [TS] **arch**
ARIMA and ARMAX, [TS] **arima**
bivariate probit regression, [R] **biprobit**
competing risks, [ST] **stcrreg**
complementary log-log regression, [R] **cloglog**
conditional logistic regression, [R] **asclogit**,
[R] **clogit**
generalized negative binomial regression, [R] **nbreg**
heckman selection model, [R] **heckman**
heteroskedastic probit model, [R] **hetprob**
interval regression, [R] **intreg**
linear regression, [R] **cnsreg**
logistic regression, [R] **logistic**, [R] **logit**

constrained estimation, *continued*
  logit and probit estimation for grouped data,
    [R] **glogit**
  multinomial logistic regression, [R] **mlogit**
  multinomial probit regression, [R] **mprobit**
  multivariate regression, [R] **mvreg**
  negative binomial regression, [R] **nbreg**
  nested logit regression, [R] **nlogit**
  ordered logistic regression, [R] **ologit**
  ordered probit regression, [R] **oprobit**
  parametric survival models, [ST] **streg**
  Poisson regression, [R] **poisson**
  probit model with
    endogenous regressors, [R] **ivprobit**
    selection, [R] **heckprob**
  probit regression, [R] **probit**
  random- and fixed-effects logit models, [XT] **xtlogit**
  random- and fixed-effects negative binomial models,
    [XT] **xtnbreg**
  random- and fixed-effects Poisson models,
    [XT] **xtpoisson**
  random-effects cloglog model, [XT] **xtcloglog**
  random-effects interval-data regression models,
    [XT] **xtintreg**
  random-effects probit model, [XT] **xtprobit**
  random-effects tobit models, [XT] **xttobit**
  seemingly unrelated regression, [R] **sureg**
  stereotype logistic regression, [R] **slogit**
  stochastic frontier models, [R] **frontier**
  stochastic frontier models for panel data,
    [XT] **xtfrontier**
  structural vector autoregressive models, [TS] **var
    svar**
  three-stage least squares, [R] **reg3**
  tobit model with endogenous regressors, [R] **ivtobit**
  treatment-effects model, [R] **treatreg**
  vector autoregressive models, [TS] **var**
  vector error-correction models, [TS] **vec**
  zero-inflated negative binomial regression, [R] **zinb**
  zero-inflated Poisson regression, [R] **zip**
  zero-truncated negative binomial regression, [R] **ztnb**
  zero-truncated Poisson regression, [R] **ztp**
constrained estimation, programming, [P] **makecns**
constrained linear regression, [SVY] **svy estimation**
constrained optimization, [TS] **arch**, [TS] **arima**,
    [TS] **dfactor**, [TS] **dvech**, [TS] **sspace**, [TS] **var**
constraint
  command, [R] **constraint**
  macro extended function, [P] **macro**
constraint matrix, creating and displaying, [P] **makecns**
constructor, [M-2] **class**
containers, [M-5] **asarray( )**
contents of data, [D] **describe**; [D] **codebook**,
    [D] **labelbook**
context, class, [P] **class**
contingency tables, [MV] **ca**, [R] **table**, [R] **tabulate
    twoway**, [ST] **epitab**, [SVY] **svy: tabulate
    twoway**

_continue, display directive, [P] **display**
continue command, [P] **continue**
continuous variable imputation, *see* imputation,
    continuous
contract command, [D] **contract**
contrast or contrasts, [MV] **Glossary**
control charts, [R] **qc**
convergence criteria, [R] **maximize**
conversion, file, [D] **filefilter**
convert, mi subcommand, [MI] **mi convert**
converting between styles, [MI] **mi convert**
convolve( ) function, [M-5] **fft( )**
Cook–Weisberg test for heteroskedasticity, [R] **regress
    postestimation**
Cook's D, [R] **predict**, [R] **regress postestimation**
coordinates, estat subcommand, [MV] **ca
    postestimation**, [MV] **mca postestimation**
copy,
  graph subcommand, [G] **graph copy**
  label subcommand, [D] **label**
  mi subcommand, [MI] **mi copy**, [MI] **styles**
  ssc subcommand, [R] **ssc**
copy and paste, [D] **edit**
.copy built-in class function, [P] **class**
copy command, [D] **copy**
copy graph, [G] **graph copy**
copy macro extended function, [P] **macro**
copycolor, set subcommand, [R] **set**
copying variables, [D] **clonevar**
copyright
  lapack, [R] **copyright lapack**
  scintilla, [R] **copyright scintilla**
  symbol, [G] *text*
  ttf2pt1, [R] **copyright ttf2pt1**
copyright command, [R] **copyright**
copysource, [M-1] **source**
Cornfield confidence intervals, [ST] **epitab**
Corr( ) function, [M-5] **fft( )**
_corr( ) function, [M-5] **corr( )**
corr( ) function, [D] **functions**, [M-5] **corr( )**,
    [P] **matrix define**
corr2data command, [D] **corr2data**
correcting data, *see* editing data
correlate command, [R] **correlate**
correlated error, *see* robust
correlated errors, [SVY] **variance estimation**, *see*
    robust, Huber/White/sandwich estimator of
    variance
correlation, [R] **correlate**; [M-5] **corr( )**, [M-5] **fft( )**
    [M-5] **mean( )**,
  compound symmetric, [MV] **mvtest correlations**
  data generation, [D] **corr2data**, [D] **drawnorm**
  for binary variables, [R] **tetrachoric**
  intracluster, [R] **loneway**
  Kendall's rank, [R] **spearman**
  matrices, [MV] **mvtest correlations**, [R] **correlate**,
    [R] **estat**

correlation, *continued*
>   pairwise, [R] **correlate**
>   partial, [R] **pcorr**
>   Pearson's product-moment, [R] **correlate**
>   Spearman's rank, [R] **spearman**
>   testing equality, [MV] **mvtest correlations**
>   tetrachoric, [R] **tetrachoric**
correlation( ) function, [M-5] **mean( )**
correlation, estat subcommand, [R] **asmprobit postestimation**, [R] **asroprobit postestimation**
correlation matrix, anti-image, [MV] **factor postestimation**, [MV] **pca postestimation**
correlation structure, [XT] **Glossary**
correlation,
>   canonical, [MV] **canon**
>   factoring of, [MV] **factor**
>   principal components of, [MV] **pca**
>   similarity measure, [MV] *measure_option*
correlation, matrices, [P] **matrix define**
correlations, estat subcommand, [MV] **canon postestimation**, [MV] **discrim lda postestimation**, [MV] **discrim qda postestimation**, [MV] **mds postestimation**
correlations, mvtest subcommand, [MV] **mvtest correlations**
correlogram, [G] **graph other**, [TS] **corrgram**, [TS] **Glossary**
correspondence analysis, [MV] **ca**, [MV] **Glossary**, [MV] **mca**
correspondence analysis projection, [MV] **Glossary**
corrgram command, [TS] **corrgram**
cos( ) function, [D] **functions**, [M-5] **sin( )**
cosh( ) function, [D] **functions**, [M-5] **sin( )**
cosine function, [D] **functions**
cost frontier model, [R] **frontier**, [XT] **xtfrontier**
costs, [MV] **Glossary**
count( ), egen function, [D] **egen**
count, ml subcommand, [R] **ml**
count command, [D] **count**
count data, [R] **expoisson**, [R] **glm**, [R] **nbreg**, [R] **poisson**, [R] **zinb**, [R] **zip**, [R] **ztnb**, [R] **ztp**
count-time data, [ST] **ct**, [ST] **Glossary**; [ST] **ctset**, [ST] **cttost**, [ST] **sttoct**, [SVY] **svy estimation**
counts, making dataset of, [D] **collapse**
courses about Stata, [U] **3.7.1 NetCourses**
covariance, estat subcommand, [MV] **discrim lda postestimation**, [MV] **discrim qda postestimation**, [R] **asmprobit postestimation**, [R] **asroprobit postestimation**
covariance, principal components of, [MV] **pca**
covariance matrix
>   anti-image, [MV] **factor postestimation**, [MV] **pca postestimation**
>   block diagonal, [MV] **mvtest covariances**
>   of estimators, [P] **ereturn**, [P] **matrix get**, [R] **correlate**, [R] **estat**, [R] **estimates store**
>   spherical, [MV] **mvtest covariances**
>   testing equality, [MV] **mvtest covariances**

covariance stationarity, [TS] **Glossary**
covariances, mvtest subcommand, [MV] **mvtest covariances**
covariate class, [D] **duplicates**
covariate patterns, [R] **logistic postestimation**, [R] **logit postestimation**, [R] **probit postestimation**
covariates, [ST] **Glossary**
covarimin rotation, [MV] **Glossary**, [MV] **rotate**, [MV] **rotatemat**
COVRATIO, [R] **regress postestimation**
cox, stpower subcommand, [ST] **stpower cox**
Cox proportional hazards model, [ST] **stcox**, [SVY] **svy estimation**
>   power, [ST] **stpower cox**
>   sample size, [ST] **stpower cox**
>   test of assumption, [ST] **stcox**, [ST] **stcox PH-assumption tests**, [ST] **stcox postestimation**, [ST] **stsplit**
>   Wald test, power, [ST] **stpower cox**
Cox–Snell residual, [ST] **stcox postestimation**, [ST] **streg postestimation**
cprplot command, [R] **regress postestimation**
Cramér's V, [R] **tabulate twoway**
Crawford–Ferguson rotation, [MV] **Glossary**, [MV] **rotate**, [MV] **rotatemat**
create, irf subcommand, [TS] **irf create**
create, serset subcommand, [P] **serset**
create_cspline, serset subcommand, [P] **serset**
create_xmedians, serset subcommand, [P] **serset**
creturn list command, [P] **creturn**
crexternal( ) function, [M-5] **findexternal( )**
Cronbach's alpha, [R] **alpha**
cross( ) function, [M-5] **cross( )**
cross command, [D] **cross**
cross-correlation function, [TS] **Glossary**, [TS] **xcorr**
cross-correlogram, [G] **graph other**, [TS] **xcorr**
crossdev( ) function, [M-5] **crossdev( )**
crossed-effects model, [XT] **Glossary**
crossed variables, [MV] **Glossary**
crossover designs, [R] **pk**, [R] **pkcross**, [R] **pkshape**
cross product, [M-5] **cross( )**, [M-5] **crossdev( )**, [M-5] **quadcross( )**
cross-product matrices, [P] **matrix accum**
cross-sectional data, [XT] **Glossary**
cross-sectional studies, [ST] **Glossary**
cross-sectional time-series
>   data, [XT] **Glossary**
>   postestimation, [XT] **xtregar postestimation**
>   regression, [XT] **xtregar**
cross-tabulations, *see* tables
crude estimate, [ST] **epitab**, [ST] **Glossary**
cs command, [ST] **epitab**
csi command, [ST] **epitab**
ct data, [ST] **Glossary**, *also see* count-time data
ctable, irf subcommand, [TS] **irf ctable**
ctset command, [ST] **ctset**
cttost command, [ST] **cttost**
cubic natural splines, [M-5] **spline3( )**

cumsp command, [TS] **cumsp**

cumul command, [R] **cumul**

cumulative distribution, empirical, [R] **cumul**

cumulative distribution functions, [D] **functions**

cumulative hazard function, [ST] **stcurve**, [ST] **sts**, [ST] **sts generate**, [ST] **sts graph**, [ST] **sts list**

cumulative hazard ratio, *see* hazard ratio

cumulative incidence
> data, [R] **poisson**, [ST] **epitab**, [SVY] **svy estimation**
>
> estimator, [ST] **Glossary**, [ST] **stcrreg**
>
> function, [ST] **Glossary**, [ST] **stcrreg**, [ST] **stcurve**

cumulative spectral distribution, empirical, [TS] **cumsp**

cumulative subhazard, [ST] **Glossary**, [ST] **stcrreg**, [ST] **stcurve**

curse of dimensionality, [MV] **Glossary**

custom prediction equations, [MI] **mi impute monotone**

cusum command, [R] **cusum**

cusum plots, [G] **graph other**

cut(), egen function, [D] **egen**

cutil, *see* classutil

cvpermute( ) function, [M-5] **cvpermute( )**

cvpermutesetup( ) function, [M-5] **cvpermute( )**

D

DA, *see* data augmentation

dashed lines, [G] *linepatternstyle*

data, [D] **data types**, [U] **12 Data**
> appending, *see* appending data
>
> autocorrelated, *see* autocorrelation
>
> case–cohort, *see* case–cohort data
>
> case–control, *see* case–control data
>
> categorical, *see* categorical data
>
> certifying, *see* certifying data
>
> characteristics of, *see* characteristics
>
> checksums of, *see* checksums of data
>
> combining, *see* combining datasets
>
> contents of, *see* contents of data
>
> count-time, *see* count-time data
>
> current, [P] **creturn**
>
> discrete survival, *see* discrete survival data
>
> displaying, *see* displaying data
>
> documenting, *see* documenting data
>
> editing, *see* editing data
>
> entering, *see* inputting data interactively; reading data from disk
>
> exporting, *see* exporting data
>
> extended missing values, *see* missing values
>
> flong, *see* flong
>
> flongsep, *see* flongsep
>
> generating, *see* generating data
>
> importing, *see* importing data
>
> inputting, *see* importing data, *see* reading data from disk
>
> labeling, *see* labeling data
>
> large, dealing with, *see* memory
>
> listing, *see* listing data

data, *continued*
> loading, *see* inputting data interactively; reading data from disk
>
> matched case–control, *see* matched case–control data
>
> missing values, *see* missing values
>
> mlong, *see* mlong
>
> multiple-record st data, *see* multiple-record st data
>
> nested case–control, *see* nested case–control data
>
> preserving, [P] **preserve**
>
> range of, *see* range of data
>
> ranking, *see* ranking data
>
> reading, *see* reading data from disk
>
> recoding, *see* recoding data
>
> rectangularizing, *see* rectangularize dataset
>
> reordering, *see* reordering data
>
> reorganizing, *see* reorganizing data
>
> restoring, *see* restoring data
>
> sampling, *see* sampling
>
> saving, *see* saving data
>
> stacking, *see* stacking data
>
> strings, *see* string variables
>
> summarizing, *see* summarizing data
>
> survey, *see* survey data
>
> survival-time, *see* survival analysis
>
> time-series, *see* time-series analysis
>
> time-span, *see* time-span data
>
> transposing, *see* transposing data
>
> verifying, *see* certifying data
>
> wide, *see* wide

data, label subcommand, [D] **label**

data augmentation, [MI] **Glossary**, [MI] **mi impute mvn**, [MI] **mi impute**

Data Editor, [D] **edit**
> copy and paste, [D] **edit**

data entry, [D] **infile (fixed format)**, [D] **infile (free format)**, [D] **input**, *also see* reading data from disk

data label macro extended function, [P] **macro**

data management, [MI] **mi add**, [MI] **mi append**, [MI] **mi expand**, [MI] **mi extract**, [MI] **mi merge**, [MI] **mi rename**, [MI] **mi replace0**, [MI] **mi reset**, [MI] **mi reshape**

data manipulation, [R] **fvrevar**, [R] **fvset**, [TS] **tsappend**, [TS] **tsfill**, [TS] **tsreport**, [TS] **tsrevar**, [TS] **tsset**, [XT] **xtset**

data matrix, [M-5] **st_data( )**, [M-5] **st_view( )**, [M-6] **Glossary**

data reduction, [MV] **ca**, [MV] **canon**, [MV] **factor**, [MV] **mds**, [MV] **pca**

data signature, [D] **datasignature**, [P] **_datasignature**, [P] **signestimationsample**

data transfer, [D] **infile (fixed format)**, [D] **infile (free format)**

data types, [I] **data types**

database, reading data from other software, [D] **odbc**, [U] **21.4 Transfer programs**

dataset,
    adding notes to, [D] **notes**
    comparing, [D] **cf**
    creating, [D] **corr2data**, [D] **drawnorm**
    rectangularize, [D] **fillin**
dataset labels, [D] **label**, [D] **label language**, [D] **notes**
    determining, [D] **codebook**, [D] **describe**
datasets, example, [U] **1.2.2 Example datasets**
datasignature
    clear command, [D] **datasignature**
    command, [D] **datasignature**
    confirm command, [D] **datasignature**
    report command, [D] **datasignature**
    set command, [D] **datasignature**
_datasignature command, [P] **_datasignature**
date
    and time stamp, [D] **describe**, [M-5] **c( )**,
      [P] **creturn**
    formats, [U] **24.3 Displaying dates and times**;
      [U] **12.5.3 Date and time formats**
    functions, [D] **dates and times**, [D] **functions**,
      [M-5] **date( )**, [U] **24.5 Extracting components**
      of dates and times
    variables, [U] **24 Working with dates and times**
date( ) function, [D] **dates and times**, [D] **functions**,
    [M-5] **date( )**
datelist, [U] **11.1.9 datelist**
dates,
    displaying, [U] **24.3 Displaying dates and times**;
      [U] **12.5.3 Date and time formats**
    inputting, [U] **24.2 Inputting dates and times**
dates and times, [D] **dates and times**
Davidon–Fletcher–Powell algorithm,
    [M-5] **moptimize( )**, [M-5] **optimize( )**, [R] **ml**
day( ) function, [D] **dates and times**, [D] **functions**,
    [M-5] **date( )**, [U] **24.5 Extracting components**
    of dates and times
db command, [R] **db**
dBASE, reading data from, [U] **21.4 Transfer**
    programs
DBETAs, [R] **regress postestimation**
.dct filename suffix, [D] **infile**, [U] **11.6 File-naming**
    conventions
debugging, [P] **trace**; [P] **discard**, [P] **pause**
decimal symbol, setting, [D] **format**
declarations, [M-2] **declarations**, [M-6] **Glossary**
declare, class, [P] **class**
.Declare built-in class modifier, [P] **class**
decode command, [D] **encode**
decomposition, [M-5] **cholesky( )**, [M-5] **fullsvd( )**,
    [M-5] **ghessenbergd( )**, [M-5] **gschurd( )**,
    [M-5] **hessenbergd( )**, [M-5] **lud( )**, [M-5] **qrd( )**,
    [M-5] **schurd( )**, [M-5] **svd( )**
deconvolve( ) function, [M-5] **fft( )**
decrement operator, [M-2] **op_increment**
default settings of system parameters, [R] **query**,
    [R] **set_defaults**
defective matrix, [M-6] **Glossary**

DEFF, see design effects
define,
    char subcommand, [P] **char**
    label subcommand, [D] **label**
    matrix subcommand, [P] **matrix define**
    program subcommand, [P] **program**, [P] **program**
      properties
    transmap subcommand, [R] **translate**
DEFT, see design effects
degrees of freedom, [MI] **mi estimate**
    complete, see complete degrees of freedom
    for coefficients, also see estimation
degree-to-radian conversion, [D] **functions**
delete, [M-5] **unlink( )**
delete, cluster subcommand, [MV] **cluster**
    programming utilities
deleting
    casewise, [D] **egen**
    files, [D] **erase**
    variables or observations, [D] **drop**
#delimit, [M-2] **semicolons**, [P] **#delimit**
delimiter
    for comments, [P] **comments**
    for lines, [P] **#delimit**
delta beta influence statistic, [R] **logistic**
    postestimation, [R] **logit postestimation**,
    [R] **regress postestimation**
delta chi-squared influence statistic, [R] **logistic**
    postestimation, [R] **logit postestimation**
delta deviance influence statistic, [R] **logistic**
    postestimation, [R] **logit postestimation**
delta method, [R] **nlcom**, [R] **predictnl**, [R] **testnl**,
    [SVY] **variance estimation**
dendrogram, [G] **graph other**, [MV] **cluster**,
    [MV] **cluster dendrogram**, [MV] **Glossary**
dendrogram, cluster subcommand, [MV] **cluster**
    dendrogram
density estimation, kernel, [R] **kdensity**
density functions, [M-5] **normal( )**
density option, [G] **graph twoway histogram**
density smoothing, [G] **graph other**
density-distribution sunflower plot, [R] **sunflower**
dereferencing, [M-2] **pointers**, [M-2] **ftof**
_deriv( ) function, [M-5] **deriv( )**
deriv( ) function, [M-5] **deriv( )**
derivative of incomplete gamma function, [D] **functions**
derivatives, [M-5] **deriv( )**
derivatives, numeric, [R] **dydx**, [R] **testnl**
derived plottypes, [G] *advanced_options*
deriv_init_*( ) functions, [M-5] **deriv( )**
deriv_query( ) function, [M-5] **deriv( )**
deriv_result_*( ) functions, [M-5] **deriv( )**
describe,
    ado subcommand, [R] **net**
    classutil subcommand, [P] **classutil**
    estimates subcommand, [R] **estimates describe**
    graph subcommand, [G] **graph describe**
    irf subcommand, [TS] **irf describe**

describe, *continued*
    mi subcommand, [MI] **mi describe**
    net subcommand, [R] **net**
    odbc subcommand, [D] **odbc**
    ssc subcommand, [R] **ssc**
describe, [M-3] **mata describe**
describe command, [D] **describe**, [U] **12.6 Dataset,
    variable, and value labels**
describing graph, [G] **graph describe**
describing mi data, [MI] **mi describe**
descriptive statistics,
    creating dataset containing, [D] **collapse**
    creating variables containing, [D] **egen**
    displaying, [D] **codebook**, [D] **pctile**, [R] **ameans**,
        [R] **lv**, [R] **mean**, [R] **proportion**, [R] **ratio**,
        [R] **summarize**, [R] **table**, [R] **tabstat**,
        [R] **tabulate, summarize()**, [R] **total**,
        [XT] **xtsum**, [XT] **xttab**
design effects, [R] **loneway**, [SVY] **estat**,
    [SVY] **Glossary**, [SVY] **svy: tabulate oneway**,
    [SVY] **svy: tabulate twoway**
design matrix, [M-5] **designmatrix()**, [M-5] **I()**
designmatrix() function, [M-5] **designmatrix()**
destring command, [D] **destring**
destroy() class function, [M-2] **class**
destructor, [M-2] **class**
destructors, class, [P] **class**
det() function, [D] **functions**, [M-5] **det()**, [P] **matrix
    define**
determinant of matrix, [M-5] **det()**, [P] **matrix define**
dettriangular() function, [M-5] **det()**
deviance residual, [R] **binreg postestimation**,
    [R] **fracpoly postestimation**, [R] **glm
    postestimation**, [R] **logistic postestimation**,
    [R] **logit postestimation**, [R] **mfp
    postestimation**, [ST] **stcox postestimation**,
    [ST] **streg postestimation**
deviation cross product, [M-5] **crossdev()**,
    [M-5] **quadcross()**
dexponential, tssmooth subcommand,
    [TS] **tssmooth dexponential**
dfactor command, [TS] **dfactor**
DFBETA, [ST] **Glossary**, [ST] **stcox postestimation**
dfbeta command, [R] **regress postestimation**
DFBETAs, [R] **regress postestimation**
dfgls command, [TS] **dfgls**
DFITS, [R] **regress postestimation**
DFP algorithm, *see* Davidon–Fletcher–Powell algorithm
dfuller command, [TS] **dfuller**
dgammapda() function, [D] **functions**, [M-5] **normal()**
dgammapdada() function, [D] **functions**,
    [M-5] **normal()**
dgammapdadx() function, [D] **functions**,
    [M-5] **normal()**
dgammapdx() function, [D] **functions**, [M-5] **normal()**
dgammapdxdx() function, [D] **functions**,
    [M-5] **normal()**

dhms() function, [D] **dates and times**, [D] **functions**,
    [M-5] **date()**
_diag() function, [M-5] **_diag()**
diag() function, [D] **functions**, [M-5] **diag()**,
    [P] **matrix define**
diag0cnt() function, [D] **functions**, [M-5] **diag0cnt()**,
    [P] **matrix define**
diagnostic codes, [D] **icd9**
diagnostic plots, [G] **graph other**, [R] **diagnostic
    plots**, [R] **logistic postestimation**, [R] **regress
    postestimation**
diagnostics, regression, [R] **regress postestimation**
diagonal, [M-5] **diagonal()**, [M-6] **Glossary**
diagonal() function, [M-5] **diagonal()**
diagonal matrix, [M-5] **_diag()**, [M-5] **diag()**,
    [M-5] **diagonal()**, [M-5] **isdiagonal()**,
    [M-6] **Glossary**
diagonal vech GARCH model, [TS] **dvech**
diagonals of matrices, [P] **matrix define**
dialog
    box, [P] **dialog programming**, [P] **window
    programming**, [R] **db**
    programming, [P] **dialog programming**, [P] **window
    programming**
Dice coefficient similarity measure,
    [MV] *measure_option*
dichotomous outcome model, [R] **exlogistic**,
    [R] **logistic**; [R] **asclogit**, [R] **binreg**,
    [R] **biprobit**, [R] **brier**, [R] **clogit**, [R] **cloglog**,
    [R] **cusum**, [R] **glm**, [R] **glogit**, [R] **heckprob**,
    [R] **hetprob**, [R] **ivprobit**, [R] **logit**, [R] **probit**,
    [R] **rocfit**, [R] **scobit**, [XT] **xtcloglog**,
    [XT] **xtgee**, [XT] **xtlogit**, [XT] **xtprobit**
Dickey–Fuller test, [TS] **dfgls**, [TS] **dfuller**
dictionaries, [M-5] **asarray()**; [D] **infile**, [D] **infile
    (fixed format)**, [D] **infix (fixed format)**,
    [D] **outfile**
diff(), egen function, [D] **egen**
difference of estimated coefficients, *see* linear
    combinations of estimators
difference operator, [TS] **Glossary**, [U] **11.4.4 Time-
    series varlists**
differentiation, [M-5] **deriv()**
difficult option, [R] **maximize**
digamma() function, [D] **functions**, [M-5] **factorial()**
digits, controlling the number displayed, [D] **format**,
    [U] **12.5 Formats: Controlling how data are
    displayed**
dilation, [MV] **Glossary**
dimension, [MV] **Glossary**
dir() function, [M-5] **dir()**
dir,
    ado subcommand, [R] **net**
    classutil subcommand, [P] **classutil**
    cluster subcommand, [MV] **cluster utility**
    _estimates subcommand, [P] **_estimates**
    estimates subcommand, [R] **estimates store**
    graph subcommand, [G] **graph dir**
    label subcommand, [D] **label**

dir, *continued*
    macro subcommand, [P] **macro**
    matrix subcommand, [P] **matrix utility**
    postutil subcommand, [P] **postfile**
    program subcommand, [P] **program**
    _return subcommand, [P] **_return**
    serset subcommand, [P] **serset**
dir command, [D] **dir**
dir macro extended function, [P] **macro**
direct standardization, [R] **dstdize**, [R] **mean**,
    [R] **proportion**, [R] **ratio**, [SVY] **direct**
    standardization, [SVY] **Glossary**
directories, [M-5] **chdir( )**, [M-5] **dir( )**,
    [M-5] **direxists( )**, [U] **11.6 File-naming**
    conventions, [U] **18.3.11 Constructing Windows**
    filenames by using macros
    changing, [D] **cd**
    creating, [D] **mkdir**
    listing, [D] **dir**
    location of ado-files, [U] **17.5 Where does Stata**
    look for ado-files?
    removing, [D] **rmdir**
directories and paths, [P] **creturn**
directory, class, [P] **classutil**
direxists( ) function, [M-5] **direxists( )**
direxternal( ) function, [M-5] **direxternal( )**
discard
    command, [P] **discard**, [U] **18.11.3 Debugging**
    ado-files
discard, relationship to graph drop, [G] **graph drop**
discrete option, [G] **graph twoway histogram**
discrete survival data, [ST] **discrete**
discrim
    knn command, [MV] **discrim**, [MV] **discrim knn**
    lda command, [MV] **discrim**, [MV] **discrim lda**
    logistic command, [MV] **discrim**, [MV] **discrim**
    logistic
    qda command, [MV] **discrim**, [MV] **discrim qda**
discriminant analysis, [MV] **candisc**, [MV] **discrim**,
    [MV] **discrim knn**, [MV] **discrim lda**,
    [MV] **discrim logistic**, [MV] **discrim qda**,
    [MV] **Glossary**
    loading plot, [MV] **scoreplot**
    score plot, [MV] **scoreplot**
discriminant function, [MV] **Glossary**
discriminating variables, [MV] **Glossary**
disparity, [MV] **Glossary**
dispersion, measures of, [D] **pctile**, [R] **summarize**,
    [R] **table**, [XT] **xtsum**; [R] **centile**, [R] **lv**
display
    as error, [M-5] **displayas( )**, [M-5] **errprintf( )**
    as text, as result, etc., [M-5] **displayas( )**
    formats, [D] **describe**, [D] **format**, [P] **macro**,
    [U] **12.5 Formats: Controlling how data are**
    displayed, [U] **24.3 Displaying dates and times**
    graph, [G] **graph display**
    saved results, [R] **saved results**
    width and length, [R] **log**

display( ) function, [M-5] **display( )**
display,
    ereturn subcommand, [P] **ereturn**
    graph subcommand, [G] **graph display**
    ml subcommand, [R] **ml**
display command, [P] **display**, [U] **19.1.2 A list of**
    the immediate commands; [P] **macro**
    as a calculator, [R] **display**
display macro extended function, [P] **display**
displayas( ) function, [M-5] **displayas( )**
displayflush( ) function, [M-5] **displayflush( )**
displaying, *also see* printing, logs (output)
    contents, [D] **describe**
    data, [D] **edit**, [D] **list**
    macros, [P] **macro**
    matrix, [P] **matrix utility**
    named graphs, [G] **graph display**, [G] **graph use**
    output, [P] **display**, [P] **quietly**, [P] **smcl**,
    [P] **tabdisp**
    previously typed lines, [R] **#review**
    saved results, [R] **saved results**
    scalar expressions, [P] **display**, [P] **scalar**
dissimilarity, [MV] **Glossary**
    matrices, [MV] **Glossary**, [MV] **matrix**
    dissimilarity, [P] **matrix dissimilarity**
    measures,
    [MV] **cluster**, [MV] **cluster programming**
    utilities, [MV] **matrix dissimilarity**, [MV] **mds**,
    [MV] *measure_option*, [P] **matrix dissimilarity**
    absolute value, [MV] *measure_option*
    Bray and Curtis, [MV] **clustermat**
    Canberra, [MV] *measure_option*
    Euclidean, [MV] *measure_option*
    gower, [MV] *measure_option*
    maximum value, [MV] *measure_option*
    Minkowski, [MV] *measure_option*
dissimilarity, matrix subcommand, [MV] **matrix**
    dissimilarity, [P] **matrix dissimilarity**
distance matrices, [MV] **matrix dissimilarity**,
    [P] **matrix dissimilarity**
distances, *see* dissimilarity measures
distances, estat subcommand, [MV] **ca**
    postestimation
distribution functions, [M-5] **normal( )**
distributional diagnostic plots, [G] **graph other**
distributions,
    diagnostic plots, [R] **diagnostic plots**
    examining, [D] **pctile**, [R] **ameans**, [R] **centile**,
    [R] **kdensity**, [R] **lv**, [R] **mean**, [R] **stem**,
    [R] **summarize**, [R] **total**
    testing equality of, [R] **ksmirnov**, [R] **kwallis**,
    [R] **ranksum**, [R] **signrank**
    testing for normality, [MV] **mvtest normality**,
    [R] **sktest**, [R] **swilk**
disturbance term, [XT] **Glossary**
division, [M-2] **op_arith**, [M-2] **op_colon**
division operator, *see* arithmetic operators
divisive hierarchical clustering methods, [MV] **Glossary**

DLL, [P] **plugin**

Dmatrix() function, [M-5] **Dmatrix()**

do command, [R] **do**, [U] **16 Do-files**

.do filename suffix, [U] **11.6 File-naming conventions**

do . . . while, [M-2] **do**, [M-2] **continue**, [M-2] **break**

dockable, set subcommand, [R] **set**

dockingguides, set subcommand, [R] **set**

documentation, [U] **1 Read this—it will help**

keyword search on, [U] **4 Stata's help and search facilities**

documentation, keyword search on, [R] **search**

documenting data, [D] **codebook**, [D] **labelbook**, [D] **notes**

doedit command, [R] **doedit**

dofC() function, [D] **dates and times**, [D] **functions**, [M-5] **date()**

dofc() function, [D] **dates and times**, [D] **functions**, [M-5] **date()**

dofh() function, [D] **dates and times**, [D] **functions**, [M-5] **date()**

do-files, [P] **break**, [P] **include**, [P] **version**, [R] **do**, [U] **16 Do-files**, [U] **18.2 Relationship between a program and a do-file**

adding comments to, [P] **comments**

editing, [R] **doedit**

long lines, [P] **#delimit**, [U] **18.11.2 Comments and long lines in ado-files**

dofm() function, [D] **dates and times**, [D] **functions**, [M-5] **date()**

dofq() function, [D] **dates and times**, [D] **functions**, [M-5] **date()**

dofw() function, [D] **dates and times**, [D] **functions**, [M-5] **date()**

dofy() function, [D] **dates and times**, [D] **functions**, [M-5] **date()**

domain sampling, [R] **alpha**

Doornik–Hansen normality test, [MV] **mvtest normality**

dose–response models, [R] **binreg**, [R] **glm**, [R] **logistic**

dot,

graph subcommand, [G] **graph dot**

graph twoway subcommand, [G] **graph twoway dot**

dot plots, [G] *area_options*, [G] **graph dot**, [G] **graph twoway dot**, [G] *line_options*

dotplot command, [R] **dotplot**

dotted lines, [G] *linepatternstyle*

double, [D] **data types**, [I] **data types**, [U] **12.2.2 Numeric storage types**

double quotes, [P] **macro**

doublebuffer, set subcommand, [R] **set**

double-exponential smoothing, [TS] **tssmooth dexponential**

double-precision floating point number, [U] **12.2.2 Numeric storage types**

dow() function, [D] **dates and times**, [D] **functions**, [M-5] **date()**, [U] **24.5 Extracting components of dates and times**

doy() function, [D] **dates and times**, [D] **functions**, [M-5] **date()**

dp, set subcommand, [D] **format**, [R] **set**

drawnorm command, [D] **drawnorm**

drop,

classutil subcommand, [P] **classutil**

cluster subcommand, [MV] **cluster utility**

duplicates subcommand, [D] **duplicates**

label subcommand, [D] **label**

notes subcommand, [D] **notes**

_estimates subcommand, [P] **_estimates**

estimates subcommand, [R] **estimates store**

graph subcommand, [G] **graph drop**

irf subcommand, [TS] **irf drop**

macro subcommand, [P] **macro**

matrix subcommand, [P] **matrix utility**

program subcommand, [P] **program**

_return subcommand, [P] **_return**

serset subcommand, [P] **serset**

drop, [M-3] **mata drop**

drop command, [D] **drop**

dropline, graph twoway subcommand, [G] **graph twoway dropline**

dropping

graphs, [G] **graph drop**

programs, [P] **discard**

variables and observations, [D] **drop**

dsign() function, [M-5] **dsign()**, [M-5] **sign()**

dstdize command, [R] **dstdize**

.dta file extension, [U] **11.6 File-naming conventions**

technical description, [P] **file formats .dta**

.dtasig filename suffix, [U] **11.6 File-naming conventions**

dual scaling, [MV] **ca**

Duda and Hart index stopping rules, [MV] **cluster stop**

dummy variables, *see* indicator variables

_dup(#), display directive, [P] **display**

duplicate observations,

dropping, [D] **duplicates**

identifying, [D] **duplicates**

duplicates

drop command, [D] **duplicates**

examples command, [D] **duplicates**

list command, [D] **duplicates**

report command, [D] **duplicates**

tag command, [D] **duplicates**

duplicating

clustered observations, [D] **expandcl**

observations, [D] **expand**

duplication matrix, [M-5] **Dmatrix()**

durbinalt, estat subcommand, [R] **regress postestimation time series**

Durbin's alternative test, [R] **regress postestimation time series**

Durbin–Watson statistic, [R] **regress postestimation time series**, [TS] **prais**

dvech command, [TS] **dvech**

dwatson, estat subcommand, [R] **regress**
postestimation time series
dydx command, [R] **dydx**
dynamic forecast, [TS] **fcast compute**, [TS] **fcast**
graph, [TS] **Glossary**
dynamic model, [XT] **Glossary**
dynamic panel-data regression, [XT] **xtabond**,
[XT] **xtdpd**, [XT] **xtdpdsys**
dynamic regression model, [TS] **arima**, [TS] **var**
dynamic structural simultaneous equations, [TS] **var**
svar
dynamic-factor model, [TS] **dfactor**, [TS] **dfactor**
postestimation, [TS] **sspace**
dynamic-multiplier function, [TS] **Glossary**, [TS] **irf**,
[TS] **irf cgraph**, [TS] **irf create**, [TS] **irf ctable**,
[TS] **irf ograph**, [TS] **irf table**, [TS] **var intro**
.dynamicmv built-in class function, [P] **class**

E

e()
function, [M-5] **e( )**
scalars, macros, matrices, functions, [D] **functions**,
[P] **ereturn**, [P] **_estimates**, [P] **return**,
[R] **saved results**, [U] **18.8 Accessing**
results calculated by other programs,
[U] **18.9 Accessing results calculated by**
estimation commands, [U] **18.10.2 Saving**
results in e()
e(functions) macro extended function, [P] **macro**
e(macros) macro extended function, [P] **macro**
e(matrices) macro extended function, [P] **macro**
e(sample) function, [D] **functions**, [P] **ereturn**,
[P] **return**
resetting, [R] **estimates save**
e(scalars) macro extended function, [P] **macro**
e-class command, [P] **program**, [P] **return**, [R] **saved**
results, [U] **18.8 Accessing results calculated by**
other programs
economist scheme, [G] **scheme economist**;
[G] *axis_options*
edit, graphs, [G] **graph editor**
edit command, [D] **edit**
editing
ado-files and do-files, [R] **doedit**
commands, [U] **10 Keyboard use**
data, [D] **edit**, [D] **generate**, [D] **merge**, [D] **recode**
files while in Stata, [R] **doedit**
output, [U] **15 Saving and printing output—log**
files
_editmissing() function, [M-5] **editmissing( )**
editmissing() function, [M-5] **editmissing( )**
_edittoint() function, [M-5] **edittoint( )**
edittoint() function, [M-5] **edittoint( )**
_edittointtol() function, [M-5] **edittoint( )**
edittointtol() function, [M-5] **edittoint( )**
_edittozero() function, [M-5] **edittozero( )**
edittozero() function, [M-5] **edittozero( )**
_edittozerotol() function, [M-5] **edittozero( )**

edittozerotol() function, [M-5] **edittozero( )**
_editvalue() function, [M-5] **editvalue( )**
editvalue() function, [M-5] **editvalue( )**
effect size, [ST] **Glossary**, [ST] **stpower**, [ST] **stpower**
cox, [ST] **stpower exponential**, [ST] **stpower**
logrank
effects, estat subcommand, [SVY] **estat**
eform_option, [R] *eform_option*
EGARCH, [TS] **arch**
egen command, [D] **egen**, [MI] **mi passive**, [MI] **mi**
xeq
EGLS (estimated generalized least squares), [XT] **xtgls**,
[XT] **xtivreg**, [XT] **xtreg**
_eigen_la() function, [M-5] **eigensystem( )**
_eigensystem() function, [M-5] **eigensystem( )**
eigensystem() function, [M-5] **eigensystem( )**
_eigensystem_select() functions,
[M-5] **eigensystemselect( )**
_eigensystemselect*() functions,
[M-5] **eigensystemselect( )**
eigensystemselect*() functions,
[M-5] **eigensystemselect( )**
eigenvalue stability condition, [TS] **varstable**,
[TS] **vecstable**
eigenvalues, [M-5] **eigensystem( )**, [M-6] **Glossary**,
[MV] **factor**, [MV] **factor postestimation**,
[MV] **Glossary**, [MV] **pca**, [MV] **rotate**,
[MV] **rotatemat**, [MV] **screeplot**, [P] **matrix**
eigenvalues, [P] **matrix symeigen**
eigenvalues, matrix subcommand, [P] **matrix**
eigenvalues
_eigenvalues() function, [M-5] **eigensystem( )**
eigenvalues() function, [M-5] **eigensystem( )**
eigenvalues and eigenvectors, [P] **matrix svd**,
[P] **matrix symeigen**
eigenvectors, [M-5] **eigensystem( )**, [M-6] **Glossary**,
[MV] **factor**, [MV] **factor postestimation**,
[MV] **Glossary**, [MV] **pca**, [MV] **rotate**,
[MV] **rotatemat**, [MV] **scoreplot**
eivreg command, [R] **eivreg**, [R] **eivreg**
postestimation, *also see* postestimation command
el() matrix function, [D] **functions**, [P] **matrix define**
elimination matrix, [M-5] **Lmatrix( )**
ellipsis, [G] *text*
else command, [P] **if**
eltype() function, [M-5] **eltype( )**
eltype, [M-2] **declarations**, [M-6] **Glossary**
EM, [MI] **Glossary**, [MI] **mi impute mvn**
parameter trace files, [MI] **mi ptrace**
empirical cumulative distribution function, [R] **cumul**
emptycells, set subcommand, [R] **set**, [R] **set**
emptycells
Encapsulated PostScript, [G] **graph export**, [G] *text*;
[G] *eps_options*
encode command, [D] **encode**, [U] **23.2 Categorical**
string variables
end, [M-3] **end**
ending a Stata session, [P] **exit**, [R] **exit**
endless loop, *see* loop, endless

end-of-line characters, [D] **changeeol**

endogenous

 regressors, [R] **gmm**, [R] **ivprobit**, [R] **ivregress**,
  [R] **ivtobit**

 treatment, [R] **treatreg**

 variables, [R] **gmm**, [R] **ivprobit**, [R] **ivregress**,
  [R] **ivtobit**, [R] **reg3**, [SVY] **svy estimation**
  [TS] **Glossary**, [XT] **Glossary**

endogenous, estat subcommand, [R] **ivregress**
  postestimation

ends(), egen function, [D] **egen**

Engle's LM test, [R] **regress postestimation time series**

Enhanced Metafile, [G] **graph export**

ensuring mi data are consistent, [MI] **mi update**

entering data, *see* inputting data interactively; reading
  data from disk, *see* reading data from disk

entropy rotation, [MV] **rotate**, [MV] **rotatemat**

environment macro extended function, [P] **macro**

environment variables (Unix), [P] **macro**

eolchar, set subcommand, [R] **set**

Epanechnikov kernel density function, [R] **kdensity**

epidemiological tables, [R] **dstdize**, [R] **symmetry**,
  [R] **tabulate twoway**, [ST] **epitab**

epsdouble() function, [D] **functions**

epsfloat() function, [D] **functions**

epsilon, [M-6] **Glossary**

epsilon() function, [M-5] **epsilon( )**

equal FMI test, [MI] **Glossary**, [MI] **mi estimate**,
  [MI] **mi estimate postestimation**

equality of means tests, [MV] **hotelling**, [MV] **manova**

equality operator, [U] **13.2.3 Relational operators**

equality test, survivor functions, [ST] **sts test**

equality tests of

 binomial proportions, [R] **bitest**

 coefficients, [R] **test**, [R] **testnl**

 distributions, [R] **ksmirnov**, [R] **kwallis**,
  [R] **ranksum**, [R] **signrank**

 means, [R] **ttest**

 medians, [R] **ranksum**

 proportions, [R] **bitest**, [R] **prtest**

 variances, [R] **sdtest**

equamax rotation, [MV] **Glossary**, [MV] **rotate**,
  [MV] **rotatemat**

equation names of matrix, [P] **matrix rownames**,
  [U] **14.2 Row and column names**; [P] **ereturn**,
  [P] **matrix define**

_equilc() function, [M-5] **_equilrc( )**

equilibration, [M-5] **_equilrc( )**

_equilr() function, [M-5] **_equilrc( )**

_equilrc() function, [M-5] **_equilrc( )**

equivalence tests, [R] **pk**, [R] **pkequiv**

erase, [M-5] **unlink( )**

erase,

 mi subcommand, [MI] **mi erase**, [MI] **styles**

 snapshot subcommand, [D] **snapshot**

erase command, [D] **erase**

erasing files, [D] **erase**

erasing graph files, [G] **graph drop**

ereturn

 clear command, [P] **ereturn**; [P] **return**

 display command, [P] **ereturn**

 list command, [P] **ereturn**, [R] **saved results**

 local command, [P] **ereturn**; [P] **return**

 matrix command, [P] **ereturn**; [P] **return**

 post command, [P] **ereturn**, [P] **makecns**;
  [P] **return**

 repost command, [P] **ereturn**; [P] **return**

 scalar command, [P] **ereturn**; [P] **return**

error

 handling, [P] **capture**, [P] **confirm**, [P] **error**,
  [U] **16.1.4 Error handling in do-files**

 messages and return codes, [P] **error**, [P] **rmsg**,
  [U] **4.9.5 Return codes**, [U] **8 Error messages
  and return codes**, *also see* error handling

_error() function, [M-5] **error( )**

error() function, [M-5] **error( )**

error checking, [D] **assert**

error codes, [M-2] **errors**

error command, [P] **error**

error messages and return codes, [R] **error messages**

error-bar charts, [R] **serrbar**

error-components model, [XT] **Glossary**,
  [XT] **xthtaylor**

errorrate, estat subcommand, [MV] **discrim
  estat**, [MV] **discrim knn postestimation**,
  [MV] **discrim lda postestimation**, [MV] **discrim
  logistic postestimation**, [MV] **discrim qda
  postestimation**

errors-in-variables regression, [R] **eivreg**

errprintf() function, [M-5] **errprintf( )**

esample, estimates subcommand, [R] **estimates save**

estat, [P] **estat programming**

 abond command, [XT] **xtabond postestimation**,
  [XT] **xtdpd postestimation**, [XT] **xtdpdsys
  postestimation**

 alternatives command, [R] **asclogit
  postestimation**, [R] **asmprobit postestimation**,
  [R] **asroprobit postestimation**

 anova command, [MV] **discrim lda postestimation**

 anti command, [MV] **factor postestimation**,
  [MV] **pca postestimation**

 archlm command, [R] **regress postestimation time
  series**

 bgodfrey command, [R] **regress postestimation
  time series**

 bootstrap command, [R] **bootstrap postestimation**

 canontest command, [MV] **discrim lda
  postestimation**

 classfunctions command, [MV] **discrim lda
  postestimation**

 classification command, [R] **logistic
  postestimation**

 classtable command, [MV] **discrim
  estat**, [MV] **discrim knn postestimation**,
  [MV] **discrim lda postestimation**, [MV] **discrim
  logistic postestimation**, [MV] **discrim qda
  postestimation**

estat, *continued*

common command, [MV] **factor postestimation**

compare command, [MV] **procrustes postestimation**

concordance command, [ST] **stcox postestimation**

config command, [MV] **mds postestimation**

coordinates command, [MV] **ca postestimation**, [MV] **mca postestimation**

correlation command, [R] **asmprobit postestimation**, [R] **asroprobit postestimation**

correlations command, [MV] **canon postestimation**, [MV] **discrim lda postestimation**, [MV] **discrim qda postestimation**, [MV] **mds postestimation**

covariance command, [MV] **discrim lda postestimation**, [MV] **discrim qda postestimation**, [R] **asmprobit postestimation**, [R] **asroprobit postestimation**

distances command, [MV] **ca postestimation**

durbinalt command, [R] **regress postestimation time series**

dwatson command, [R] **regress postestimation time series**

effects command, [SVY] **estat**

endogenous command, [R] **ivregress postestimation**

errorrate command, [MV] **discrim estat**, [MV] **discrim knn postestimation**, [MV] **discrim lda postestimation**, [MV] **discrim logistic postestimation**, [MV] **discrim qda postestimation**

factors command, [MV] **factor postestimation**

facweights command, [R] **asmprobit postestimation**, [R] **asroprobit postestimation**

firststage command, [R] **ivregress postestimation**

gof command, [R] **logistic postestimation**, [R] **poisson postestimation**

grdistances command, [MV] **discrim lda postestimation**, [MV] **discrim qda postestimation**

grmeans command, [MV] **discrim lda postestimation**

group command, [XT] **xtmelogit**, [XT] **xtmelogit postestimation**, [XT] **xtmepoisson**, [XT] **xtmepoisson postestimation**, [XT] **xtmixed**, [XT] **xtmixed postestimation**

grsummarize command, [MV] **discrim estat**, [MV] **discrim knn postestimation**, [MV] **discrim lda postestimation**, [MV] **discrim logistic postestimation**, [MV] **discrim qda postestimation**

hettest command, [R] **regress postestimation**

ic command, [R] **estat**

imtest command, [R] **regress postestimation**

inertia command, [MV] **ca postestimation**

kmo command, [MV] **factor postestimation**, [MV] **pca postestimation**

lceffects command, [SVY] **estat**

estat, *continued*

list command, [MV] **discrim estat**, [MV] **discrim knn postestimation**, [MV] **discrim lda postestimation**, [MV] **discrim logistic postestimation**, [MV] **discrim qda postestimation**

loadings command, [MV] **canon postestimation**, [MV] **discrim lda postestimation**, [MV] **pca postestimation**

manova command, [MV] **discrim lda postestimation**

mfx command, [R] **asclogit postestimation**, [R] **asmprobit postestimation**, [R] **asroprobit postestimation**

mvreg command, [MV] **procrustes postestimation**

overid command, [R] **gmm postestimation**, [R] **ivregress postestimation**

ovtest command, [R] **regress postestimation**

pairwise command, [MV] **mds postestimation**

phtest command, [ST] **stcox PH-assumption tests**

predict command, [R] **exlogistic postestimation**

profiles command, [MV] **ca postestimation**

quantiles command, [MV] **mds postestimation**

recovariance command, [XT] **xtmelogit**, [XT] **xtmelogit postestimation**, [XT] **xtmepoisson**, [XT] **xtmepoisson postestimation**, [XT] **xtmixed**, [XT] **xtmixed postestimation**

residuals command, [MV] **factor postestimation**, [MV] **pca postestimation**

rotate command, [MV] **canon postestimation**

rotatecompare command, [MV] **canon postestimation**, [MV] **factor postestimation**, [MV] **pca postestimation**

sargan command, [XT] **xtabond postestimation**, [XT] **xtdpd postestimation**, [XT] **xtdpdsys postestimation**

sd command, [SVY] **estat**

se command, [R] **exlogistic postestimation**, [R] **expoisson postestimation**

size command, [SVY] **estat**

smc command, [MV] **factor postestimation**, [MV] **pca postestimation**

strata command, [SVY] **estat**

stress command, [MV] **mds postestimation**

structure command, [MV] **discrim lda postestimation**, [MV] **factor postestimation**

subinertia command, [MV] **mca postestimation**

summarize command, [MV] **ca postestimation**, [MV] **discrim estat**, [MV] **discrim knn postestimation**, [MV] **discrim lda postestimation**, [MV] **discrim logistic postestimation**, [MV] **discrim qda postestimation**, [MV] **factor postestimation**, [MV] **mca postestimation**, [MV] **mds postestimation**, [MV] **pca postestimation**, [MV] **procrustes postestimation**, [R] **estat**

svyset command, [SVY] **estat**

szroeter command, [R] **regress postestimation**

estat, *continued*
> table command, [MV] **ca postestimation**
> vce command, [R] **estat**, [SVY] **estat**
> vif command, [R] **regress postestimation**
> wcorrelation command, [XT] **xtgee**
>> **postestimation**

estimable, [R] **margins**

estimate, mi subcommand, [MI] **mi estimate**, [MI] **mi estimate using**

estimate linear combinations of coefficients, *see* linear combinations of estimators

estimated marginal means, [R] **margins**

_estimates
> clear command, [P] **_estimates**
> dir command, [P] **_estimates**
> drop command, [P] **_estimates**
> hold command, [P] **_estimates**
> unhold command, [P] **_estimates**

estimates
> clear command, [R] **estimates store**
> command, [R] **suest**
>> introduction, [R] **estimates**
> describe command, [R] **estimates describe**
> dir command, [R] **estimates store**
> drop command, [R] **estimates store**
> esample command, [R] **estimates save**
> for command, [R] **estimates for**
> notes command, [R] **estimates notes**
> query command, [R] **estimates store**
> replay command, [R] **estimates replay**
> restore command, [R] **estimates store**
> save command, [R] **estimates save**
> stats command, [R] **estimates stats**
> store command, [R] **estimates store**
> table command, [R] **estimates table**
> title command, [R] **estimates title**
> use command, [R] **estimates save**

estimates command, [SVY] **svy postestimation**

estimation
> allowed estimation commands, [MI] **estimation**
> commands, [I] **estimation commands**, [P] **ereturn**, [P] **_estimates**, [U] **18.9 Accessing results calculated by estimation commands**, [U] **26 Overview of Stata estimation commands**
>> accessing stored information from, [P] **matrix get**
>> allowing constraints in, [P] **makecns**
>> eliminating stored information from, [P] **discard**
>> obtaining predictions after, [P] **_predict**
>> obtaining robust estimates, [P] **_robust**
>> saving results from, [P] **_estimates**
> degrees of freedom for coefficients, [MI] **mi estimate**
> options, [R] **estimation options**
> posting VCE, [MI] **mi estimate**
> results,
>> clearing, [R] **estimates store**; [P] **ereturn**, [P] **_estimates**

estimation, results, *continued*
>> listing, [P] **ereturn**, [P] **_estimates**
>> saving, [P] **ereturn**, [P] **_estimates**
>> storing and restoring, [R] **estimates store**
>> tables of, [R] **estimates table**
> sample, summarizing, [R] **estat**
> tests after, [MI] **mi estimate**, [MI] **mi estimate postestimation**, [SVY] **estat**, [SVY] **svy postestimation**

estimators,
> covariance matrix of, [P] **ereturn**, [P] **matrix get**, [R] **correlate**, [R] **estat**, [U] **20.8 Obtaining the variance–covariance matrix**
> linear combinations, [U] **20.12 Obtaining linear combinations of coefficients**
> linear combinations of, [R] **lincom**
> nonlinear combinations of, [R] **nlcom**

etiologic fraction, [ST] **epitab**

Euclidean dissimilarity measure, [MV] *measure_option*

Euclidean distance, [MV] **Glossary**

event, [ST] **Glossary**, [ST] **stpower logrank**

event of interest, [ST] **Glossary**

exact
> logistic regression model, [R] **exlogistic**
> Poisson regression model, [R] **expoisson**
> test, Fisher's, [R] **tabulate twoway**

example datasets, [U] **1.2.2 Example datasets**

examples, duplicates subcommand, [D] **duplicates**

Excel, Microsoft, reading data from, [D] **odbc**, [D] **xmlsave**, *also see* spreadsheets, transferring

excess fraction, [ST] **epitab**

exec(), odbc subcommand, [D] **odbc**

executable, update subcommand, [R] **update**

existence, confirm subcommand, [P] **confirm**

exit() function, [M-5] **exit( )**

exit, class subcommand, [P] **class exit**

exit class program, [P] **class exit**

exit command, [P] **capture**, [P] **exit**, [R] **exit**, [U] **16.1.4 Error handling in do-files**

exit mata, [M-3] **end**

exiting Stata, *see* exit command

exlogistic command, [R] **exlogistic**, [R] **exlogistic postestimation**, *also see* postestimation command

exogenous variable, [TS] **Glossary**, [XT] **Glossary**

exp, [M-2] **exp**, [M-6] **Glossary**

=exp, [U] **11 Language syntax**

exp() function, [D] **functions**, [M-5] **exp( )**

exp_list, [SVY] **svy brr**, [SVY] **svy jackknife**

exp_list, [TS] **rolling**

expand, mi subcommand, [MI] **mi expand**

expand command, [D] **expand**

expand factor varlists, [P] **fvexpand**

expand for mi data, [MI] **mi expand**

expandcl command, [D] **expandcl**

expectation-maximization algorithm, *see* EM

exploded logit model, [R] **rologit**

expoisson command, [R] **expoisson**, [R] **expoisson postestimation**, *also see* postestimation command
exponential,
    stpower subcommand, [ST] **stpower exponential**
    tssmooth subcommand, [TS] **tssmooth exponential**
exponential
    distribution, [ST] **streg**
    function, [D] **functions**
    model, stochastic frontier, [R] **frontier**
    notation, [U] **12.2 Numbers**
    smoothing, [TS] **Glossary**, [TS] **tssmooth**, [TS] **tssmooth exponential**
    survival
        power, [ST] **stpower exponential**
        regression, [ST] **streg**
        sample size, [ST] **stpower exponential**
    test, [ST] **Glossary**
        power, [ST] **stpower exponential**
        sample size, [ST] **stpower exponential**
exponentiated coefficients, [R] *eform_option*
exponentiation, [M-5] **exp( )**, [M-5] **matexpsym( )**
export, graph subcommand, [G] **graph export**
export, mi subcommand, [MI] **mi export**, [MI] **mi export ice**, [MI] **mi export nhanes1**
exporting
    data, [D] **outfile**, [D] **outsheet**, [MI] **mi export**, [MI] **mi export ice**, [MI] **mi export nhanes1**, [U] **21.4 Transfer programs**
    graphs, [G] *eps_options*, [G] **graph export**, [G] **graph set**, [G] *png_options*, [G] *ps_options*, [G] **text**, [G] *tif_options*
exposure variable, [ST] **Glossary**
expressions, [P] **matrix define**, [U] **13 Functions and expressions**
extended macro functions, [P] **char**, [P] **display**, [P] **macro**, [P] **macro lists**, [P] **serset**
extended memory, [D] **memory**
external, [M-2] **declarations**
externals, [M-2] **declarations**, [M-5] **direxternal( )**, [M-5] **findexternal( )**, [M-5] **valofexternal( )**, [M-6] **Glossary**
extract, mi subcommand, [MI] **mi extract**, [MI] **mi replace0**
extract diagonal, [M-5] **diag( )**, [M-5] **diagonal( )**
extracting $m=\#$ data from mi data, [MI] **mi extract**, [MI] **mi select**
extracting original data from mi data, [MI] **mi extract**
extrapolation, [D] **ipolate**

F

F( ) function, [D] **functions**, [M-5] **normal( )**
F density,
    central, [D] **functions**
    noncentral, [D] **functions**
    distribution,
        cumulative, [D] **functions**
        inverse cumulative, [D] **functions**

F density, distribution, *continued*
        inverse reverse cumulative, [D] **functions**
        inverse reverse cumulative noncentral, [D] **functions**
        reverse cumulative, [D] **functions**
        reverse cumulative noncentral, [D] **functions**
factor, [MV] **Glossary**
    analysis, [MV] **canon**, [MV] **factor**, [MV] **factor postestimation**, [MV] **Glossary**, [R] **alpha**
    model, [TS] **dfactor**
    loading plot, [MV] **Glossary**, [MV] **scoreplot**
    loadings, [MV] **Glossary**
    parsimony, [MV] **Glossary**
    score plot, [MV] **scoreplot**
    scores, [MV] **factor postestimation**, [MV] **Glossary**
    variables, [U] **11.4.3 Factor variables**, [U] **13.8 Indicator values for levels of factor variables**, [U] **14.2.2 Two-part names**, [U] **20.10 Accessing estimated coefficients**, [U] **25 Working with categorical data and factor variables**; [P] **fvexpand**, [P] **intro**, [P] **matrix rownames**, [P] **_rmcoll**, [P] **syntax**, [P] **unab**, [R] **fvrevar**, [R] **fvset**
factor command, [MV] **factor**
factorial, [U] **11.4.3 Factor variables**
    design, [MV] **manova**, [R] **anova**
    function, [D] **functions**
factorial( ) function, [M-5] **factorial( )**
factormat command, [MV] **factor**
factors, estat subcommand, [MV] **factor postestimation**
factor-variable settings, [R] **fvset**
facweights, estat subcommand, [R] **asmprobit postestimation**, [R] **asroprobit postestimation**
failure event, [ST] **Glossary**
failure tables, [ST] **ltable**
failure-time model, *see* survival analysis
false-negative result, *see* type II error
false-positive result, *see* type I error
FAQs, [U] **3.2 The Stata web site (www.stata.com)**
    search, [R] **search**, [U] **4.9.4 FAQ searches**
fastscroll, set subcommand, [R] **set**
favorspeed( ) function, [M-5] **favorspeed( )**
fbufget( ) function, [M-5] **bufio( )**
fbufput( ) function, [M-5] **bufio( )**
fcast compute command, [TS] **fcast compute**
fcast graph command, [TS] **fcast graph**
_fclose( ) function, [M-5] **fopen( )**
fclose( ) function, [M-5] **fopen( )**
FDA (SAS XPORT) format, [D] **fdasave**
fdadescribe command, [D] **fdasave**
fdasave command, [D] **fdasave**; [D] **infile**
fdause command, [D] **fdasave**; [D] **infile**
Fden( ) function, [D] **functions**, [M-5] **normal( )**
feasible generalized least squares, *see* FGLS
fences, [R] **lv**
ferrortext( ) function, [M-5] **ferrortext( )**

FEVD, [TS] **Glossary**, [TS] **irf**, [TS] **irf create**,
 [TS] **irf ograph**, [TS] **irf table**, [TS] **var intro**,
 [TS] **varbasic**, [TS] **vec intro**
_fft( ) function, [M-5] **fft( )**
fft( ) function, [M-5] **fft( )**
_fget( ) function, [M-5] **fopen( )**
fget( ) function, [M-5] **fopen( )**
_fgetmatrix( ) function, [M-5] **fopen( )**
fgetmatrix( ) function, [M-5] **fopen( )**
_fgetnl( ) function, [M-5] **fopen( )**
fgetnl( ) function, [M-5] **fopen( )**
FGLS, [R] **reg3**, [TS] **dfgls**, [TS] **prais**, [TS] **var**,
 [XT] **xtgls**, [XT] **xtivreg**, [XT] **xtreg**
 postestimation, [XT] **xtgls postestimation**
file
 conversion, [D] **filefilter**
 extensions, [I] **file extensions**
 format, Stata, [P] **file formats .dta**
 modification, [D] **filefilter**
 processing, [M-5] **bufio( )**; [M-4] **io**,
  [M-5] **findfile( )**, [M-5] **fileexists( )**,
  [M-5] **ferrortext( )**, [M-5] **fopen( )**, [M-5] **cat( )**,
  [M-5] **unlink( )**
 translatation, [D] **filefilter**
file
 close command, [P] **file**
 open command, [P] **file**
 query command, [P] **file**
 read command, [P] **file**
 seek command, [P] **file**
 sersetread command, [P] **serset**
 sersetwrite command, [P] **serset**
 set command, [P] **file**
 write command, [P] **file**
file, confirm subcommand, [P] **confirm**
file, find in path, [P] **findfile**
fileexists( ) function, [M-5] **fileexists( )**
filefilter command, [D] **filefilter**
filename manipulation, [M-5] **adosubdir( )**,
 [M-5] **pathjoin( )**
filenames, displaying, [D] **dir**
files,
 checksum of, [D] **checksum**
 comparison, [D] **cf**
 compress, [D] **zipfile**
 copying and appending, [D] **copy**
 display contents of, [D] **type**
 downloading, [D] **checksum**, [R] **adoupdate**,
  [R] **net**, [R] **sj**, [R] **ssc**, [R] **update** [U] **28 Using
  the Internet to keep up to date**
 erasing, [D] **erase**
 exporting, *see* exporting data
 extensions, [U] **11.6 File-naming conventions**
 importing, *see* importing data
 names, [U] **11.6 File-naming conventions**,
  [U] **18.3.11 Constructing Windows filenames
  by using macros**
 opening, [P] **window programming**

files, *continued*
 reading ASCII text or binary, [P] **file**
 saving, [D] **fdasave**, [D] **save**, [P] **window
  programming**
 temporary, [P] **macro**, [P] **preserve**, [P] **scalar**
 uncompress, [D] **zipfile**
 writing ASCII text or binary, [P] **file**
fill
 areas, dimming and brightening, [G] *colorstyle*,
  [G] **graph twoway histogram**, [G] **graph
  twoway kdensity**
 color, setting, [G] *region_options*
fill( ), egen function, [D] **egen**
fillin command, [D] **fillin**
filling in values, [ST] **stfill**
_fillmissing( ) function, [M-5] **_fillmissing( )**
filters, [TS] **tssmooth** [TS] **tssmooth dexponential**,
 [TS] **tssmooth exponential**, [TS] **tssmooth
 hwinters**, [TS] **tssmooth ma**, [TS] **tssmooth nl**,
 [TS] **tssmooth shwinters**,
final, [M-2] **class**
findexternal( ) function, [M-5] **findexternal( )**
findfile( ) function, [M-5] **findfile( )**
findfile command, [P] **findfile**
finding file in path, [P] **findfile**
finding variables, [D] **lookfor**
findit command, [R] **search**
first-differenced estimator, [XT] **xtivreg**
firststage, estat subcommand, [R] **ivregress
 postestimation**
fisher, xtunitroot subcommand, [XT] **xtunitroot**
Fisher's exact test, [R] **tabulate twoway**, [ST] **epitab**
Fisher-type tests, [XT] **xtunitroot**
fits, adding, [G] **graph twoway fpfit**, [G] **graph
 twoway fpfitci**, [G] **graph twoway lfit**,
 [G] **graph twoway lfitci**, [G] **graph twoway qfit**,
 [G] **graph twoway qfitci**
fixed-effects model, [R] **anova**, [R] **areg**, [R] **asclogit**,
 [R] **clogit**, [R] **xi**, [XT] **Glossary**, [XT] **xtabond**,
 [XT] **xtdpd**, [XT] **xtdpdsys**, [XT] **xtivreg**,
 [XT] **xtlogit**, [XT] **xtnbreg**, [XT] **xtpoisson**,
 [XT] **xtreg**, [XT] **xtregar**
F-keys, [U] **10 Keyboard use**
flexible functional form, [R] **boxcox**, [R] **fracpoly**,
 [R] **mfp**
flist command, [D] **list**
float, [D] **data types**, [I] **data types**,
 [M-5] **floatround( )**, [U] **13.11 Precision and
 problems therein**, [U] **12.2.2 Numeric storage
 types**
float( ) function, [D] **functions**, [U] **13.11 Precision
 and problems therein**
floatresults, set subcommand, [R] **set**
floatround( ) function, [M-5] **floatround( )**
floatwindows, set subcommand, [R] **set**
flong
 data style, [MI] **Glossary**, [MI] **styles**
 technical description, [MI] **technical**

flongsep
data style, [MI] **Glossary**, [MI] **mi xeq**, [MI] **styles**
estimating memory requirements, [MI] **mi convert**
style, [MI] **mi copy**, [MI] **mi erase**
technical description, [MI] **technical**
floor() function, [D] **functions**, [M-5] **trunc( )**
_flopin function, [M-5] **lapack( )**
_flopout function, [M-5] **lapack( )**
FMI, *see* fraction missing information
%fmts, [D] **format**, [U] **12.5 Formats: Controlling**
how data are displayed
fmtwidth() function, [D] **functions**, [M-5] **fmtwidth( )**
folders, *see* directories
follow-up period, [ST] **Glossary**
follow-up studies, *see* incidence studies
fonts, in graphs, [G] *text*
footnote, ml subcommand, [R] **ml**
_fopen() function, [M-5] **fopen( )**
fopen() function, [M-5] **fopen( )**
fopen, window subcommand, [P] **window**
programming
for, [M-2] **break**, [M-2] **continue**, [M-2] **for**,
[M-2] **semicolons**
for, estimates subcommand, [R] **estimates for**
foreach command, [P] **foreach**
forecast, *see* forecasting
standard error of, [R] **predict**, [R] **regress**
postestimation
forecast-error variance decomposition, *see* FEVD
forecast-error variance decompositions (FEVD),
[G] **graph other**
forecasting, [G] **graph other**, [TS] **arch**, [TS] **arima**,
[TS] **dfactor postestimation**, [TS] **dvech**
postestimation, [TS] **fcast compute**,
[TS] **fcast graph**, [TS] **irf create**, [TS] **sspace**
postestimation, [TS] **tsappend**, [TS] **tssmooth**,
[TS] **tssmooth dexponential**, [TS] **tssmooth**
exponential, [TS] **tssmooth hwinters**,
[TS] **tssmooth ma**, [TS] **tssmooth shwinters**,
[TS] **var**, [TS] **var intro**, [TS] **vec**, [TS] **vec**
intro
foreground color, [G] **schemes intro**
format, confirm subcommand, [P] **confirm**
format command, [D] **format**
format macro extended function, [P] **macro**
format width, [M-5] **fmtwidth( )**
formating, setting, [D] **varmanage**
formats, [D] **dates and times**, [D] **describe**,
[D] **format**, [U] **12.5 Formats: Controlling how**
data are displayed; [U] **24.3 Displaying dates**
and times
formatted data, reading, [U] **21 Inputting data**;
[D] **infile**, [D] **infile (fixed format)**, [D] **infile**
(free format), [D] **infix (fixed format)**,
[D] **insheet**
formatting contents of macros, [P] **macro**
formatting statistical output, [D] **format**
FORTRAN, [M-2] **goto**, [M-5] **dsign( )**

forvalues command, [P] **forvalues**
forward operator, [TS] **Glossary**
fourfold tables, [ST] **epitab**
Fourier transform, [M-5] **fft( )**
FoxPro, reading data from, [U] **21.4 Transfer programs**
FPC, [SVY] **Glossary**, [SVY] **survey**, [SVY] **svy**
estimation, [SVY] **svyset**, [SVY] **variance**
estimation
fpfit, graph twoway subcommand, [G] **graph**
twoway fpfit
fpfitci, graph twoway subcommand, [G] **graph**
twoway fpfitci
_fput() function, [M-5] **fopen( )**
fput() function, [M-5] **fopen( )**
_fputmatrix() function, [M-5] **fopen( )**
fputmatrix() function, [M-5] **fopen( )**
fracgen command, [R] **fracpoly**
fracplot command, [R] **fracpoly postestimation**
fracpoly prefix command, [R] **fracpoly**, [R] **fracpoly**
postestimation, *also see* postestimation command
fracpred command, [R] **fracpoly postestimation**
fraction defective, [R] **qc**
fraction missing information, [MI] **Glossary**, [MI] **mi**
estimate, [MI] **mi estimate postestimation**
fraction option, [G] **graph twoway histogram**
fractional polynomial regression, [R] **fracpoly**
multivariable, [R] **mfp**
frailty, [ST] **Glossary**
frailty model, [ST] **streg**; [ST] **stcox**, [ST] **stcurve**
_fread() function, [M-5] **fopen( )**
fread() function, [M-5] **fopen( )**
frequencies,
creating dataset of, [D] **collapse**, [D] **contract**
graphical representation, [R] **histogram**,
[R] **kdensity**
table of, [R] **table**, [R] **tabstat**, [R] **tabulate**
oneway, [R] **tabulate, summarize()**, [R] **tabulate**
twoway, [SVY] **svy: tabulate oneway**,
[SVY] **svy: tabulate twoway**
frequency option, [G] **graph twoway histogram**
frequency table, [XT] **xttab**
frequency weights, [U] **11.1.6 weight**,
[U] **20.18.1 Frequency weights**
frequency-domain analysis, [TS] **cumsp**, [TS] **Glossary**,
[TS] **pergram**
[frequency=*exp*] modifier, [U] **11.1.6 weight**,
[U] **20.18.1 Frequency weights**
frequentist concepts, [MI] **intro substantive**
freturncode() function, [M-5] **ferrortext( )**
from() option, [R] **maximize**
from,
net subcommand, [R] **net**
update subcommand, [R] **update**
frombase() function, [M-5] **inbase( )**
frontier command, [R] **frontier**, [R] **frontier**
postestimation, *also see* postestimation command
frontier model, stochastic, [XT] **xtfrontier**
frontier models, [R] **frontier**

fsave, window subcommand, [P] **window programming**

_fseek() function, [M-5] **fopen( )**

fseek() function, [M-5] **fopen( )**

fstatus() function, [M-5] **fopen( )**

Ftail() function, [D] **functions**, [M-5] **normal( )**

_ftell() function, [M-5] **fopen( )**

ftell() function, [M-5] **fopen( )**

ftfreqs() function, [M-5] **fft( )**

ftpad() function, [M-5] **fft( )**

ftperiodogram() function, [M-5] **fft( )**

ftretime() function, [M-5] **fft( )**

_ftruncate() function, [M-5] **fopen( )**

ftruncate() function, [M-5] **fopen( )**

ftunwrap() function, [M-5] **fft( )**

ftwrap() function, [M-5] **fft( )**

full factorial, [U] **11.4.3 Factor variables**

fullsdiag() function, [M-5] **fullsvd( )**

_fullsvd() function, [M-5] **fullsvd( )**

fullsvd() function, [M-5] **fullsvd( )**

fully factorial, [R] **margins**

function, [M-2] **declarations**, [M-6] **Glossary**

function, graph twoway subcommand, [G] **graph twoway function**

function arguments, [M-1] **returnedargs**, *also see* arguments

function naming convention, [M-1] **naming**

functions, [D] **functions**, [U] **13.3 Functions**

    adding, [MV] **cluster programming subroutines**

    aggregate, [D] **egen**

    combinatorial, [D] **functions**

    creating dataset of, [D] **collapse**, [D] **obs**

    date, [U] **24.5 Extracting components of dates and times**

    date and time, [D] **functions**

    extended macro, [P] **char**, [P] **display**, [P] **macro**, [P] **macro lists**, [P] **serset**

    graphing, [D] **range**, [G] **graph twoway function**

    link, [R] **glm**

    mathematical, [D] **functions**

    matrix, [D] **functions**, [P] **matrix define**, [U] **14.8 Matrix functions**

    passing to functions, [M-2] **ftof**

    piecewise cubic, [R] **mkspline**

    piecewise linear, [R] **mkspline**

    programming, [D] **functions**

    random number, [D] **generate**

    statistical, [D] **functions**

    string, [D] **functions**

    time-series, [D] **functions**

    underscore, [M-6] **Glossary**

future history, [ST] **Glossary**, [ST] **stset**

fvexpand command, [P] **fvexpand**

fvrevar command, [R] **fvrevar**

fvset, mi subcommand, [MI] **mi XXXset**

fvset command, [R] **fvset**

fvset command for mi data, [MI] **mi XXXset**

fvunab command, [P] **unab**

[fweight=*exp*] modifier, [U] **11.1.6 weight**, [U] **20.18.1 Frequency weights**

_fwrite() function, [M-5] **fopen( )**

fwrite() function, [M-5] **fopen( )**

fxsize() option, [G] **graph combine**

fysize() option, [G] **graph combine**

G

g2 inverse of matrix, [P] **matrix define**, [P] **matrix svd**

gamma

    density function, [D] **functions**

        incomplete, [D] **functions**

    distribution

        cumulative, [D] **functions**

        inverse cumulative, [D] **functions**

        inverse reverse cumulative, [D] **functions**

        reverse cumulative, [D] **functions**

gamma() function, [M-5] **factorial( )**

gammaden() function, [D] **functions**, [M-5] **normal( )**

gammap() function, [D] **functions**, [M-5] **normal( )**

gammaptail() function, [D] **functions**, [M-5] **normal( )**

gap() option, [G] **graph twoway histogram**

gaps, [ST] **Glossary**

GARCH, [TS] **arch**, [TS] **dvech**, [TS] **Glossary**

garch command, [TS] **arch**

Gauss, reading data from, [U] **21.4 Transfer programs**

Gauss–Hermite quadrature, [XT] **quadchk**

GEE (generalized estimating equations), [XT] **Glossary**, [XT] **xtgee**

    postestimation, [XT] **xtgee postestimation**

_geigen_la() function, [M-5] **geigensystem( )**

_geigenselect*_la() functions, [M-5] **geigensystem( )**

geigensystem() function, [M-5] **geigensystem( )**

_geigensystem_la() function, [M-5] **geigensystem( )**

geigensystemselect*() functions, [M-5] **geigensystem( )**

generalized

    autoregressive conditional heteroskedasticity, [TS] **arch**, [TS] **dvech**

    eigensystem, [M-5] **geigensystem( )**

    eigenvalues, [M-6] **Glossary**

    gamma survival regression, [ST] **streg**

    Hessenberg decomposition, [M-5] **ghessenbergd( )**

    inverse, [M-5] **invsym( )**, [M-5] **pinv( )**, [M-5] **qrinv( )**

    inverse of matrix, [P] **matrix define**, [P] **matrix svd**

    least squares, *see* FGLS

    least-squares estimator, [TS] **Glossary**

    linear latent and mixed models, *see* GLLAMM

    linear models, *see* GLM

    method of moments, *see* gmm command

    method of moments (GMM), [P] **matrix accum**, [XT] **xtabond**, [XT] **xtdpd**, [XT] **xtdpdsys**

generalized, *continued*
  negative binomial regression, [R] **nbreg**, [SVY] **svy estimation**
  Schur decomposition, [M-5] **gschurd( )**
generate,
  cluster subcommand, [MV] **cluster generate**
  icd9 subcommand, [D] **icd9**
  icd9p subcommand, [D] **icd9**
  sts subcommand, [ST] **sts generate**
generate command, [D] **generate**, [MI] **mi passive**, [MI] **mi xeq**; [D] **egen**
generate functions, adding, [MV] **cluster programming subroutines**
generating
  data, [D] **generate**; [D] **egen**
  variables, [ST] **stgen**, [ST] **sts generate**
get( ) matrix function, [D] **functions**, [P] **matrix define**, [P] **matrix get**
get, net subcommand, [R] **net**
getting started, [U] **1 Read this—it will help**
Getting Started with Stata manuals, [U] **1.1 Getting Started with Stata**
  keyword search of, [U] **4 Stata's help and search facilities**
gettoken command, [P] **gettoken**
Geweke–Hajivassiliou–Keane multivariate normal simulator, [M-5] **ghk( )**, [M-5] **ghkfast( )**
ghalton( ) function, [M-5] **halton( )**
_ghessenbergd( ) function, [M-5] **ghessenbergd( )**
ghessenbergd( ) function, [M-5] **ghessenbergd( )**
_ghessenbergd_la( ) function, [M-5] **ghessenbergd( )**
ghk( ) function, [M-5] **ghk( )**
ghkfast( ) function, [M-5] **ghkfast( )**
ghkfast_init( ) function, [M-5] **ghkfast( )**
ghkfast_init_*( ) function, [M-5] **ghkfast( )**
ghkfast_query_*( ) function, [M-5] **ghkfast( )**
ghk_init( ) function, [M-5] **ghk( )**
ghk_init_*( ) function, [M-5] **ghk( )**
ghk_query_npts( ) function, [M-5] **ghk( )**
GJR, [TS] **arch**
gladder command, [R] **ladder**
GLLAMM, [R] **gllamm**
gllamm command, [R] **gllamm**
GLM, [R] **binreg**, [R] **glm**, [SVY] **svy estimation**, [XT] **Glossary**
glm command, [R] **glm**, [R] **glm postestimation**, *also see* postestimation command
Global, class prefix operator, [P] **class**
global command, [P] **macro**, [U] **18.3.2 Global macros**, [U] **18.3.10 Advanced global macro manipulation**
global variable, [M-2] **declarations**, [M-5] **direxternal( )**, [M-5] **findexternal( )**, [M-5] **valofexternal( )**, [M-6] **Glossary**
glogit command, [R] **glogit**, [R] **glogit postestimation**, *also see* postestimation command
glsaccum, matrix subcommand, [P] **matrix accum**

gmm command, [R] **gmm**, [R] **gmm postestimation**, *also see* postestimation command
gnbreg command, [R] **nbreg**, [R] **nbreg postestimation**, *also see* postestimation command
gof, estat subcommand, [R] **logistic postestimation**, [R] **poisson postestimation**
Gompertz survival regression, [ST] **streg**
Goodman and Kruskal's gamma, [R] **tabulate twoway**
goodness-of-fit tests, [R] **brier**, [R] **diagnostic plots**, [R] **ksmirnov**, [R] **logistic postestimation**, [R] **poisson postestimation**, [R] **regress postestimation**, [R] **swilk**
goto, [M-2] **goto**
Gower coefficient similarity measure, [MV] *measure_option*
gph files, [G] **concept: gph files**, [G] **graph manipulation**
  describing contents, [G] **graph describe**
.gph files, [U] **11.6 File-naming conventions**
gprobit command, [R] **glogit**, [R] **glogit postestimation**, *also see* postestimation command
gradient option, [R] **maximize**
grammar, [M-2] **syntax**
Granger causality, [TS] **Glossary**, [TS] **vargranger**
graph
  bar command, [G] **graph bar**
  box command, [G] **graph box**
  combine command, [G] **graph combine**
  command, [G] **graph**
  copy command, [G] **graph copy**
  describe, [G] **graph describe**
  dir command, [G] **graph dir**
  display command, [G] **graph display**
  dot command, [G] **graph dot**; [G] *area_options*, [G] *line_options*
  drop command, [G] **graph drop**
  export command, [G] **graph export**
  hbar command, [G] **graph bar**
  hbox command, [G] **graph box**
  matrix command, [G] **graph matrix**
  pie command, [G] **graph pie**
  play command, [G] **graph play**
  print command, [G] **graph print**, [G] *pr_options*; [G] **graph set**
  query command, [G] **graph query**
  rename command, [G] **graph rename**
  save command, [G] **graph save**
  set command, [G] **graph set**
  twoway area command, [G] **graph twoway area**
  twoway bar command, [G] **graph twoway bar**
  twoway command, [G] **graph twoway**
  twoway connected command, [G] **graph twoway connected**
  twoway dot command, [G] **graph twoway dot**
  twoway dropline command, [G] **graph twoway dropline**
  twoway fpfit command, [G] **graph twoway fpfit**

graph, *continued*

twoway fpfitci command, [G] **graph twoway fpfitci**

twoway function command, [G] **graph twoway function**

twoway histogram command, [G] **graph twoway histogram**

twoway kdensity command, [G] **graph twoway kdensity**

twoway lfit command, [G] **graph twoway lfit**

twoway lfitci command, [G] **graph twoway lfitci**

twoway line command, [G] **graph twoway line**

twoway lowess command, [G] **graph twoway lowess**

twoway lpoly command, [G] **graph twoway lpoly**

twoway lpolyci command, [G] **graph twoway lpolyci**

twoway mband command, [G] **graph twoway mband**

twoway mspline command, [G] **graph twoway mspline**

twoway pcarrow command, [G] **graph twoway pcarrow**

twoway pcarrowi command, [G] **graph twoway pcarrowi**

twoway pcbarrow command, [G] **graph twoway pcarrow**

twoway pccapsym command, [G] **graph twoway pccapsym**

twoway pci command, [G] **graph twoway pci**

twoway pcscatter command, [G] **graph twoway pcscatter**

twoway pcspike command, [G] **graph twoway pcspike**

twoway qfit command, [G] **graph twoway qfit**

twoway qfitci command, [G] **graph twoway qfitci**

twoway rarea command, [G] **graph twoway rarea**

twoway rbar command, [G] **graph twoway rbar**

twoway rcap command, [G] **graph twoway rcap**

twoway rcapsym command, [G] **graph twoway rcapsym**

twoway rconnected command, [G] **graph twoway rconnected**

twoway rline command, [G] **graph twoway rline**

twoway rscatter command, [G] **graph twoway rscatter**

twoway rspike command, [G] **graph twoway rspike**

twoway scatter command, [G] **graph twoway scatter**

twoway scatteri command, [G] **graph twoway scatteri**

twoway spike command, [G] **graph twoway spike**

twoway tsline command, [G] **graph twoway tsline**

twoway tsrline command, [G] **graph twoway tsline**

use command, [G] **graph use**

graph,

adjusted Kaplan–Meier survivor curves, [ST] **sts**

baseline hazard and survivor, [ST] **stcox**, [ST] **sts**

cumulative hazard function, [ST] **stcurve**, [ST] **sts graph**

hazard function, [ST] **ltable**, [ST] **stcurve**, [ST] **sts graph**

Kaplan–Meier survivor curves, [ST] **stcox PH-assumption tests**, [ST] **sts**

log-log curve, [ST] **stcox PH-assumption tests**

survivor function, [ST] **stcurve**, [ST] **sts graph**

graph,

fcast subcommand, [TS] **fcast graph**

irf subcommand, [TS] **irf graph**

ml subcommand, [R] **ml**

sts subcommand, [ST] **sts graph**

Graph Editor, [G] **graph editor**

graph region, [G] *region_options*

graph text, [G] *text*

graphical user interface, [P] **dialog programming**

graphics, set subcommand, [G] **set graphics**, [R] **set**

graphregion() option, [G] *region_options*

graphs,

added-variable plot, [R] **regress postestimation**

adjusted partial residual plot, [R] **regress postestimation**

augmented component-plus-residual plot, [R] **regress postestimation**

augmented partial residual plot, [R] **regress postestimation**

autocorrelations, [TS] **corrgram**

binary variable cumulative sum, [R] **cusum**

biplot, [MV] **biplot**, [MV] **ca postestimation**

CA dimension projection, [MV] **ca postestimation**

cluster tree, [MV] **cluster dendrogram**

component-plus-residual, [R] **regress postestimation**

correlogram, [TS] **corrgram**

cross-correlogram, [TS] **xcorr**

cross-sectional time-series data, [XT] **xtdata**, [XT] **xtline**

cumulative distribution, [R] **cumul**

cumulative spectral density, [TS] **cumsp**

dendrogram, [MV] **cluster dendrogram**

density, [R] **kdensity**

density-distribution sunflower, [R] **sunflower**

derivatives, [R] **dydx**, [R] **testnl**

describing contents, [G] **graph describe**

diagnostic, [R] **diagnostic plots**

dotplot, [R] **dotplot**

eigenvalue

after factor, [MV] **screeplot**

after pca, [MV] **screeplot**

error-bar charts, [R] **serrbar**

forecasts, [TS] **fcast graph**

fractional polynomial, [R] **fracpoly**

functions, [D] **obs**, [D] **range**

histograms, [R] **histogram**, [R] **kdensity**

graphs, *continued*
 impulse–response functions, [TS] **irf**, [TS] **irf cgraph**, [TS] **irf graph**, [TS] **irf ograph**
 integrals, [R] **dydx**
 ladder-of-power histograms, [R] **ladder**
 leverage-versus-(squared)-residual, [R] **regress postestimation**
 loading
  after candisc, [MV] **scoreplot**
  after discrim lda, [MV] **scoreplot**
  after factor, [MV] **scoreplot**
  after pca, [MV] **scoreplot**
 logistic diagnostic, [R] **logistic postestimation**
 lowess smoothing, [R] **lowess**
 MDS configuration, [MV] **mds postestimation**
 means and medians, [R] **grmeanby**
 normal probability, [R] **diagnostic plots**
 overall look of, [G] **schemes intro**
 parameterized curves, [D] **range**
 partial correlogram, [TS] **corrgram**
 partial residual, [R] **regress postestimation**
 partial-regression leverage, [R] **regress postestimation**
 periodogram, [TS] **pergram**
 procrustes overlay, [MV] **procrustes postestimation**
 quality control, [R] **qc**
 quantile, [R] **diagnostic plots**
 quantile–normal, [R] **diagnostic plots**
 quantile–quantile, [R] **diagnostic plots**
 regression diagnostic, [R] **regress postestimation**
 residual versus fitted, [R] **regress postestimation**
 residual versus predictor, [R] **regress postestimation**
 ROC curve, [R] **logistic postestimation**, [R] **roc**, [R] **rocfit postestimation**
 rootograms, [R] **spikeplot**
 saving, [G] **graph save**, [G] *saving_option*
 score
  after candisc, [MV] **scoreplot**
  after discrim lda, [MV] **scoreplot**
  after factor, [MV] **scoreplot**
  after pca, [MV] **scoreplot**
 Shepard diagram, [MV] **mds postestimation**
 smoothing, [R] **kdensity**, [R] **lowess**, [R] **lpoly**
 spike plot, [R] **spikeplot**
 stem-and-leaf, [R] **stem**
 sunflower, [R] **sunflower**
 suppressing, [G] *nodraw_option*
 symmetry, [R] **diagnostic plots**
 time-versus-concentration curve, [R] **pk**, [R] **pkexamine**
 white-noise test, [TS] **wntestb**
grdistances, estat subcommand, [MV] **discrim lda postestimation**, [MV] **discrim qda postestimation**
greater than (or equal) operator, [U] **13.2.3 Relational operators**
.grec files, [U] **11.6 File-naming conventions**

Greek letters, [G] *text*
Greenhouse–Geisser epsilon, [R] **anova**
Greenwood confidence intervals, [ST] **sts**
grid
 definition, [G] *gridstyle*
 lines, [G] *axis_label_options*
  without ticks, [G] *tickstyle*
gridstyle, [G] *gridstyle*
grmeanby command, [R] **grmeanby**
grmeans, estat subcommand, [MV] **discrim lda postestimation**
group(), egen function, [D] **egen**
group, estat subcommand, [XT] **xtmelogit postestimation**, [XT] **xtmepoisson postestimation**, [XT] **xtmixed postestimation**
group-data regression, [R] **glogit**, [R] **intreg**
grouping variables, generating, [MV] **cluster generate**
groups, graphs by, [G] *by_option*
grsummarize, estat subcommand, [MV] **discrim estat**, [MV] **discrim knn postestimation**, [MV] **discrim lda postestimation**, [MV] **discrim logistic postestimation**, [MV] **discrim qda postestimation**
gs1 print color mapping, [G] **set printcolor**
gs2 print color mapping, [G] **set printcolor**
gs3 print color mapping, [G] **set printcolor**
_gschurd() function, [M-5] **gschurd( )**
gschurd() function, [M-5] **gschurd( )**
_gschurdgroupby() function, [M-5] **gschurd( )**
gschurdgroupby() function, [M-5] **gschurd( )**
_gschurdgroupby_la() function, [M-5] **gschurd( )**
_gschurd_la() function, [M-5] **gschurd( )**
gsort command, [D] **gsort**
GUI, examples of, [U] **2 A brief description of Stata**
GUI programming, [P] **dialog programming**

H

hadamard() matrix function, [D] **functions**, [P] **matrix define**
Hadamard matrix, [SVY] **Glossary**, [SVY] **svy brr**
hadri, xtunitroot subcommand, [XT] **xtunitroot**
Hadri Lagrange multiplier stationarity test, [XT] **xtunitroot**
half option, [G] **graph matrix**
half-normal model, stochastic frontier, [R] **frontier**
halfyear() function, [D] **dates and times**, [D] **functions**, [M-5] **date( )**
halfyearly() function, [D] **dates and times**, [D] **functions**, [M-5] **date( )**
_halton() function, [M-5] **halton( )**
halton() function, [M-5] **halton( )**
Halton set, [M-5] **halton( )**
Hamann coefficient similarity measure, [MV] *measure_option*
Hammersley set, [M-5] **halton( )**
hard missing value, [MI] **Glossary**
harmonic mean, [R] **ameans**

Harrell's C, [ST] **stcox postestimation**

Harris–Tzavalis test, [XT] **xtunitroot**

has_eprop() function, [D] **functions**

hash functions, [M-5] **hash1( )**

hash tables, [M-5] **asarray( )**

hash1() function, [M-5] **hash1( )**

hasmissing() function, [M-5] **missing( )**

hat matrix, *see* projection matrix, diagonal elements of

hausman command, [R] **hausman**

Hausman specification test, [R] **hausman**, [XT] **xtreg postestimation**

Hausman–Taylor
  estimator, [XT] **xthtaylor**
  postestimation, [XT] **xthtaylor postestimation**

Haver Analytics, [TS] **haver**

haver command, [D] **infile**, [TS] **haver**

hazard
  constributions, [ST] **Glossary**
  function, [G] **graph other**, [ST] **sts**, [ST] **sts generate**, [ST] **sts list**
    graph of, [ST] **ltable**, [ST] **stcurve**, [ST] **sts graph**
  ratio, [ST] **Glossary**
    minimal detectable difference, [ST] **stpower**
    minimal effect size, [ST] **stpower**

hazard tables, [ST] **ltable**

hbar, graph subcommand, [G] **graph bar**

hbox, graph subcommand, [G] **graph box**

headlabel option, [G] **graph twoway pccapsym**, [G] **graph twoway pcscatter**

health ratios, [R] **binreg**

heart attack data, [MI] **intro substantive**, [MI] **mi estimate**, [MI] **mi impute logit**, [MI] **mi impute mlogit**, [MI] **mi impute monotone**, [MI] **mi impute mvn**, [MI] **mi impute ologit**, [MI] **mi impute pmm**, [MI] **mi impute regress**

heckman command, [R] **heckman**, [R] **heckman postestimation**, *also see* postestimation command

Heckman selection model, [R] **heckman**, [R] **heckprob**, [SVY] **svy estimation**

heckprob command, [R] **heckprob**, [R] **heckprob postestimation**, *also see* postestimation command

height() textbox option, [G] *added_text_options*

help, [M-1] **help**, [M-3] **mata help**

help, view subcommand, [R] **view**

help command, [R] **help**, [U] **4 Stata's help and search facilities**, [U] **7 –more– conditions**
  writing your own, [U] **18.11.6 Writing online help**

help file search, [R] **hsearch**

help—I don't know what to do, [U] **3 Resources for learning and using Stata**

help system, searching, [R] **hsearch**

help_d, view subcommand, [R] **view**

Henze–Zirkler normality test, [MV] **mvtest normality**

Hermitian
  adjoin, [M-2] **op_transpose**, [M-5] **conj( )**
  matrices, [M-5] **issymmetric( )**, [M-5] **makesymmetric( )**, [M-6] **Glossary**

Hermitian, *continued*
  transpose, [M-2] **op_transpose**, [M-5] **conj( )**

Hessenberg
  decomposition, [M-5] **hessenbergd( )**, [M-6] **Glossary**
  form, [M-6] **Glossary**

_hessenbergd() function, [M-5] **hessenbergd( )**

hessenbergd() function, [M-5] **hessenbergd( )**

_hessenbergd_la() function, [M-5] **hessenbergd( )**

hessian option, [R] **maximize**

heterogeneity test, [ST] **epitab**

heteroskedastic probit
  model, [SVY] **svy estimation**
  regression, [R] **hetprob**

heteroskedasticity, [XT] **xtgls**
  conditional, [R] **regress postestimation time series**, [TS] **arch**
  multiplicative heteroskedastic regression, [TS] **arch**
  robust variances, *see* robust, Huber/White/sandwich estimator of variance
  stochastic frontier, [R] **frontier**
  test for, [R] **regress postestimation**, [R] **regress postestimation time series**

hetprob command, [R] **hetprob**, [R] **hetprob postestimation**, *also see* postestimation command

hettest, estat subcommand, [R] **regress postestimation**

hexadecimal report, [D] **hexdump**

hexdump command, [D] **hexdump**

Heywood case, [MV] **Glossary**

Heywood solution, [MV] **Glossary**

hh() function, [D] **dates and times**, [D] **functions**, [M-5] **date( )**

hhC() function, [D] **dates and times**, [D] **functions**, [M-5] **date( )**

hierarchical
  cluster analysis, [MV] **cluster**, [MV] **cluster linkage**, [MV] **clustermat**
  clustering, [MV] **Glossary**
  clustering methods, [MV] **Glossary**
  model, [XT] **Glossary**, [XT] **xtmelogit**, [XT] **xtmepoisson**, [XT] **xtmixed**
  regression, [R] **areg**, [R] **nestreg**
  samples, [R] **anova**, [R] **loneway**; [R] **areg**

high-low charts, [G] **graph twoway rbar**, [G] **graph twoway rcap**, [G] **graph twoway rspike**

Hilbert() function, [M-5] **Hilbert( )**

Hildreth–Lu regression, [TS] **prais**

HILO, [M-5] **byteorder( )**

histogram, graph twoway subcommand, [G] **graph twoway histogram**

histogram command, [R] **histogram**

histograms, [G] **graph twoway histogram**, [R] **histogram**
  dotplots, [R] **dotplot**
  kernel density estimator, [R] **kdensity**
  ladder-of-powers, [R] **ladder**

histograms, *continued*
    of categorical variables, [R] **histogram**
    stem-and-leaf, [R] **stem**
histories, [G] *by_option*, [G] **graph bar**, [G] **graph
    box**, [G] **graph matrix**, [G] **graph pie**,
    [G] **graph twoway histogram**
hms( ) function, [D] **dates and times**, [D] **functions**,
    [M-5] **date( )**
hofd( ) function, [D] **dates and times**, [D] **functions**,
    [M-5] **date( )**
hold,
    _estimates subcommand, [P] **_estimates**
    _return subcommand, [P] **_return**
Holm adjustment, [R] **test**
Holt–Winters smoothing, [TS] **Glossary**,
    [TS] **tssmooth**, [TS] **tssmooth dexponential**,
    [TS] **tssmooth exponential**, [TS] **tssmooth
    hwinters**, [TS] **tssmooth shwinters**
home resale-price data, [MI] **mi estimate**, [MI] **mi
    estimate postestimation**, [MI] **mi estimate using**
homogeneity of variances, [R] **oneway**, [R] **sdtest**
homogeneity test, [ST] **epitab**
homoskedasticity tests, [R] **regress postestimation**
Horst normalization, *see* Kaiser normalization
Hosmer–Lemeshow goodness-of-fit test, [R] **logistic
    postestimation**
hot, ssc subcommand, [R] **ssc**
hotelling command, [MV] **hotelling**
Hotelling's
    generalized *T*-squared statistic, [MV] **manova**
    T-squared, [MV] **Glossary**, [MV] **hotelling**,
        [MV] **mvtest means**
hours( ) function, [D] **dates and times**, [D] **functions**,
    [M-5] **date( )**
_hqrd( ) function, [M-5] **qrd( )**
hqrd( ) function, [M-5] **qrd( )**
hqrdmultq( ) function, [M-5] **qrd( )**
hqrdmultq1t( ) function, [M-5] **qrd( )**
_hqrdp( ) function, [M-5] **qrd( )**
hqrdp( ) function, [M-5] **qrd( )**
_hqrdp_la( ) function, [M-5] **qrd( )**
hqrdq( ) function, [M-5] **qrd( )**
hqrdq1( ) function, [M-5] **qrd( )**
hqrdr( ) function, [M-5] **qrd( )**
hqrdr1( ) function, [M-5] **qrd( )**
hsearch command, [R] **hsearch**, [U] **4 Stata's help
    and search facilities**
ht, xtunitroot subcommand, [XT] **xtunitroot**
http://www.stata.com, [U] **3.2 The Stata web site
    (www.stata.com)**
httpproxy, set subcommand, [R] **netio**, [R] **set**
httpproxyauth, set subcommand, [R] **netio**, [R] **set**
httpproxyhost, set subcommand, [R] **netio**, [R] **set**
httpproxyport, set subcommand, [R] **netio**, [R] **set**
httpproxypw, set subcommand, [R] **netio**, [R] **set**
httpproxyuser, set subcommand, [R] **netio**, [R] **set**
Huber weighting, [R] **rreg**

Huber/White/sandwich estimator of variance,
    [SVY] **variance estimation**
Huynh–Feldt epsilon, [R] **anova**
hwinters, tssmooth subcommand, [TS] **tssmooth
    hwinters**
hyperbolic functions, [M-5] **sin( )**
hypergeometric( ) function, [D] **functions**,
    [M-5] **normal( )**
hypergeometric,
    cumulative distribution, [D] **functions**
    probability mass function, [D] **functions**
hypergeometricp( ) function, [D] **functions**,
    [M-5] **normal( )**
hypertext help, [R] **help**, [U] **4 Stata's help and search
    facilities**, [U] **18.11.6 Writing online help**
hypothesis tests, *see* tests

I

I( ) function, [D] **functions**, [M-5] **I( )**, [P] **matrix
    define**
ibeta( ) function, [D] **functions**, [M-5] **normal( )**
ibetatail( ) function, [D] **functions**, [M-5] **normal( )**
ic, estat subcommand, [R] **estat**
icd9
    check command, [D] **icd9**
    clean command, [D] **icd9**
    generate command, [D] **icd9**
    lookup command, [D] **icd9**
    query command, [D] **icd9**
    search command, [D] **icd9**
icd9p
    check command, [D] **icd9**
    clean command, [D] **icd9**
    generate command, [D] **icd9**
    lookup command, [D] **icd9**
    query command, [D] **icd9**
    search command, [D] **icd9**
ice command, [MI] **mi export ice**, [MI] **mi import ice**
ID variable, [ST] **Glossary**
identifier, class, [P] **class**
identifier, unique, [D] **isid**
identity matrix, [M-5] **I( )**, [P] **matrix define**
idiosyncratic error term, [XT] **Glossary**
if *exp*, [M-2] **if**, [P] **syntax**, [U] **11 Language syntax**
if programming command, [P] **if**
ignorable missing-data mechanism, [MI] **intro
    substantive**
IIA, [R] **asclogit**, [R] **asmprobit**, [R] **clogit**,
    [R] **hausman**, [R] **nlogit**
Im( ) function, [M-5] **Re( )**
imaginary part, [M-5] **Re( )**
immediate commands, [I] **immediate commands**,
    [I] **prefix commands**, [P] **display**, [R] **bitest**,
    [R] **ci**, [R] **prtest**, [R] **sampsi**, [R] **sdtest**,
    [R] **symmetry**, [R] **tabulate twoway**,
    [R] **ttest**, [U] **19 Immediate commands**;
    [U] **18.4.5 Parsing immediate commands**

Im–Pesaran–Shin test, [XT] **xtunitroot**

implied context, class, [P] **class**

import, mi subcommand, [MI] **mi import**, [MI] **mi import flong**, [MI] **mi import flongsep**, [MI] **mi import ice**, [MI] **mi import nhanes1**, [MI] **mi import wide**

importance weights, [U] **11.1.6 weight**, [U] **20.18.4 Importance weights**

importing data, [D] **fdasave**, [D] **infile**, [D] **infile (fixed format)**, [D] **infile (free format)**, [D] **infix (fixed format)**, [D] **insheet**, [D] **odbc**, [D] **xmlsave** [MI] **mi import**, [MI] **mi import flong**, [MI] **mi import flongsep**, [MI] **mi import ice**, [MI] **mi import nhanes1**, [MI] **mi import wide**, [U] **21.4 Transfer programs**;

impulse–response functions, *see* IRF

imputation,

    binary, [MI] **mi impute logit**

    categorical, [MI] **mi impute mlogit**, [MI] **mi impute ologit**

    continuous, [MI] **mi impute pmm**, [MI] **mi impute regress**

    linear regression, [MI] **mi impute regress**

    logistic regression, [MI] **mi impute logit**

    modeling, [MI] **mi impute**

    monotone, [MI] **mi impute**, [MI] **mi impute monotone**

    multinomial logistic regression, [MI] **mi impute mlogit**

    multiple, [MI] **intro substantive**

    multivariate, [MI] **mi impute monotone**, [MI] **mi impute mvn**

    multivariate monotone, [MI] **mi impute**, [MI] **mi impute logit**, [MI] **mi impute mlogit**, [MI] **mi impute ologit**, [MI] **mi impute pmm**, [MI] **mi impute regress**

    multivariate normal, [MI] **mi impute mvn**

    on subsamples, [MI] **mi impute**

    ordered logistic regression, [MI] **mi impute ologit**

    passive variables, [MI] **mi impute regress**

    predictive mean matching, [MI] **mi impute**, [MI] **mi impute pmm**

    regression, [MI] **mi impute**, [MI] **mi impute regress**

    semiparameteric, [MI] **mi impute pmm**

    step, [MI] **intro substantive**, [MI] **mi estimate**

    transformations, [MI] **mi impute**

    univariate, [MI] **mi impute logit**, [MI] **mi impute mlogit**, [MI] **mi impute ologit**, [MI] **mi impute pmm**, [MI] **mi impute regress**

imputation method, [MI] **mi impute**, [MI] **mi impute logit**, [MI] **mi impute mlogit**, [MI] **mi impute ologit**, [MI] **mi impute pmm**, [MI] **mi impute regress**

    proper, [MI] **intro substantive**, [MI] **mi impute monotone**, [MI] **mi impute mvn**

imputations, recommended number of, [MI] **intro substantive**

impute, mi subcommand, [MI] **mi impute**, [MI] **mi impute logit**, [MI] **mi impute mlogit**, [MI] **mi impute monotone**, [MI] **mi impute mvn**, [MI] **mi impute ologit**, [MI] **mi impute pmm**, [MI] **mi impute regress**,

imputed variables, *see* variables, imputed

imtest, estat subcommand, [R] **regress postestimation**

in *range* modifier, [P] **syntax**, [U] **11 Language syntax**

in smcl, display directive, [P] **display**

inbase( ) function, [M-5] **inbase( )**

incidence, [ST] **Glossary**

incidence-rate ratio, [R] **poisson**, [R] **zip**, [ST] **epitab**, [ST] **stci**, [ST] **stir**, [ST] **stptime**, [ST] **stsum** differences, [R] **lincom**, [R] **nlcom**

incidence-rate ratio and rate ratio, [SVY] **svy estimation**, [XT] **xtgee**, [XT] **xtnbreg**, [XT] **xtpoisson**

incidence studies, [ST] **Glossary**

include command, [P] **include**

income distributions, [R] **inequality**

income tax rate function, [D] **egen**

incomplete

    beta function, [D] **functions**, [M-5] **normal( )**

    gamma function, [D] **functions**, [M-5] **normal( )**

    observations, [MI] **Glossary**

increment operator, [M-2] **op_increment**

independence of irrelevant alternatives, *see* IIA

independence tests, *see* tests

independent and identically distributed, [TS] **Glossary**

index,

    mathematical functions, [M-4] **statistical**

    matrix functions, [M-4] **utility**

    statistical functions, [M-4] **statistical**

    stopping rules, *see* stopping rules

    utility functions, [M-4] **utility**

index of probit and logit, [R] **logit postestimation**, [R] **predict**, [R] **probit postestimation**

index search, [R] **search**, [U] **4 Stata's help and search facilities**

indexnot( ) function, [D] **functions**, [M-5] **indexnot( )**

indicator variables, [R] **anova**, [R] **areg**, [R] **xi**

indicators, [U] **11.4.3 Factor variables**

indirect standardization, [R] **dstdize**

ineligible missing values, [MI] **Glossary**

inequality measures, [R] **inequality**

inertia, [MV] **Glossary**

inertia, estat subcommand, [MV] **ca postestimation**

infile command, [D] **infile (fixed format)**, [D] **infile (free format)**; [D] **infile**

infix command, [D] **infix (fixed format)**; [D] **infile**

influence statistics, [R] **logistic postestimation**, [R] **predict**, [R] **regress postestimation**

%infmt, [D] **infile (fixed format)**

information

    criteria, *see* AIC, BIC

    matrix, [P] **matrix get**, [R] **correlate**, [R] **maximize**

    matrix test, [R] **regress postestimation**

Informix, reading data from, [U] **21.4 Transfer programs**
inheritance, [M-2] **class**, [P] **class**
init, ml subcommand, [R] **ml**
initialization, class, [P] **class**
inlist() function, [D] **functions**
inner fence, [R] **lv**
innovation accounting, [TS] **irf**
input, matrix subcommand, [P] **matrix define**
input, obtaining from console in programs, *see* console, obtaining input from
input command, [D] **input**
input/output functions, [M-4] **io**
inputting data
    from a file, *see* reading data from disk
    interactively, [D] **edit**, [D] **input**, *also see* editing data; reading data from disk
inrange() function, [D] **functions**
insert, odbc subcommand, [D] **odbc**
insheet command, [D] **insheet**; [D] **infile**
inspect command, [D] **inspect**
install,
    net subcommand, [R] **net**
    ssc subcommand, [R] **ssc**
installation
    of official updates, [R] **update**, [U] **28 Using the Internet to keep up to date**
    of SJ and STB, [R] **net**, [R] **sj**, [U] **3.6 Updating and adding features from the web**, [U] **17.6 How do I install an addition?**
instance, class, [P] **class**
.instancemv built-in class function, [P] **class**
instance-specific variable, [P] **class**
instrumental-variables, [XT] **Glossary**
    estimator, [XT] **Glossary**
    regression, [R] **gmm**, [R] **ivprobit**, [R] **ivregress**, [R] **ivtobit**, [R] **nlsur**, [SVY] **svy estimation**, [XT] **xtabond**, [XT] **xtdpd**, [XT] **xtdpdsys**, [XT] **xthtaylor**, [XT] **xtivreg**, [XT] **xtivreg postestimation**
int, [D] **data types**, [I] **data types**
int (storage type), [U] **12.2.2 Numeric storage types**
int() function, [D] **functions**
integ command, [R] **dydx**
integer truncation function, [D] **functions**
integers, [M-5] **trunc()**
integrals, numeric, [R] **dydx**
integrated process, [TS] **Glossary**
intensity, color, adjustment, [G] *colorstyle*, [G] **graph twoway histogram**, [G] **graph twoway kdensity**
intensitystyle, [G] *intensitystyle*
interaction, [R] **margins**
interaction expansion, [R] **xi**
interactions, [U] **11.4.3 Factor variables**, [U] **13.5.3 Factor variables and time-series operators**, [U] **25.2 Estimation with factor variables**
internal consistency, test for, [R] **alpha**

Internet, [U] **3.2 The Stata web site (www.stata.com)**
    commands to control connections to, [R] **netio**
    installation of updates from, [R] **adoupdate**, [R] **net**, [R] **sj**, [R] **update** [U] **28 Using the Internet to keep up to date**
    search, [R] **net search**
interpolation, [D] **ipolate**
interquantile range, [R] **qreg**
interquartile range, [R] **lv**
    generating variable containing, [D] **egen**
    making dataset of, [D] **collapse**
    reporting, [R] **table**, [R] **tabstat**
    summarizing, [D] **pctile**
interrater agreement, [R] **kappa**
interrupting command execution, [U] **10 Keyboard use**
interval
    data, [XT] **Glossary**
    regression, [R] **intreg**, [SVY] **svy estimation**
    regression, random-effects, [XT] **xtintreg** postestimation, [XT] **xtintreg postestimation**
intracluster correlation, [R] **loneway**
intreg command, [R] **intreg**, [R] **intreg postestimation**, *also see* postestimation command
inv() matrix function, [D] **functions**, [P] **matrix define**
invbinomial() function, [D] **functions**, [M-5] **normal()**
invbinomialtail() function, [D] **functions**, [M-5] **normal()**
invchi2() function, [D] **functions**, [M-5] **normal()**
invchi2tail() function, [D] **functions**, [M-5] **normal()**
invcloglog() function, [D] **functions**, [M-5] **logit()**
inverse
    cumulative
        beta distribution, [D] **functions**
        binomial function, [D] **functions**
        chi-squared distribution function, [D] **functions**
        F distribution function, [D] **functions**
        incomplete gamma function, [D] **functions**
    matrix, [M-4] **solvers**, [M-5] **invsym()**, [M-5] **cholinv()**, [M-5] **luinv()**, [M-5] **qrinv()**, [M-5] **pinv()**, [M-5] **solve_tol()**
    noncentral
        beta distribution, [D] **functions**
        chi-squared distribution function, [D] **functions**
        F distribution, [D] **functions**
    normal distribution function, [D] **functions**
    reverse cumulative
        beta distribution, [D] **functions**
        binomial function, [D] **functions**
        chi-squared distribution function, [D] **functions**
        F distribution function, [D] **functions**
        incomplete gamma function, [D] **functions**
        t distribution function, [D] **functions**
inverse of matrix, [P] **matrix define**, [P] **matrix svd**
invF() function, [D] **functions**, [M-5] **normal()**
_invfft() function, [M-5] **fft()**

invfft() function, [M-5] **fft( )**
invFtail() function, [D] **functions**, [M-5] **normal( )**
invgammap() function, [D] **functions**, [M-5] **normal( )**
invgammaptail() function, [D] **functions**,
    [M-5] **normal( )**
invHilbert() function, [M-5] **Hilbert( )**
invibeta() function, [D] **functions**, [M-5] **normal( )**
invibetatail() function, [D] **functions**,
    [M-5] **normal( )**
invlogit() function, [D] **functions**, [M-5] **logit( )**
invnbinomial() function, [D] **functions**,
    [M-5] **normal( )**
invnbinomialtail() function, [D] **functions**,
    [M-5] **normal( )**
invnchi2() function, [D] **functions**, [M-5] **normal( )**
invnFtail() function, [D] **functions**, [M-5] **normal( )**
invnibeta() function, [D] **functions**, [M-5] **normal( )**
invnormal() function, [D] **functions**, [M-5] **normal( )**
invorder() function, [M-5] **invorder( )**
invpoisson() function, [D] **functions**,
    [M-5] **normal( )**
invpoissontail() function, [D] **functions**,
    [M-5] **normal( )**
_invsym() function, [M-5] **invsym( )**
invsym() function, [D] **functions**, [M-5] **invsym( )**,
    [P] **matrix define**
invtokens() function, [M-5] **invtokens( )**
invttail() function, [D] **functions**, [M-5] **normal( )**
invvech() function, [M-5] **vec( )**
I/O functions, [M-4] **io**
ipolate command, [D] **ipolate**
ips, xtunitroot subcommand, [XT] **xtunitroot**
iqr(), egen function, [D] **egen**
IQR, *see* interquartile range
iqreg command, [R] **qreg**, [R] **qreg postestimation**,
    also see postestimation command
ir command, [ST] **epitab**
irecode() function, [D] **functions**
irf
    add command, [TS] **irf add**
    cgraph command, [TS] **irf cgraph**
    commands, introduction, [TS] **irf**
    create command, [TS] **irf create**
    ctable command, [TS] **irf ctable**
    describe command, [TS] **irf describe**
    drop command, [TS] **irf drop**
    graph command, [TS] **irf graph**
    ograph command, [TS] **irf ograph**
    rename command, [TS] **irf rename**
    set command, [TS] **irf set**
    table command, [TS] **irf table**
IRF, [G] **graph other**, [TS] **Glossary**, [TS] **irf**, [TS] **irf**
    add, [TS] **irf cgraph**, [TS] **irf create**, [TS] **irf**
    ctable, [TS] **irf describe**, [TS] **irf drop**, [TS] **irf**
    graph, [TS] **irf ograph**, [TS] **irf rename**,
    [TS] **irf set**, [TS] **irf table**, [TS] **var intro**,
    [TS] **varbasic**, [TS] **vec intro**

IRF, *continued*
    cumulative impulse–response functions, [TS] **irf**
    create
    cumulative orthogonalized impulse–response
    functions, [TS] **irf create**
    orthogonalized impulse–response functions, [TS] **irf**
    create
.irf files, [U] **11.6 File-naming conventions**
iri command, [ST] **epitab**
IRLS, [R] **glm**, [R] **reg3**
.isa built-in class function, [P] **class**
iscale() option, [G] **graph matrix**
iscomplex() function, [M-5] **isreal( )**
isdiagonal() function, [M-5] **isdiagonal( )**
isfleeting() function, [M-5] **isfleeting( )**
isid command, [D] **isid**
.isofclass built-in class function, [P] **class**
ispointer() function, [M-5] **isreal( )**
isreal() function, [M-5] **isreal( )**
isrealvalues() function, [M-5] **isrealvalues( )**
isstring() function, [M-5] **isreal( )**
issymmetric() function, [D] **functions**,
    [M-5] **issymmetric( )**, [P] **matrix define**
issymmetriconly() function, [M-5] **issymmetric( )**
istdize command, [R] **dstdize**
istmt, [M-1] **how**, [M-6] **Glossary**
isview() function, [M-5] **isview( )**
italics, [G] *text*
iterate() option, [R] **maximize**
iterated least squares, [R] **reg3**, [R] **sureg**
iterated principal factor method, [MV] **Glossary**
iterations, controlling the maximum number,
    [R] **maximize**
itrim() string function, [D] **functions**
ivprobit command, [R] **ivprobit**, [R] **ivprobit**
    postestimation, *also see* postestimation command
ivregress command, [R] **ivregress**, [R] **ivregress**
    postestimation, *also see* postestimation command
ivtobit command, [R] **ivtobit**, [R] **ivtobit**
    postestimation, *also see* postestimation command
[iweight=*exp*] modifier, [U] **11.1.6 weight**,
    [U] **20.18.4 Importance weights**

J

J() function, [D] **functions**, [M-5] **J( )**, [M-2] **void**,
    [M-6] **Glossary**, [P] **matrix define**
Jaccard coefficient similarity measure,
    [MV] *measure_option*
jackknife
    estimation, [R] **jackknife**, [SVY] **Glossary**,
        [SVY] *jackknife_options*, [SVY] **svy jackknife**,
        [SVY] **variance estimation**
    standard errors, [R] *vce_option*, [XT] *vce_options*
jackknife_options, [SVY] *jackknife_options*
jackknife prefix command, [R] **jackknife**,
    [R] **jackknife postestimation**, *also see*
    postestimation command

jackknifed residuals, [R] **predict**, [R] **regress postestimation**
Jarque–Bera statistic, [TS] **varnorm**, [TS] **vecnorm**
JCA, [MV] **Glossary**
Jeffreys noninformative prior, [MI] **mi impute mvn**
jitter() option, [G] **graph matrix**, [G] **graph twoway scatter**
jitterseed() option, [G] **graph matrix**, [G] **graph twoway scatter**
join operator, [M-2] **op_join**
joinby command, [D] **joinby**, [U] **22 Combining datasets**
joining
    datasets, *see* combining datasets
    time-span records, [ST] **stsplit**
joint correspondence analysis, [MV] **Glossary**
_jumble() function, [M-5] **sort( )**
jumble() function, [M-5] **sort( )**
justification of text, [G] ***textbox_options***
justificationstyle, [G] ***justificationstyle***

K

Kaiser normalization, [MV] **factor postestimation**, [MV] **pca postestimation**, [MV] **rotate**, [MV] **rotatemat**
Kaiser–Meyer–Olkin sampling adequacy, [MV] **factor postestimation**, [MV] **Glossary**, [MV] **pca postestimation**
Kalman
    filter, [TS] **arima**, [TS] **dfactor**, [TS] **dfactor postestimation**, [TS] **Glossary**, [TS] **sspace** [TS] **sspace postestimation**,
    forecast, [TS] **dfactor postestimation**, [TS] **sspace postestimation**
    smoothing, [TS] **dfactor postestimation**, [TS] **sspace postestimation**
kap command, [R] **kappa**
Kaplan–Meier
    product-limit estimate, [ST] **Glossary**, [ST] **sts** [ST] **sts generate**, [ST] **sts graph**, [ST] **sts list**, [ST] **sts test**,
    survivor function, [ST] **sts**; [ST] **ltable**, [ST] **stcox PH-assumption tests**
kappa command, [R] **kappa**
kapwgt command, [R] **kappa**
kdensity, graph twoway subcommand, [G] **graph twoway kdensity**
kdensity command, [R] **kdensity**
keep command, [D] **drop**
keeping variables or observations, [D] **drop**
Kendall's tau, [R] **spearman**, [R] **tabulate twoway**
kernel-density
    estimator, [R] **kdensity**
    smoothing, [G] **graph other**
kernel-weighted local polynomial estimator, [R] **lpoly**
keyboard
    entry, [U] **10 Keyboard use**
    search, [U] **4 Stata's help and search facilities**

keys, [G] ***legend_option***
Kish design effects, [R] **loneway**, [SVY] **estat**
Kmatrix() function, [M-5] **Kmatrix( )**
kmeans, [MV] **Glossary**
kmeans, cluster subcommand, [MV] **cluster kmeans and kmedians**
kmeans clustering, [MV] **cluster**, [MV] **cluster kmeans and kmedians**
kmedians, [MV] **Glossary**
kmedians, cluster subcommand, [MV] **cluster kmeans and kmedians**
kmedians clustering, [MV] **cluster**, [MV] **cluster kmeans and kmedians**
kmo, estat subcommand, [MV] **factor postestimation**, [MV] **pca postestimation**
KNN, [MV] **Glossary**
knn, discrim subcommand, [MV] **discrim knn**
Kolmogorov–Smirnov test, [R] **ksmirnov**
KR-20, [R] **alpha**
Kronecker
    direct product, [M-2] **op_kronecker**
    product, [D] **cross**, [P] **matrix define**
Kruskal stress, [MV] **Glossary**, [MV] **mds postestimation**
Kruskal–Wallis test, [R] **kwallis**
ksmirnov command, [R] **ksmirnov**
ktau command, [R] **spearman**
*k*th-nearest neighbor, [MV] **discrim knn**, [MV] **Glossary**
Kuder–Richardson Formula 20, [R] **alpha**
Kulczynski coefficient similarity measure, [MV] ***measure_option***
kurt(), egen function, [D] **egen**
kurtosis, [MV] **mvtest normality**, [R] **lv**, [R] **regress postestimation**, [R] **sktest**, [R] **summarize**, [R] **tabstat**, [TS] **varnorm**, [TS] **vecnorm**
kwallis command, [R] **kwallis**

L

L1-norm models, [R] **qreg**
l1title() option, [G] ***title_options***
l2title() option, [G] ***title_options***
label, snapshot subcommand, [D] **snapshot**
label
    copy command, [D] **label**
    command, [U] **12.6 Dataset, variable, and value labels**
    data command, [D] **label**
    define command, [D] **label**
    dir command, [D] **label**
    drop command, [D] **label**
    language command, [D] **label language**
    list command, [D] **label**
    macro extended function, [P] **macro**
    save command, [D] **label**
    values command, [D] **label**
    variable command, [D] **label**

label values, [P] **macro**, [U] **12.6 Dataset, variable, and value labels**; [U] **13.10 Label values**
labelbook command, [D] **labelbook**
labeling data, [U] **12.6 Dataset, variable, and value labels**; [D] **describe**, [D] **label**, [D] **label language**, [D] **notes**
  in other languages, [U] **12.6.4 Labels in other languages**
labels,
  axis, [G] *axis_label_options*
  creating, [D] **varmanage**
  editing, [D] **varmanage**
  marker, [G] *marker_label_options*
LAD regression, [R] **qreg**
ladder command, [R] **ladder**
ladder-of-powers plots, [G] **graph other**, [R] **ladder**
lag-exclusion statistics, [TS] **varwle**
lag operator, [TS] **Glossary**, [U] **11.4.4 Time-series varlists**
lag-order selection statistics, [TS] **varsoc**; [TS] **var**, [TS] **var intro**, [TS] **var svar**, [TS] **vec intro**
lagged values, [U] **13.7 Explicit subscripting**, [U] **13.9.1 Generating lags, leads, and differences**
Lagrange-multiplier test, [R] **regress postestimation time series**, [TS] **varlmar**, [TS] **veclmar**
Lance and Williams' formula, [MV] **cluster**
language syntax, [P] **syntax**, [U] **11 Language syntax**
languages, multiple, [D] **label language**
LAPACK, [M-1] **LAPACK**, [M-5] **cholesky( )**, [M-5] **cholinv( )**, [M-5] **cholsolve( )**, [M-5] **eigensystem( )**, [M-5] **eigensystemselect( )**, [M-5] **fullsvd( )**, [M-5] **ghessenbergd( )**, [M-5] **lapack( )**, [M-5] **lud( )**, [M-5] **luinv( )**, [M-5] **lusolve( )**, [M-5] **qrd( )**, [M-5] **qrinv( )**, [M-5] **qrsolve( )**, [M-5] **svd( )**, [M-5] **svsolve( )**, [M-6] **Glossary**
latent roots, [M-5] **eigensystem( )**
Latin-square designs, [MV] **manova**, [R] **anova**, [R] **pkshape**
launch dialog box, [R] **db**
LAV regression, [R] **qreg**
Lawley–Hotelling trace statistic, [MV] **canon**, [MV] **Glossary**, [MV] **manova**, [MV] **mvtest means**
lceffects, estat subcommand, [SVY] **estat**
lcolor( ) option, [G] *connect_options*, [G] *rspike_options*
LDA, [MV] **discrim lda**, [MV] **Glossary**
lda, discrim subcommand, [MV] **discrim lda**
lead
  operator, [U] **11.4.4 Time-series varlists**
  values, *see* lagged values
leap seconds, [TS] **tsset**
least absolute
  deviations, [R] **qreg**
  residuals, [R] **qreg**
  value regression, [R] **qreg**

least squared deviations, [R] **regress**, [R] **regress postestimation**; [R] **areg**, [R] **cnsreg**, [R] **nl**
least squares, *see* linear regression
  generalized, *see* FGLS
least-squares means, [R] **margins** [U] **20.14.1 Obtaining expected marginal means**
left eigenvectors, [M-5] **eigensystem( )**
left suboption, [G] *justificationstyle*
_lefteigensystem( ) function, [M-5] **eigensystem( )**
lefteigensystem( ) function, [M-5] **eigensystem( )**
lefteigensystemselect*( ) functions, [M-5] **eigensystemselect( )**
leftgeigensystem( ) function, [M-5] **geigensystem( )**
leftgeigensystemselect*( ) function, [M-5] **geigensystem( )**
legend( ) option, [G] *legend_option*
legends, [G] *legend_option*
  problems, [G] *legend_option*
  use with by( ), [G] *by_option*, [G] *legend_option*
legendstyle, [G] *legendstyle*
length, [M-5] **abs( )**, [M-5] **rows( )**, [M-5] **strlen( )**
length( ) function, [D] **functions**, [M-5] **rows( )**
length macro extended function, [P] **macro**
length of string function, [D] **functions**
less than (or equal) operator, [U] **13.2.3 Relational operators**
letter values, [R] **lv**
level, set subcommand, [R] **level**, [R] **set**
level command and value, [P] **macro**
levels, [U] **11.4.3 Factor variables**
levelsof command, [P] **levelsof**
Levene's robust test statistic, [R] **sdtest**
leverage, [R] **logistic postestimation**, [R] **predict**, [R] **regress postestimation**
  obtaining with weighted data, [R] **predict**
leverage plots, [G] **graph other**
leverage-versus-(squared)-residual plot, [R] **regress postestimation**
Levin–Lin–Chu test, [XT] **xtunitroot**
lexis command, [ST] **stsplit**
lexis diagram, [ST] **stsplit**
lfit, graph twoway subcommand, [G] **graph twoway lfit**
lfitci, graph twoway subcommand, [G] **graph twoway lfitci**
libraries, [M-1] **how**, [M-3] **mata mlib**, [M-3] **mata which**
license, [R] **about**
life tables, [ST] **Glossary**, [ST] **ltable**
likelihood, *see* maximum likelihood estimation
likelihood displacement value, [ST] **Glossary**, [ST] **stcox postestimation**
likelihood-ratio
  chi-squared of association, [R] **tabulate twoway**
  test, [R] **lrtest**, [U] **20.11.3 Likelihood-ratio tests**
Likert summative scales, [R] **alpha**

limits, [D] **describe**, [D] **memory**, [M-1] **limits**,
  [R] **matsize**, [U] **6 Setting the size of memory**
  numerical and string, [P] **creturn**
  system, [P] **creturn**
lincom command, [R] **lincom**, [R] **test**, [SVY] **svy**
  postestimation
line, definition, [G] *linestyle*
line, graph twoway subcommand, [G] **graph twoway**
  line
linear
  combinations, [SVY] **estat**, [SVY] **svy**
    postestimation
    forming, [P] **matrix score**
    of estimators, [R] **lincom**, [U] **20.12 Obtaining**
      linear combinations of coefficients
  discriminant analysis, [MV] **candisc**, [MV] **discrim**
    lda, [MV] **Glossary**
  interpolation and extrapolation, [D] **ipolate**
  mixed models, [XT] **xtmixed**
    postestimation, [XT] **xtmelogit postestimation**,
      [XT] **xtmepoisson postestimation**,
      [XT] **xtmixed postestimation**
  regression, [R] **anova**, [R] **areg**, [R] **cnsreg**,
    [R] **eivreg**, [R] **frontier**, [R] **glm**, [R] **gmm**,
    [R] **heckman**, [R] **intreg**, [R] **ivregress**,
    [R] **ivtobit**, [R] **mvreg**, [R] **qreg**, [R] **reg3**,
    [R] **regress**, [R] **rreg**, [R] **sureg**, [R] **tobit**,
    [R] **vwls**, [SVY] **svy estimation**, [TS] **newey**,
    [TS] **prais**, [TS] **var**, [TS] **var intro**, [TS] **var**
    svar, [TS] **varbasic**, [XT] **xtabond**, [XT] **xtdpd**,
    [XT] **xtdpdsys**, [XT] **xtfrontier**, [XT] **xtgee**,
    [XT] **xtgls**, [XT] **xthtaylor**, [XT] **xtivreg**,
    [XT] **xtmixed**, [XT] **xtpcse**, [XT] **xtrc**,
    [XT] **xtreg**, [XT] **xtregar**
  regression imputation, *see* imputation, regression
  splines, [R] **mkspline**
  tests, *see* estimation, tests after
linearized variance estimator, [SVY] **Glossary**,
  [SVY] **variance estimation**
linegap, set subcommand, [R] **set**
linepalette, palette subcommand, [G] **palette**
linepatternstyle, [G] *linepatternstyle*
lines, [G] **concept: lines**
  adding, [G] *added_line_options*, [G] **graph**
    twoway lfit, *also see* fits, adding
  connecting points, [G] *connect_options*,
    [G] *connectstyle*
  dashed, [G] *linepatternstyle*
  dotted, [G] *linepatternstyle*
  grid, [G] *axis_label_options*, [G] *linestyle*
  long, in do-files and ado-files, [P] **#delimit**,
    [U] **18.11.2 Comments and long lines in ado-**
    files
  look of, [G] *fcline_options*, [G] *line_options*,
    [G] *linestyle*
  patterns, [G] *linepatternstyle*
  suppressing, [G] *linestyle*
  thickness, [G] *linewidthstyle*
linesize, set subcommand, [R] **log**, [R] **set**

linestyle, [G] *linestyle*
  added, [G] *addedlinestyle*
linewidthstyle, [G] *linewidthstyle*
link, net subcommand, [R] **net**
link function, [R] **glm**, [XT] **Glossary**, [XT] **xtgee**
linkage, [MV] **Glossary**
linktest command, [R] **linktest**
list
  manipulation, [P] **macro lists**
  subscripts, *see* subscripts
list,
  cluster subcommand, [MV] **cluster utility**
  estat subcommand, [MV] **discrim estat**,
    [MV] **discrim knn postestimation**,
    [MV] **discrim lda postestimation**, [MV] **discrim**
    logistic postestimation, [MV] **discrim qda**
    postestimation
  char subcommand, [P] **char**
  creturn subcommand, [P] **creturn**
  duplicates subcommand, [D] **duplicates**
  ereturn subcommand, [P] **ereturn**, [R] **saved**
    results
  label subcommand, [D] **label**
  macro subcommand, [P] **macro**
  matrix subcommand, [P] **matrix utility**
  notes subcommand, [D] **notes**
  odbc subcommand, [D] **odbc**
  program subcommand, [P] **program**
  return subcommand, [R] **saved results**
  snapshot subcommand, [D] **snapshot**
  sreturn subcommand, [R] **saved results**
  sts subcommand, [ST] **sts list**
  sysdir subcommand, [P] **sysdir**
  timer subcommand, [P] **timer**
list command, [D] **list**; [D] **format**
list macro extended function, [P] **macro lists**
listing
  data, [D] **edit**, [D] **list**
  estimation results, [P] **ereturn**, [P] **_estimates**
  macro expanded functions, [P] **macro lists**
  values of a variable, [P] **levelsof**
listserver, [U] **3.4 The Stata listserver**
liststruct() function, [M-5] **liststruct( )**
listwise deletion, [MI] **Glossary**, [MI] **intro**
  substantive, [MI] **mi estimate**
llc, xtunitroot subcommand, [XT] **xtunitroot**
Lmatrix() function, [M-5] **Lmatrix( )**
LMAX value, [ST] **Glossary**, [ST] **stcox**
  postestimation
ln() function, [D] **functions**, [M-5] **exp( )**
lnfactorial() function, [D] **functions**,
  [M-5] **factorial( )**
lngamma() function, [D] **functions**, [M-5] **factorial( )**
lnnormal() function, [D] **functions**, [M-5] **normal( )**
lnnormalden() function, [D] **functions**,
  [M-5] **normal( )**
lnskew0 command, [R] **lnskew0**
load, odbc subcommand, [D] **odbc**

loading, [MV] **Glossary**
   data, *see* inputting data interactively; reading data
      from disk, *see* reading data from disk
   plot, [MV] **Glossary**, [MV] **scoreplot**
   saved data, [D] **use**
loadingplot command, [MV] **discrim lda**
   postestimation, [MV] **factor postestimation**,
   [MV] **pca postestimation**, [MV] **scoreplot**
loadings, estat subcommand, [MV] **canon**
   postestimation, [MV] **discrim lda**
   postestimation, [MV] **pca postestimation**
local
   ++ command, [P] **macro**
   − command, [P] **macro**
   command, [P] **macro**
local,
   ereturn subcommand, [P] **ereturn**
   return subcommand, [P] **return**
   sreturn subcommand, [P] **return**
Local, class prefix operator, [P] **class**
local command, [U] **18.3.1 Local macros**,
   [U] **18.3.9 Advanced local macro manipulation**
local linear, [R] **lpoly**
local polynomial smoothing, [G] **graph other**,
   [G] **graph twoway lpoly**, [G] **graph twoway**
   lpolyci, [R] **lpoly**
locally weighted smoothing, [R] **lowess**
location,
   measures of, [R] **lv**, [R] **summarize**, [R] **table**
   specifying, [G] *clockposstyle*, [G] *compassdirstyle*,
   [G] *ringposstyle*
locksplitters, set subcommand, [R] **set**
log() function, [D] **functions**, [M-5] **exp( )**
log10() function, [D] **functions**, [M-5] **exp( )**
log command, [R] **log**, [R] **view**, [U] **15 Saving and**
   printing output—log files, [U] **16.1.2 Comments**
   and blank lines in do-files
.log filename suffix, [U] **11.6 File-naming conventions**
log files, *see* log command
   printing, [R] **translate**, *also see* log command
log-linear model, [R] **glm**, [R] **poisson**, [R] **zip**
log or nolog option, [R] **maximize**
log scales, [G] *axis_scale_options*
log transformations, [R] **boxcox**, [R] **lnskew0**
logarithms, [M-5] **exp( )**, [M-5] **matexpsym( )**
logical operators, [M-2] **op_logical**, [U] **13.2.4 Logical**
   operators
logistic, discrim subcommand, [MV] **discrim**
   logistic
logistic and logit regression, [R] **logistic**, [R] **logit**,
   [SVY] **svy estimation**
   complementary log-log, [R] **cloglog**
   conditional, [R] **asclogit**, [R] **clogit**, [R] **rologit**
   exact, [R] **exlogistic**
   fixed-effects, [R] **asclogit**, [R] **clogit**, [XT] **xtlogit**
   generalized estimating equations, [XT] **xtgee**
   generalized linear model, [R] **glm**
   mixed-effects, [XT] **xtmelogit**

logistic and logit regression, *continued*
   multinomial, [R] **asclogit**, [R] **clogit**, [R] **mlogit**
   nested, [R] **nlogit**
   ordered, [R] **ologit**
   polytomous, [R] **mlogit**
   population-averaged, [XT] **xtlogit**; [XT] **xtgee**
   postestimation, [XT] **xtlogit postestimation**
   random-effects, [XT] **xtlogit**
   rank-ordered, [R] **rologit**
   skewed, [R] **scobit**
   stereotype, [R] **slogit**
   with grouped data, [R] **glogit**
logistic command, [R] **logistic**, [R] **logistic**
   postestimation, *also see* postestimation command
logistic discriminant analysis, [MV] **discrim logistic**,
   [MV] **Glossary**
logistic regression imputation, *see* imputation, logistic
   regression
logit command, [R] **logit**, [R] **logit postestimation**,
   also see postestimation command
logit function, [D] **functions**, [M-5] **logit( )**
logit regression, *see* logistic and logit regression
log-linear model, [SVY] **svy estimation**
log-log plot, [ST] **stcox PH-assumption tests**
loglogistic survival regression, [ST] **streg**
lognormal distribution, [R] **ameans**
lognormal survival regression, [ST] **streg**
log-rank,
   power, [ST] **stpower logrank**
   sample size, [ST] **stpower logrank**
   test, [ST] **stpower logrank**, [ST] **sts test**
logrank, stpower subcommand, [ST] **stpower**
   logrank
logtype, set subcommand, [R] **log**, [R] **set**
LOHI, [M-5] **byteorder( )**
loneway command, [R] **loneway**
long, [D] **data types**, [I] **data types**,
   [U] **12.2.2 Numeric storage types**
long lines in ado-files and do-files, [P] **#delimit**,
   [U] **18.11.2 Comments and long lines in ado-**
   files
longitudinal
   data, [XT] **Glossary**, *see* panel data
   studies, *see* incidence studies
   survey data, [SVY] **svy estimation**
LOO, [MV] **Glossary**
look of areas, [G] *area_options*, [G] *fitarea_options*
lookfor command, [D] **lookfor**
lookup,
   icd9 subcommand, [D] **icd9**
   icd9p subcommand, [D] **icd9**
loop, endless, *see* endless loop
looping, [P] **continue**, [P] **foreach**, [P] **forvalues**,
   [P] **while**
Lorenz curve, [R] **inequality**
loss, [MV] **Glossary**
loss to follow-up, [ST] **Glossary**

Lotus 1-2-3, reading data from, *see* spreadsheets, transferring, *see* spreadsheets

lower( ) string function, [D] **functions**

lowercase, [M-5] **strupper( )**

lowercase-string function, [D] **functions**

_lowertriangle( ) function, [M-5] **lowertriangle( )**

lowertriangle( ) function, [M-5] **lowertriangle( )**

lower-triangular matrix, *see* triangular matrix

lowess, *see* locally weighted smoothing

lowess, graph twoway subcommand, [G] **graph twoway lowess**

lowess command, [R] **lowess**

lowess smoothing, [G] **graph other**

lpattern( ) option, [G] *connect_options*, [G] *rspike_options*

lpoly command, [R] **lpoly**, *also see* smooth command

L-R plots, [G] **graph other**, [R] **regress postestimation**

LRECLs, [D] **infile (fixed format)**

lroc command, [R] **logistic postestimation**

lrtest command, [R] **lrtest**

ls command, [D] **dir**

lsens command, [R] **logistic postestimation**

lstat command, *see* estat classification command

lstyle( ) option, [G] *rspike_options*

ltable command, [ST] **ltable**

ltolerance( ) option, [R] **maximize**

ltrim( ) string function, [D] **functions**

LU decomposition, [M-5] **lud( )**

_lud( ) function, [M-5] **lud( )**

lud( ) function, [M-5] **lud( )**

_lud_la( ) function, [M-5] **lud( )**

_luinv( ) function, [M-5] **luinv( )**

luinv( ) function, [M-5] **luinv( )**

_luinv_la( ) function, [M-5] **luinv( )**

_lusolve( ) function, [M-5] **lusolve( )**

lusolve( ) function, [M-5] **lusolve( )**

_lusolve_la( ) function, [M-5] **lusolve( )**

lv command, [R] **lv**

lval, [M-2] **op_assignment**, [M-6] **Glossary**

lvalue, class, [P] **class**

lvr2plot command, [R] **regress postestimation**

lwidth( ) option, [G] *connect_options*, [G] *rspike_options*

M

M, [MI] **Glossary**, [MI] **mi impute**
size recommendations, [MI] **intro substantive**

m, [MI] **Glossary**

MA, [TS] **arch**, [TS] **arima**, [TS] **dfactor**, [TS] **sspace**

ma, tssmooth subcommand, [TS] **tssmooth ma**

Mac,
keyboard use, [U] **10 Keyboard use**
pause, [P] **sleep**
specifying filenames, [U] **11.6 File-naming conventions**

machine precision, [M-5] **epsilon( )**, [M-6] **Glossary**

macro
dir command, [P] **macro**
drop command, [P] **macro**
list command, [P] **macro**
shift command, [P] **macro**

macro substitution, [P] **macro**
class, [P] **class**

macros, [P] **macro**, [U] **18.3 Macros**; [P] **creturn**, [P] **scalar**, [P] **syntax**, *also see* e( )

macval( ) macro expansion function, [P] **macro**

mad( ), egen function, [D] **egen**

MAD regression, [R] **qreg**

Mahalanobis
distance, [MV] **Glossary**
transformation, [MV] **Glossary**

main effects, [MV] **manova**, [R] **anova**

makecns command, [P] **makecns**

_makesymmetric( ) function, [M-5] **makesymmetric( )**

makesymmetric( ) function, [M-5] **makesymmetric( )**

man command, [R] **help**

manage, window subcommand, [P] **window programming**

MANCOVA, [MV] **Glossary**, [MV] **manova**

mangle option, [G] **graph twoway pcarrow**

manipulation commands, [G] **graph manipulation**

Mann–Whitney two-sample statistics, [R] **ranksum**

MANOVA, [MV] **Glossary**, [MV] **manova**

manova, estat subcommand, [MV] **discrim lda postestimation**

manova command, [MV] **manova**

manovatest command, [MV] **manova postestimation**

Mantel–Cox method, [ST] **strate**

Mantel–Haenszel
method, [ST] **strate**
test, [ST] **epitab**, [ST] **stir**

mapping strings to numbers, [D] **encode**, [D] **label**

maps, [M-5] **asarray( )**

MAR, [MI] **intro substantive**

marginal
effects, [R] **margins**, [U] **20.14 Obtaining marginal means, adjusted predictions, and predictive margins**
homogeneity, test of, [R] **symmetry**
means, [R] **margins**, [U] **20.14 Obtaining marginal means, adjusted predictions, and predictive margins**
tax rate egen function, [D] **egen**

margins, size of, [G] *marginstyle*

margins command, [R] **margins**, [R] **margins postestimation**, [SVY] **svy postestimation**, [U] **20.14 Obtaining marginal means, adjusted predictions, and predictive margins**

marginstyle, [G] *marginstyle*, [G] *region_options*, [G] *textbox_options*

mark command, [P] **mark**

marker labels, [G] *marker_label_options*, [G] *markerlabelstyle*

markerlabelstyle, [G] *markerlabelstyle*

markers, [G] *marker_options*, *also see* marker labels
  color, [G] *colorstyle*
  resizing, [G] *scale_option*
  shape of, [G] *symbolstyle*
  size of, [G] *markersizestyle*
markersizestyle, [G] *markersizestyle*
markerstyle, [G] *markerstyle*
markin command, [P] **mark**
marking observations, [P] **mark**
markout command, [P] **mark**
Markov chain Monte Carlo, *see* MCMC
marksample command, [P] **mark**
Marquardt algorithm, [M-5] **moptimize( )**,
  [M-5] **optimize( )**
martingale residual, [ST] **stcox postestimation**,
  [ST] **streg postestimation**
mass, [MV] **Glossary**
Mata
  commands, [M-3] **intro**
  error messages, [M-5] **error( )**, *also see* traceback
    log
mata, [M-3] **mata clear**, [M-3] **mata describe**,
  [M-3] **mata drop**, [M-3] **mata help**, [M-3] **mata
  matsave**, [M-3] **mata memory**, [M-3] **mata
  mlib**, [M-3] **mata mosave**, [M-3] **mata rename**,
  [M-3] **mata set**, [M-3] **mata stata**, [M-3] **mata
  which**, [M-3] **namelists**
mata, clear subcommand, [D] **clear**
.mata file, [M-1] **source**, [M-3] **mata mlib**,
  [M-6] **Glossary**
.mata filename suffix, [U] **11.6 File-naming
  conventions**
mata invocation command, [M-3] **mata**
matacache, [M-3] **mata set**
matacache, set subcommand, [R] **set**
matafavor, [M-3] **mata set**, [M-5] **favorspeed( )**
matafavor, set subcommand, [R] **set**
matalibs, [M-3] **mata set**
matalibs, set subcommand, [R] **set**
matalnum, [M-3] **mata set**
matalnum, set subcommand, [R] **set**
matamofirst, [M-3] **mata set**
matamofirst, set subcommand, [R] **set**
mataoptimize, [M-3] **mata set**
mataoptimize, set subcommand, [R] **set**
matastrict, [M-3] **mata set**, [M-2] **declarations**,
  [M-1] **ado**
matastrict, set subcommand, [R] **set**
matched case–control data, [R] **asclogit**, [R] **clogit**,
  [R] **symmetry**, [ST] **epitab**, [ST] **Glossary**,
  [ST] **sttocc**
matched-pairs tests, [R] **signrank**, [R] **ttest**
matching coefficient, [MV] **Glossary**
  similarity measure, [MV] *measure_option*
matching configuration, [MV] **Glossary**
matcproc command, [P] **makecns**
matdescribe, [M-3] **mata matsave**
_matexpsym( ) function, [M-5] **matexpsym( )**

matexpsym( ) function, [M-5] **matexpsym( )**
math symbols, [G] *text*
mathematical functions, [M-4] **mathematical**,
  [M-4] **matrix**, [M-4] **scalar**, [M-4] **solvers**,
  [M-4] **standard**
  and expressions, [D] **functions**, [P] **matrix define**,
  [U] **13.3 Functions**
Matlab, reading data from, [U] **21.4 Transfer programs**
matlist command, [P] **matlist**
_matlogsym( ) function, [M-5] **matexpsym( )**
matlogsym( ) function, [M-5] **matexpsym( )**
matmissing( ) matrix function, [D] **functions**,
  [P] **matrix define**
matname command, [P] **matrix mkmat**
_matpowersym( ) function, [M-5] **matpowersym( )**
matpowersym( ) function, [M-5] **matpowersym( )**
mat_put_rr command, [P] **matrix get**
matrices, [P] **matrix**, [U] **14 Matrix expressions**
  accessing internal, [P] **matrix get**
  accumulating, [P] **matrix accum**
  appending rows and columns, [P] **matrix define**
  Cholesky decomposition, [P] **matrix define**
  coefficient matrices, [P] **ereturn**
  column names, *see* matrices, row and column names
  constrained estimation, [P] **makecns**
  copying, [P] **matrix define**, [P] **matrix get**,
    [P] **matrix mkmat**
  correlation, [MV] **pca**, [P] **matrix define**
  covariance matrix of estimators, [MV] **pca**,
    [P] **ereturn**, [P] **matrix get**
  cross-product, [P] **matrix accum**
  determinant, [P] **matrix define**
  diagonals, [P] **matrix define**
  displaying, [P] **matlist**, [P] **matrix utility**
  dissimilarity, [MV] **matrix dissimilarity**, [P] **matrix
    dissimilarity**
  distances, [MV] **matrix dissimilarity**, [P] **matrix
    dissimilarity**
  dropping, [P] **matrix utility**
  eigenvalues, [P] **matrix eigenvalues**, [P] **matrix
    symeigen**
  eigenvectors, [P] **matrix symeigen**
  elements, [P] **matrix define**
  equation names, *see* matrices, row and column
    names
  estimation results, [P] **ereturn**, [P] **_estimates**
  functions, [D] **functions**, [P] **matrix define**
  identity, [P] **matrix define**
  input, [P] **matrix define**, [U] **14.4 Inputting
    matrices by hand**
  inversion, [P] **matrix define**, [P] **matrix svd**
  Kronecker product, [P] **matrix define**
  labeling rows and columns, *see* matrices, row and
    column names
  linear combinations with data, [P] **matrix score**
  listing, [P] **matlist**, [P] **matrix utility**
  namespace and conflicts, [P] **matrix**, [P] **matrix
    define**

matrices, *continued*
    number of rows and columns, [P] **matrix define**
    operators such as addition, [P] **matrix define**,
        [U] **14.7 Matrix operators**
    orthonormal basis, [P] **matrix svd**
    partitioned, [P] **matrix define**
    performing constrained estimation, [P] **makecns**
    posting estimation results, [P] **ereturn**,
        [P] **_estimates**
    renaming, [P] **matrix utility**
    row and column names, [P] **ereturn**, [P] **matrix
        define**, [P] **matrix mkmat**, [P] **matrix
        rownames**, [U] **14.2 Row and column names**
    rows and columns, [P] **matrix define**
    saving matrix, [P] **matrix mkmat**
    scoring, [P] **matrix score**
    similarity, [MV] **matrix dissimilarity**, [P] **matrix
        dissimilarity**
    store variables as matrix, [P] **matrix mkmat**
    submatrix extraction, [P] **matrix define**
    submatrix substitution, [P] **matrix define**
    subscripting, [P] **matrix define**,
        [U] **14.9 Subscripting**
    sweep operator, [P] **matrix define**
    temporary names, [P] **matrix**
    trace, [P] **matrix define**
    transposing, [P] **matrix define**
    variables, make into matrix, [P] **matrix mkmat**
    zero, [P] **matrix define**
matrix
    accum command, [P] **matrix accum**
    coleq command, [P] **matrix rownames**
    colnames command, [P] **matrix rownames**
    command, [MV] **matrix dissimilarity**
        introduction, [P] **matrix**
    define command, [P] **matrix define**
    dir command, [P] **matrix utility**
    dissimilarity command, [P] **matrix dissimilarity**
    drop command, [P] **matrix utility**
    eigenvalues command, [P] **matrix eigenvalues**
    glsaccum command, [P] **matrix accum**
    input command, [P] **matrix define**
    list command, [P] **matrix utility**
    opaccum command, [P] **matrix accum**
    rename command, [P] **matrix utility**
    roweq command, [P] **matrix rownames**
    rownames command, [P] **matrix rownames**
    score command, [P] **matrix score**
    svd command, [P] **matrix svd**
    symeigen command, [P] **matrix symeigen**
    vecaccum command, [P] **matrix accum**
matrix, [M-4] **intro**, [M-6] **Glossary**
    dissimilarity, [MV] **Glossary**
    functions, [M-4] **manipulation**, [M-4] **matrix**,
        [M-4] **solvers**, [M-4] **standard**
    graphs, [G] **graph matrix**
    norm, [M-5] **norm( )**

matrix, [M-2] **declarations**
matrix,
    clear subcommand, [D] **clear**
    confirm subcommand, [P] **confirm**
    ereturn subcommand, [P] **ereturn**
    graph subcommand, [G] **graph matrix**
    return subcommand, [P] **return**
matrix() function, [D] **functions**, [P] **matrix define**
matsave, [M-3] **mata matsave**
matsize, [M-1] **limits**, [P] **creturn**, [P] **macro**,
    [U] **6.2.2 Advice on setting matsize**,
    [U] **14 Matrix expressions**
matsize, set subcommand, [R] **matsize**, [R] **set**
matuniform() matrix function, [D] **functions**,
    [P] **matrix define**
matuse, [M-3] **mata matsave**
max(),
    built-in function, [D] **functions**
    egen function, [D] **egen**
    function, [M-5] **minmax( )**
maxbyte() function, [D] **functions**
maxdb, set subcommand, [R] **db**, [R] **set**
maxdouble() function, [D] **functions**,
    [M-5] **mindouble( )**
maxes() option, [G] **graph matrix**
maxfloat() function, [D] **functions**
maximization, [M-5] **moptimize( )**, [M-5] **optimize( )**
maximization technique explained, [R] **maximize**,
    [R] **ml**
maximize, ml subcommand, [R] **ml**
maximum
    function, [D] **egen**, [D] **functions**
    length of string, [M-1] **limits**
    likelihood factor method, [MV] **Glossary**
    likelihood estimation, [MV] **factor**, [R] **maximize**,
        [R] **ml**, [TS] **var**, [TS] **var intro**, [TS] **var svar**,
        [TS] **varbasic**
    number of variables and observations, [D] **describe**,
        [D] **memory**, [U] **6 Setting the size of memory**
    number of variables in a model, [R] **matsize**
    pseudolikelihood estimation, [SVY] **ml for svy**,
        [SVY] **variance estimation**
    restricted-likelihood, [XT] **xtmixed**
    size of dataset, [D] **describe**, [D] **memory**,
        [U] **6 Setting the size of memory**
    size of matrix, [M-1] **limits**
    value dissimilarity measure, [MV] *measure_option*
maximums, [M-5] **minindex( )**
maximums and minimums,
    creating dataset of, [D] **collapse**
    functions, [D] **egen**, [D] **functions**
    reporting, [R] **lv**, [R] **summarize**, [R] **table**
maxindex() function, [M-5] **minindex( )**
maxint() function, [D] **functions**
maxiter, set subcommand, [R] **maximize**, [R] **set**
maxlong() function, [D] **functions**
maxvar, set subcommand, [D] **memory**, [R] **set**

mband, graph twoway subcommand, [G] **graph twoway mband**

MCA, [MV] **Glossary**

mca command, [MV] **mca**

mcaplot command, [MV] **mca postestimation**

mcaprojection command, [MV] **mca postestimation**

MCAR, [MI] **intro substantive**

mcc command, [ST] **epitab**

mcci command, [ST] **epitab**

McFadden's choice model, [R] **asclogit**

MCMC, [MI] **Glossary**, [MI] **mi impute**, [MI] **mi impute mvn**

parameter trace files, [MI] **mi ptrace**

McNemar's chi-squared test, [R] **clogit**, [ST] **epitab**

mcolor() option, [G] *marker_options*

md command, [D] **mkdir**

mdev(), egen function, [D] **egen**

MDS, [MV] **Glossary**

configuration plot, [MV] **Glossary**

mds command, [MV] **mds**

mdsconfig, estat subcommand, [MV] **mds postestimation**

mdslong command, [MV] **mdslong**

mdsmat command, [MV] **mdsmat**

mdsshepard, estat subcommand, [MV] **mds postestimation**

mdy() function, [D] **dates and times**, [D] **functions**, [M-5] **date( )**

mdyhms() function, [D] **dates and times**, [D] **functions**, [M-5] **date( )**

mean command, [R] **mean**, [R] **mean postestimation**, *also see* postestimation command

mean(), egen function, [D] **egen**

mean() function, [M-5] **mean( )**

means, mvtest subcommand, [MV] **mvtest means**

means,

across variables, not observations, [D] **egen**

arithmetic, geometric, and harmonic, [R] **ameans**

confidence interval and standard error, [R] **ci**

creating

dataset of, [D] **collapse**

variable containing, [D] **egen**

displaying, [R] **ameans**, [R] **summarize**, [R] **table**, [R] **tabstat**, [R] **tabulate, summarize()**, [XT] **xtsum**

estimates of, [R] **mean**

graphing, [R] **grmeanby**

robust, [R] **rreg**

survey data, [SVY] **svy estimation**

testing equality of, [MV] **hotelling**, [MV] **manova**, [MV] **mvtest means**, [R] **ttest**

sample size or power, [R] **sampsi**

meanvariance() function, [M-5] **mean( )**

measure, [MV] **Glossary**

measurement error, [R] **alpha**, [R] **vwls**

measures, cluster subcommand, [MV] **cluster programming utilities**

measures of

association, [R] **tabulate twoway**

inequality, [R] **inequality**

location, [R] **lv**, [R] **summarize**

median(), egen function, [D] **egen**

median command, [R] **ranksum**

median

regression, [R] **qreg**

test, [R] **ranksum**

medianlinkage, cluster subcommand, [MV] **cluster linkage**

median-linkage clustering, [MV] **cluster**, [MV] **cluster linkage**, [MV] **clustermat**, [MV] **Glossary**

medians,

creating

dataset of, [D] **collapse**

variable containing, [D] **egen**

displaying, [D] **pctile**, [R] **centile**, [R] **lv**, [R] **summarize**, [R] **table**, [R] **tabstat**

graphing, [R] **grmeanby**

testing equality of, [R] **ranksum**

MEFF, *see* misspecification effects

MEFT, *see* misspecification effects

member

function, [M-2] **class**

programs, [P] **class**

variables, [M-2] **class**, [P] **class**

memory, set subcommand, [D] **memory**, [R] **set**

memory, [U] **6 Setting the size of memory**

clearing, [D] **clear**

determining and resetting limits, [D] **describe**, [D] **memory**

loading, [D] **use**

matsize, *see* matsize, set subcommand

reducing utilization, [D] **compress**, [D] **encode**, [P] **discard**

setting, [U] **6.2.3 Advice on setting memory**

virtual, [U] **6.5 Virtual memory and speed considerations**

memory, [M-3] **mata memory**

memory command, [U] **6 Setting the size of memory**

memory graphs, describing contents, [G] **graph describe**

memory requirements, estimating for flongsep, [MI] **mi convert**

memory settings, [P] **creturn**

memory utilization, [M-1] **limits**, [M-3] **mata memory**

memory command, [D] **memory**

menu, window subcommand, [P] **window programming**

menus, programming, [P] **dialog programming**, [P] **window programming**

merge, mi subcommand, [MI] **mi merge**

merge command, [D] **merge**, [U] **22 Combining datasets**

_merge variables, [D] **merge**

merged-explicit options, [G] **concept: repeated options**

merged-implicit options, [G] **concept: repeated options**

merging data, [MI] **mi merge**, *see* combining datasets

messages and return codes, *see* error messages and
return codes

meta-analysis, [R] **meta**

method, [M-2] **class**

metric scaling, [MV] **Glossary**

mfcolor() option, [G] *marker_options*

mfp prefix command, [R] **mfp**, [R] **mfp postestimation**,
also see postestimation command

mfx, estat subcommand, [R] **asclogit postestimation**,
[R] **asmprobit postestimation**, [R] **asroprobit
postestimation**

mhodds command, [ST] **epitab**

mi

    add command, [MI] **mi add**

    append command, [MI] **mi append**

    convert command, [MI] **mi convert**

    copy command, [MI] **mi copy**, [MI] **styles**

    describe command, [MI] **mi describe**

    erase command, [MI] **mi erase**, [MI] **styles**

    estimate command, [MI] **mi estimate**, [MI] **mi
estimate postestimation**, [MI] **mi estimate using**

    estimate postestimation, [MI] **mi estimate
postestimation**

    expand command, [MI] **mi expand**

    export command, [MI] **mi export**, [MI] **mi export
ice**, [MI] **mi export nhanes1**

    extract command, [MI] **mi extract**, [MI] **mi
replace0**

    fvset command, [MI] **mi XXXset**

    import command, [MI] **mi import**, [MI] **mi import
flong**, [MI] **mi import flongsep**, [MI] **mi import
ice**, [MI] **mi import nhanes1**, [MI] **mi import
wide**

    impute command, [MI] **mi impute**, [MI] **mi impute
logit**, [MI] **mi impute mlogit**, [MI] **mi impute
monotone**, [MI] **mi impute mvn**, [MI] **mi
impute ologit**, [MI] **mi impute pmm**, [MI] **mi
impute regress**

    merge command, [MI] **mi merge**

    misstable command, [MI] **mi misstable**

    passive command, [MI] **mi passive**

    ptrace command, [MI] **mi ptrace**

    query command, [MI] **mi describe**

    register command, [MI] **mi set**

    rename command, [MI] **mi rename**

    replace0 command, [MI] **mi replace0**

    reset command, [MI] **mi reset**

    reshape command, [MI] **mi reshape**

    select command, [MI] **mi select**

    set command, [MI] **mi set**

    st command, [MI] **mi XXXset**

    stjoin command, [MI] **mi stsplit**

    streset command, [MI] **mi XXXset**

    stset command, [MI] **mi XXXset**

    stsplit command, [MI] **mi stsplit**

    svyset command, [MI] **mi XXXset**

    test command, [MI] **mi estimate postestimation**

mi, *continued*

    testtransform command, [MI] **mi estimate
postestimation**

    tsset command, [MI] **mi XXXset**

    unregister command, [MI] **mi set**

    unset command, [MI] **mi set**

    update command, [MI] **mi update**, [MI] **noupdate
option**

    varying command, [MI] **mi varying**

    xeq command, [MI] **mi xeq**

    xtset command, [MI] **mi XXXset**

mi() function, [D] **functions**

mi command, [MI] **intro**, [MI] **styles**, [MI] **workflow**

mi data, [MI] **Glossary**

Microsoft

    Access, reading data from, [D] **odbc**,
[U] **21.4 Transfer programs**

    Excel, reading data from, [D] **odbc**

    SpreadsheetML, [D] **xmlsave**

    Windows, *see* Windows

middle suboption, [G] *alignmentstyle*

midsummaries, [R] **lv**

mild outliers, [R] **lv**

Mills' ratio, [R] **heckman**, [R] **heckman
postestimation**

min(),

    built-in function, [D] **functions**

    egen function, [D] **egen**

    function, [M-5] **minmax( )**

minbyte() function, [D] **functions**

mindouble() function, [D] **functions**,
[M-5] **mindouble( )**

minfloat() function, [D] **functions**

minimal

    detectable difference, hazard ratio, [ST] **stpower**

    effect size, hazard ratio, [ST] **stpower**

minimization, [M-5] **moptimize( )**, [M-5] **optimize( )**

minimum

    absolute deviations, [R] **qreg**

    entropy rotation, [MV] **Glossary**

    squared deviations, [R] **areg**, [R] **cnsreg**, [R] **nl**,
[R] **regress**, [R] **regress postestimation**

minimums, [M-5] **minindex( )**

minimums and maximums, *see* maximums and
minimums

minindex() function, [M-5] **minindex( )**

minint() function, [D] **functions**

Minkowski dissimilarity measure,
[MV] *measure_option*

minlong() function, [D] **functions**

minmax() function, [M-5] **minmax( )**

minutes() function, [D] **dates and times**,
[D] **functions**, [M-5] **date( )**

misclassification rate, [MV] **Glossary**

missing() function, [D] **functions**, [M-5] **missing( )**

missing at random, *see* MAR

missing completely at random, *see* MCAR

missing data, [MI] **intro substantive**
   arbitrary pattern, [MI] **Glossary**, [MI] **intro substantive**, [MI] **mi impute mvn**
   monotone pattern, [MI] **Glossary**, [MI] **intro substantive**, [MI] **mi impute monotone**, [MI] **mi impute mvn**
missing not at random, *see* MNAR
missing values, [D] **missing values**, [I] **missing values**, [R] **misstable**, [U] **12.2.1 Missing values**, [U] **13 Functions and expressions**; [M-5] **missing( )**, [M-5] **missingof( )**, [M-5] **editmissing( )**, [M-5] **_fillmissing( )**
   counting, [D] **codebook**, [D] **inspect**
   encoding and decoding, [D] **mvencode**
   extended, [D] **mvencode**
   hard and soft, [MI] **Glossary**
   ineligible, [MI] **Glossary**
   pattern of, [MI] **mi misstable**
   replacing, [D] **merge**
missingness, pattern, *see* pattern of missingness
missingof( ) function, [M-5] **missingof( )**
misspecification effects, [SVY] **estat**, [SVY] **Glossary**
misstable, mi subcommand, [MI] **mi misstable**
misstable
   nested command, [R] **misstable**
   patterns command, [R] **misstable**
   summarize command, [R] **misstable**
   tree command, [R] **misstable**
misstable for mi data, [MI] **mi misstable**
mixed designs, [MV] **manova**, [R] **anova**
mixed model, [XT] **Glossary**
_mkdir( ) function, [M-5] **chdir( )**
mkdir( ) function, [M-5] **chdir( )**
mkdir command, [D] **mkdir**
mkmat command, [P] **matrix mkmat**
mkspline command, [R] **mkspline**
ml
   check command, [R] **ml**
   clear command, [R] **ml**
   command, [SVY] **ml for svy**
   count command, [R] **ml**
   display command, [R] **ml**
   footnote command, [R] **ml**
   graph command, [R] **ml**
   init command, [R] **ml**
   maximize command, [R] **ml**
   model command, [R] **ml**
   plot command, [R] **ml**
   query command, [R] **ml**
   report command, [R] **ml**
   score command, [R] **ml**
   search command, [R] **ml**
   trace command, [R] **ml**
mlabangle( ) option, [G] *marker_label_options*
mlabcolor( ) option, [G] *marker_label_options*
mlabel( ) option, [G] *marker_label_options*
mlabgap( ) option, [G] *marker_label_options*

mlabposition( ) option, [G] *marker_label_options*
mlabsize( ) option, [G] *marker_label_options*
mlabstyle( ) option, [G] *marker_label_options*
mlabtextstyle( ) option, [G] *marker_label_options*
mlabvposition( ) option, [G] *marker_label_options*
mlcolor( ) option, [G] *marker_options*
mleval command, [R] **ml**
mlib, [M-3] **mata mlib**
.mlib filename suffix, [U] **11.6 File-naming conventions**
.mlib library files, [M-1] **how**, [M-3] **mata describe**, [M-3] **mata mlib**, [M-3] **mata set**, [M-3] **mata which**, [M-6] **Glossary**
mlmatbysum command, [R] **ml**
mlmatsum command, [R] **ml**
mlogit command, [R] **mlogit**, [R] **mlogit postestimation**, *also see* postestimation command
mlong
   data style, [MI] **Glossary**, [MI] **styles**
   technical description, [MI] **technical**
mlpattern( ) option, [G] *marker_options*
mlstyle( ) option, [G] *marker_options*
mlsum command, [R] **ml**
mlvecsum command, [R] **ml**
mlwidth( ) option, [G] *marker_options*
mm( ) function, [D] **dates and times**, [D] **functions**, [M-5] **date( )**
.mmat filename suffix, [U] **11.6 File-naming conventions**
.mmat files, [M-3] **mata matsave**
mmC( ) function, [D] **dates and times**, [D] **functions**, [M-5] **date( )**
MNAR, [MI] **intro substantive**
MNP, *see* multinomial outcome model
.mo file, [M-1] **how**, [M-3] **mata mosave**, [M-3] **mata which**, [M-6] **Glossary**
.mo filename suffix, [U] **11.6 File-naming conventions**
mod( ) function, [D] **functions**, [M-5] **mod( )**
mode( ), egen function, [D] **egen**
model, ml subcommand, [R] **ml**
model,
   maximum number of variables in, [R] **matsize**
   sensitivity, [R] **regress postestimation**, [R] **rreg**
   specification test, [R] **linktest**, [R] **regress postestimation**, [XT] **xtreg postestimation**
modern scaling, [MV] **Glossary**
modification, file, [D] **filefilter**
modifying data, [D] **generate**, *also see* editing data
modulus( ) function, [M-5] **mod( )**
modulus function, [D] **functions**
modulus transformations, [R] **boxcox**
mofd( ) function, [D] **dates and times**, [D] **functions**, [M-5] **date( )**
monotone imputation, *see* imputation, monotone
monotone missing values, [R] **misstable**
monotonicity, *see* pattern of missingness
Monte Carlo simulations, [P] **postfile**, [R] **permute**, [R] **simulate**

month( ) function, [D] **dates and times**, [D] **functions**,
  [M-5] **date( )**, [U] **24.5 Extracting components
  of dates and times**
monthly( ) function, [D] **dates and times**,
  [D] **functions**, [M-5] **date( )**
Moore–Penrose inverse, [M-5] **pinv( )**
_moptimize( ) function, [M-5] **moptimize( )**
moptimize( ) function, [M-5] **moptimize( )**
moptimize_ado_cleanup( ) function,
  [M-5] **moptimize( )**
_moptimize_evaluate( ) function,
  [M-5] **moptimize( )**
moptimize_evaluate( ) function, [M-5] **moptimize( )**
moptimize_init_*( ) functions, [M-5] **moptimize( )**
moptimize_query( ) function, [M-5] **moptimize( )**
moptimize_result_*( ) functions,
  [M-5] **moptimize( )**
moptimize_util_*( ) functions, [M-5] **moptimize( )**
more( ) function, [M-5] **more( )**
more, set subcommand, [R] **more**, [R] **set**
more command and parameter, [P] **macro**, [P] **more**,
  [R] **more**, [U] **7 –more– conditions**
more condition, [R] **query**, [U] **7 –more– conditions**,
  [U] **16.1.6 Preventing –more– conditions**
mortality table, *see* life tables
mosave, [M-3] **mata mosave**
moving average, [TS] **arch**, [TS] **arima**, [TS] **Glossary**,
  [TS] **tssmooth**, [TS] **tssmooth ma**
mprobit command, [R] **mprobit**
mreldif( ) function, [D] **functions**, [M-5] **reldif( )**,
  [P] **matrix define**
mreldifre( ) function, [M-5] **reldif( )**
mreldifsym( ) function, [M-5] **reldif( )**
msize( ) option, [G] *marker_options*,
  [G] *rcap_options*
msofhours( ) function, [D] **dates and times**,
  [D] **functions**, [M-5] **date( )**
msofminutes( ) function, [D] **dates and times**,
  [D] **functions**, [M-5] **date( )**
msofseconds( ) function, [D] **dates and times**,
  [D] **functions**, [M-5] **date( )**
mspline, graph twoway subcommand, [G] **graph
  twoway mspline**
mstyle( ) option, [G] *marker_options*
msymbol( ) option, [G] *marker_options*
mtr( ), egen function, [D] **egen**
multiarm trial, [ST] **Glossary**, [ST] **stpower**
multidimensional scaling, [MV] **Glossary**, [MV] **mds**,
  [MV] **mds postestimation**, [MV] **mdslong**,
  [MV] **mdsmat**
multilevel models, [R] **gllamm**, [XT] **xtmelogit**,
  [XT] **xtmepoisson**, [XT] **xtmixed**
multinomial logistic regression, [SVY] **svy estimation**
multinomial logistic regression, imputation, *see*
  imputation, multinomial logistic regression
multinomial outcome model, [R] **asclogit**,
  [R] **asmprobit**, [R] **asroprobit**, [R] **clogit**,
  [R] **mlogit**, [R] **mprobit**, [R] **nlogit**, [R] **slogit**,
  also see polytomous outcome model

multinomial probit regression, [SVY] **svy estimation**
multiple
  comparison tests, [R] **oneway**
  correspondence analysis, [MV] **Glossary**
  languages, [D] **label language**
  regression, *see* linear regression
  testing, [R] **regress postestimation**, [R] **test**,
    [R] **testnl**
multiple imputation, [MI] **intro**, [MI] **intro substantive**,
  [MI] **styles**, [MI] **workflow**
  analysis step, [MI] **intro substantive**, [MI] **mi
    estimate**, [MI] **mi estimate postestimation**,
    [MI] **mi estimate using**
  estimation, [MI] **estimation**
  imputation step, [MI] **intro substantive**, [MI] **mi
    impute**
  inference, [MI] **intro substantive**
  pooling step, [MI] **intro substantive**, [MI] **mi
    estimate**, [MI] **mi estimate using**
  theory, [MI] **intro substantive**
multiple-record st data, [ST] **stfill**, [ST] **stvary**
multiplication, [M-2] **op_arith**, [M-2] **op_colon**
multiplication operator, *see* arithmetic operators
multiplicative heteroskedasticity, [TS] **arch**
multistage clustered sampling, [SVY] **survey**,
  [SVY] **svydescribe**, [SVY] **svyset**
multivariable fractional polynomial regression, [R] **mfp**
multivariate analysis, [MV] **canon**, [MV] **hotelling**,
  [MV] **mvtest**, [U] **26.21 Multivariate and
  cluster analysis**
  bivariate probit, [R] **biprobit**
  of covariance, [MV] **manova**
  of variance, [MV] **manova**
  regression, [R] **mvreg**
  three-stage least squares, [R] **reg3**
  Zellner's seemingly unrelated, [R] **nlsur**, [R] **sureg**
multivariate ARCH, [TS] **dvech postestimation**
multivariate Behrens–Fisher problem, [MV] **mvtest
  means**
multivariate GARCH, [TS] **dvech** [TS] **dvech
  postestimation**,
multivariate imputation, *see* imputation, multivariate
multivariate kurtosis, [MV] **mvtest normality**
multivariate logistic variable imputation, *see* imputation,
  multivariate logistic
multivariate normal, [MV] **mvtest normality**
multivariate normal imputation, *see* imputation,
  multivariate normal
multivariate normal simulator, [M-5] **ghk( )**,
  [M-5] **ghkfast( )**
multivariate regression, [MV] **Glossary**
multivariate regression imputation, *see* imputation,
  multivariate
multivariate skewness, [MV] **mvtest normality**
multivariate time series, [TS] **dfactor**, [TS] **dvech**,
  [TS] **sspace**, [TS] **var**, [TS] **var intro**, [TS] **var
  svar**, [TS] **varbasic**, [TS] **vec**, [TS] **vec intro**,
  [TS] **xcorr**

mvdecode command, [D] **mvencode**

mvencode command, [D] **mvencode**

MVN imputation, *see* imputation, multivariate normal

mvreg, estat subcommand, [MV] **procrustes postestimation**

mvreg command, [R] **mvreg**, [R] **mvreg postestimation**, *also see* postestimation command

mvtest

    command, [MV] **mvtest**

    correlations command, [MV] **mvtest correlations**

    covariances command, [MV] **mvtest covariances**

    means command, [MV] **mvtest means**

    normality command, [MV] **mvtest normality**

N

_n and _N built-in variables, [U] **13.4 System variables (_variables)**, [U] **13.7 Explicit subscripting**

name() option, [G] *name_option*

nameexternal() function, [M-5] **findexternal( )**

namelists, [M-3] **namelists**

names, [U] **11.3 Naming conventions**

    conflicts, [P] **matrix**, [P] **matrix define**, [P] **scalar**

    matrix row and columns, [P] **ereturn**, [P] **matrix define**, [P] **matrix rownames**

names, confirm subcommand, [P] **confirm**

namespace and conflicts, matrices and scalars, [P] **matrix**, [P] **matrix define**

naming

    convention, [M-1] **naming**

    variables, [D] **rename**

NARCH, [TS] **arch**

NARCHK, [TS] **arch**

natural

    log function, [D] **functions**

    splines, [R] **mkspline**

nbetaden() function, [D] **functions**, [M-5] **normal( )**

nbinomial() function, [D] **functions**, [M-5] **normal( )**

nbinomialp() function, [D] **functions**, [M-5] **normal( )**

nbinomialtail() function, [D] **functions**, [M-5] **normal( )**

nbreg command, [R] **nbreg postestimation**, [R] **nbreg**, *also see* postestimation command

nchi2() function, [D] **functions**, [M-5] **normal( )**

n-class command, [P] **program**, [P] **return**

ndots() option, [G] **graph twoway dot**

nearest neighbor, [MI] **mi impute pmm**, [MV] **discrim knn**, [MV] **Glossary**

needle plot, [R] **spikeplot**

_negate() function, [M-5] **_negate( )**

negation, [M-2] **op_arith**, [M-5] **_negate( )**

negation operator, *see* arithmetic operators

negative binomial

    distribution,

        cumulative, [D] **functions**

        inverse cumulative, [D] **functions**

negative binomial, distribution, *continued*

        inverse reverse cumulative, [D] **functions**

        reverse cumulative, [D] **functions**

        probability mass function, [D] **functions**

    regression, [R] **nbreg**, [SVY] **svy estimation**; [R] **glm**

        fixed-effects, [XT] **xtnbreg**

        model, [XT] **Glossary**

        population-averaged, [XT] **xtnbreg**; [XT] **xtgee**

        postestimation, [XT] **xtnbreg postestimation**

        random-effects, [XT] **xtnbreg**

        zero-inflated, [R] **zinb**

        zero-truncated, [R] **ztnb**

Nelder–Mead method, [M-5] **moptimize( )**, [M-5] **optimize( )**

Nelson–Aalen cumulative hazard, [ST] **sts**, [ST] **sts generate**, [ST] **sts graph**, [ST] **sts list**

nested, misstable subcommand, [R] **misstable**

nested

    case–control data, [ST] **sttocc**

    designs, [MV] **manova**, [R] **anova**

    effects, [MV] **manova**, [R] **anova**

    logit, [R] **nlogit**

    model statistics, [R] **nestreg**

    regression, [R] **nestreg**

nestreg prefix command, [R] **nestreg**

net, view subcommand, [R] **view**

net

    cd command, [R] **net**

    describe command, [R] **net**

    from command, [R] **net**

    get command, [R] **net**

    install command, [R] **net**

    link command, [R] **net**

    query command, [R] **net**

    search command, [R] **net search**

    set ado command, [R] **net**

    set other command, [R] **net**

    sj command, [R] **net**

    stb command, [R] **net**

NetCourseNow, [U] **3.7.1 NetCourses**

NetCourses, [U] **3.7.1 NetCourses**

net_d, view subcommand, [R] **view**

new() class function, [M-2] **class**

new, ssc subcommand, [R] **ssc**

.new built-in class function, [P] **class**

new lines, data without, [D] **infile (fixed format)**

newey command, [TS] **newey**, [TS] **newey postestimation**

Newey–West

    covariance matrix, [TS] **Glossary**

    postestimation, [TS] **newey postestimation**

    regression, [TS] **newey**

    standard errors, [P] **matrix accum**, [R] **glm**

_newline(#), display directive, [P] **display**

news, view subcommand, [R] **view**

news command, [R] **news**

newsletter, [U] **3 Resources for learning and using Stata**

Newton–Raphson algorithm, [M-5] **moptimize( )**, [M-5] **optimize( )**, [R] **ml**

Neyman allocation, [SVY] **estat**

nFden( ) function, [D] **functions**, [M-5] **normal( )**

nFtail( ) function, [D] **functions**, [M-5] **normal( )**

NHANES data, [MI] **mi export nhanes1**, [MI] **mi import nhanes1**

nibeta( ) function, [D] **functions**, [M-5] **normal( )**

nl, tssmooth subcommand, [TS] **tssmooth nl**

nl command, [R] **nl**, [R] **nl postestimation**, *also see* postestimation command

nlcom command, [R] **nlcom**, [SVY] **svy postestimation**

nlogit command, [R] **nlogit**, [R] **nlogit postestimation**, *also see* postestimation command

nlogitgen command, [R] **nlogit**

nlogittree command, [R] **nlogit**

nlsur command, [R] **nlsur**, [R] **nlsur postestimation**, *also see* postestimation command

nobreak command, [P] **break**

nodraw option, [G] *nodraw_option*

noisily prefix, [P] **quietly**

nolog or log option, [R] **maximize**

noncentral

    beta density, [D] **functions**

    beta distribution, [D] **functions**

    chi-squared distribution function, [D] **functions**

    F density, [D] **functions**

    F distribution, [D] **functions**

nonconformities, quality control, [R] **qc**

nonconstant variance, *see* robust, Huber/White/sandwich estimator of variance

nonlinear

    combinations, predictions, and tests, [SVY] **svy postestimation**

    combinations of estimators, [R] **nlcom**

    estimation, [TS] **arch**

    least squares, [R] **nl**, [SVY] **svy estimation**

    predictions, [R] **predictnl**

    regression, [R] **boxcox**, [R] **nl**, [R] **nlsur**

    smoothing, [TS] **tssmooth nl**

    tests, *see* estimation, tests after

nonmetric scaling, [MV] **Glossary**

nonmissing( ) function, [M-5] **missing( )**

nonparametric estimation, [R] **lpoly**

nonparametric methods, [MV] **Glossary**

nonparametric test, equality of survivor functions, [ST] **sts test**

nonparametric tests,

    association, [R] **spearman**

    equality of

        distributions, [R] **ksmirnov**, [R] **kwallis**, [R] **ranksum**, [R] **signrank**

        medians, [R] **ranksum**

        proportions, [R] **bitest**, [R] **prtest**

    percentiles, [R] **centile**

    ROC analysis, [R] **roc**

nonparametric tests, *continued*

    serial independence, [R] **runtest**

    tables, [R] **tabulate twoway**

    trend, [R] **cusum**, [R] **nptrend**

nonrtolerance option, [R] **maximize**

nonstationary time series, [TS] **dfgls**, [TS] **dfuller**, [TS] **pperron**, [TS] **vec**, [TS] **vec intro**

nopreserve option, [P] **nopreserve option**

norm, [M-6] **Glossary**

norm( ) function, [M-5] **norm( )**

normal( ) function, [D] **functions**, [M-5] **normal( )**

normal distribution and normality,

    bivariate, [D] **functions**

    cdf, [D] **functions**

    density, [D] **functions**

    examining distributions for, [R] **diagnostic plots**, [R] **lv**

    generating multivariate data, [D] **corr2data**, [D] **drawnorm**

    inverse, [D] **functions**

    probability and quantile plots, [R] **diagnostic plots**

    test for, [R] **sktest**, [R] **swilk**

    transformations to achieve, [R] **boxcox**, [R] **ladder**, [R] **lnskew0**

normal probability plots, [G] **graph other**

normalden( ) function, [D] **functions**, [M-5] **normal( )**

normality, mvtest subcommand, [MV] **mvtest normality**

normality test

    after VAR or SVAR, [TS] **varnorm**

    after VEC, [TS] **vecnorm**

normalization, [MV] **Glossary**

normally distributed random numbers, [D] **functions**

not equal operator, [U] **13.2.3 Relational operators**

not operator, [U] **13.2.4 Logical operators**

note( ) option, [G] *title_options*

notes,

    cluster analysis, [MV] **cluster notes**

    creating, [D] **varmanage**

    editing, [D] **varmanage**

notes,

    cluster subcommand, [MV] **cluster notes**

    estimates subcommand, [R] **estimates notes**

notes command, [D] **notes**

notes on estimation results, [R] **estimates notes**

notifyuser, set subcommand, [R] **set**

noupdate option, [MI] **noupdate option**

novarabbrev command, [P] **varabbrev**

NPARCH, [TS] **arch**

npnchi2( ) function, [D] **functions**, [M-5] **normal( )**

nptrend command, [R] **nptrend**

NR algorithm, *see* Newton–Raphson algorithm

nrtolerance( ) option, [R] **maximize**

NULL, [M-2] **pointers**, [M-6] **Glossary**

nullmat( ) matrix function, [D] **functions**, [P] **matrix define**

number, confirm subcommand, [P] **confirm**

number to string conversion, *see* string functions, expressions, and operators, *see* string functions

numbered styles, [G] *linestyle*, [G] *markerlabelstyle*, [G] *markerstyle*, [G] *pstyle*

numbers, [U] **12.2 Numbers**

    formatting, [D] **format**

    mapping to strings, [D] **encode**, [D] **label**

numeric, [M-2] **declarations**, [M-6] **Glossary**

numeric

    display formats, [I] **format**

    list, [P] **numlist**, [P] **syntax**, [U] **11.1.8 numlist**

    value labels, [D] **labelbook**

numerical precision, [U] **13.11 Precision and problems therein**

numlabel command, [D] **labelbook**

numlist command, [P] **numlist**, [U] **11.1.8 numlist**

N-way

    analysis of variance, [R] **anova**

    multivariate analysis of variance, [MV] **manova**

O

object, [P] **class**

object code, [M-1] **how**, [M-6] **Glossary**

object-oriented programming, [M-2] **class**, [M-6] **Glossary**, [P] **class**

objects, size of, [G] *relativesize*

.objkey built-in class function, [P] **class**

.objtype built-in class function, [P] **class**

oblimax rotation, [MV] **Glossary**, [MV] **rotate**, [MV] **rotatemat**

oblimin rotation, [MV] **Glossary**, [MV] **rotate**, [MV] **rotatemat**

oblique rotation, [MV] **factor postestimation**, [MV] **Glossary**, [MV] **rotate**, [MV] **rotatemat**

obs, set subcommand, [D] **obs**, [R] **set**

obs parameter, [D] **obs**; [D] **describe**

observations,

    built-in counter variable, [U] **11.3 Naming conventions**

    complete and incomplete, [MI] **Glossary**

    creating dataset of, [D] **collapse**

    dropping, [D] **drop**

    dropping duplicate, [D] **duplicates**

    duplicating, [D] **expand**

    duplicating, clustered, [D] **expandcl**

    identifying duplicate, [D] **duplicates**

    increasing number of, [D] **obs**

    marking, [P] **mark**

    maximum number of, [D] **describe**, [D] **memory**, [U] **6 Setting the size of memory**

    ordering, [D] **sort**; [D] **gsort**

    transposing with variables, [D] **xpose**

observed information matrix, *see* OIM

Ochiai coefficient similarity measure, [MV] *measure_option*

odbc

    command, [D] **infile**

    describe command, [D] **odbc**

    exec() command, [D] **odbc**

    insert command, [D] **odbc**

    list command, [D] **odbc**

    load command, [D] **odbc**

    query command, [D] **odbc**

    sqlfile() command, [D] **odbc**

odbcmgr, set subcommand, [R] **set**

ODBC data source, reading data from, [D] **odbc**, [U] **21.4 Transfer programs**, [U] **21.5 ODBC sources**

odds ratio, [R] **asclogit**, [R] **asroprobit**, [R] **binreg**, [R] **clogit**, [R] **cloglog**, [R] **glm**, [R] **glogit**, [R] **logistic**, [R] **logit**, [R] **mlogit**, [R] **rologit**, [R] **scobit**, [ST] **epitab**, [ST] **Glossary**, [SVY] **svy estimation**, [XT] **xtcloglog**, [XT] **xtgee**, [XT] **xtlogit**, [XT] **xtnbreg**, [XT] **xtpoisson**, [XT] **xtprobit**

    differences, [R] **lincom**, [R] **nlcom**, [SVY] **svy postestimation**

off, timer subcommand, [P] **timer**

offset

    between axes and data, setting, [G] *region_options*

    variable, [ST] **Glossary**

ograph, irf subcommand, [TS] **irf ograph**

OIM, [R] **ml**, [R] *vce_option*, [XT] *vce_options*

OLDPLACE directory, [P] **sysdir**, [U] **17.5 Where does Stata look for ado-files?**

OLE Automation, [P] **automation**

ologit command, [R] **ologit**, [R] **ologit postestimation**, *also see* postestimation command

OLS regression, *see* linear regression

omitted variables test, [R] **regress postestimation**

on, timer subcommand, [P] **timer**

one-level model, [XT] **Glossary**

one-way analysis of variance, [R] **kwallis**, [R] **loneway**, [R] **oneway**

oneway command, [R] **oneway**

online help, [M-1] **help**, [M-3] **mata help**, [R] **help**, [R] **hsearch**, [R] **search**, [U] **4 Stata's help and search facilities**; [U] **7 –more– conditions**

opaccum, matrix subcommand, [P] **matrix accum**

open, file subcommand, [P] **file**

operating system command, [D] **cd**, [D] **copy**, [D] **dir**, [D] **erase**, [D] **mkdir**, [D] **rmdir**, [D] **shell**, [D] **type**

operator,

    difference, [U] **11.4.4 Time-series varlists**

    lag, [U] **11.4.4 Time-series varlists**

    lead, [U] **11.4.4 Time-series varlists**

    seasonal lag, [U] **11.4.4 Time-series varlists**

operators, [M-2] **op_arith**, [M-2] **op_assignment**, [M-2] **op_colon**, [M-2] **op_conditional**, [M-2] **op_increment**, [M-2] **op_join**, [M-2] **op_kronecker**, [M-2] **op_logical**, [M-2] **op_range**, [M-2] **op_transpose**, [M-6] **Glossary**, [P] **matrix define**,

operators, *continued* [U] **13.2 Operators**
 order of evaluation, [U] **13.2.5 Order of evaluation,
  all operators**
OPG, [R] **ml**, [R] *vce_option*, [XT] *vce_options*
oprobit command, [R] **oprobit**, [R] **oprobit
  postestimation**, *also see* postestimation command
optimization, [M-3] **mata set**, [M-5] **moptimize( )**,
  [M-5] **optimize( )**, [M-6] **Glossary**
_optimize() function, [M-5] **optimize( )**
optimize() function, [M-5] **optimize( )**
_optimize_evaluate() function, [M-5] **optimize( )**
optimize_evaluate() function, [M-5] **optimize( )**
optimize_init_*() functions, [M-5] **optimize( )**
optimize_query() function, [M-5] **optimize( )**
optimize_result_*() functions, [M-5] **optimize( )**
options, [U] **11 Language syntax**
 in a programming context, [P] **syntax**, [P] **unab**
 repeated, [G] **concept: repeated options**
 or operator, [U] **13.2.4 Logical operators**
ORACLE, reading data from, [D] **odbc**,
  [U] **21.4 Transfer programs**
order() function, [M-5] **sort( )**
order command, [D] **order**
order statistics, [D] **egen**, [R] **lv**
ordered
 logistic regression, [SVY] **svy estimation**
 logistic regression imputation, *see* imputation,
  ordered logistic regression
 logit, [R] **ologit**
 ordered probit regression, [SVY] **svy estimation**
 probit, [R] **oprobit**
ordering
 observations, [D] **sort**; [D] **gsort**
 variables, [D] **order**, [D] **sort**
ordinal outcome model, [R] **ologit**, [R] **oprobit**,
  [R] **rologit**, [R] **slogit**, *also see* polytomous
  outcome model
ordinary least squares, *see* linear regression
ordination, [MV] **Glossary**
orgtype, [M-2] **declarations**, [M-6] **Glossary**
orgtype() function, [M-5] **eltype( )**
orientationstyle, [G] *orientationstyle*
orthog command, [R] **orthog**
orthogonal
 matrix, [M-6] **Glossary**
 polynomials, [R] **orthog**
 rotation, [MV] **factor postestimation**,
  [MV] **Glossary**, [MV] **rotate**, [MV] **rotatemat**
 transformation, [MV] **Glossary**
orthogonalized impulse–response function,
  [TS] **Glossary**, [TS] **irf**, [TS] **var intro**, [TS] **vec**
  [TS] **vec intro**,
orthonormal basis, [P] **matrix svd**
orthpoly command, [R] **orthog**
other graph commands, [G] **graph other**
.out filename suffix, [D] **outsheet**, [U] **11.6 File-
  naming conventions**

outer
 fence, [R] **lv**
 product, [D] **cross**
 product of the gradient, *see* OPG
outfile command, [D] **outfile**
outliers, [R] **lv**, [R] **qreg**, [R] **regress postestimation**,
  [R] **rreg**
outlines, suppressing, [G] *linestyle*
outlining regions, [G] *region_options*
out-of-sample predictions, [R] **predict**, [R] **predictnl**,
  [U] **20.9.3 Making out-of-sample predictions**
output,
 controlling the scrolling of, [R] **more**
 displaying, [P] **display**, [P] **smcl**
 formatting numbers, [D] **format**
 printing, [R] **translate**, [U] **15 Saving and printing
  output—log files**
 recording, [R] **log**
 suppressing, [P] **quietly**
output, set subcommand, [P] **quietly**, [R] **set**
output settings, [P] **creturn**
outsheet command, [D] **outsheet**
outside values, [R] **lv**
over() option, [G] **graph bar**, [G] **graph box**,
  [G] **graph dot**
overid, estat subcommand, [R] **gmm postestimation**,
  [R] **ivregress postestimation**
overidentifying restrictions, [XT] **Glossary**
overloading, class program names, [P] **class**
ovtest, estat subcommand, [R] **regress
  postestimation**

P

P charts, [G] **graph other**
pac command, [TS] **corrgram**
pagesize, set subcommand, [R] **more**, [R] **set**
paging of screen output, controlling, [P] **more**,
  [R] **more**
paired-coordinate plots, [G] **graph twoway pcarrow**,
  [G] **graph twoway pccapsym**, [G] **graph
  twoway pcscatter**, [G] **graph twoway pcspike**
pairwise, estat subcommand, [MV] **mds
  postestimation**
pairwise
 combinations, [D] **cross**, [D] **joinby**
 correlation, [R] **correlate**
palette command, [G] **palette**
panel-corrected standard errors, [XT] **Glossary**
panel data, [M-5] **panelsetup( )**, [XT] **Glossary**,
  [U] **26.15 Panel-data models**
panels, variable identifying, [XT] **xtset**
panelsetup() function, [M-5] **panelsetup( )**
panelstats() function, [M-5] **panelsetup( )**
panelsubmatrix() function, [M-5] **panelsetup( )**
panelsubview() function, [M-5] **panelsetup( )**
Paradox, reading data from, [U] **21.4 Transfer
  programs**

parameter trace files, [MI] **mi impute mvn**, [MI] **mi ptrace**
parameterized curves, [D] **range**
parameters, system, *see* system parameters
parametric
  methods, [MV] **Glossary**
  survival models, [ST] **streg**, [SVY] **svy estimation**
PARCH, [TS] **arch**
parsedistance, cluster subcommand, [MV] **cluster programming utilities**
parsimax rotation, [MV] **Glossary**, [MV] **rotate**, [MV] **rotatemat**
parsing, [M-5] **tokenget( )**, [M-5] **tokens( )**, [P] **syntax**, [U] **18.4 Program arguments**; [P] **gettoken**, [P] **numlist**, [P] **tokenize**
partial
  autocorrelation function, [TS] **corrgram**, [TS] **Glossary**
  correlation, [R] **pcorr**
  DFBETA, [ST] **Glossary**
  effects, [R] **margins**
  likelihood displacement value, [ST] **Glossary**
  LMAX value, [ST] **Glossary**
  regression leverage plot, [R] **regress postestimation**
  regression plot, [R] **regress postestimation**
  residual plot, [R] **regress postestimation**
  target rotation, [MV] **rotate**, [MV] **rotatemat**
partially specified target rotation, [MV] **Glossary**
partition cluster-analysis methods, [MV] **cluster kmeans and kmedians**, [MV] **Glossary**
partition clustering, [MV] **Glossary**
partitioned matrices, [P] **matrix define**
partitioning memory, [D] **memory**, [U] **6 Setting the size of memory**
Parzen kernel density function, [R] **kdensity**
passive, mi subcommand, [MI] **mi passive**
passive variables, *see* variables, passive
past history, [ST] **Glossary**, [ST] **stset**
pathasciisuffix() function, [M-5] **pathjoin( )**
pathbasename() function, [M-5] **pathjoin( )**
pathisabs() function, [M-5] **pathjoin( )**
pathisurl() function, [M-5] **pathjoin( )**
pathjoin() function, [M-5] **pathjoin( )**
pathlist() function, [M-5] **pathjoin( )**
pathrmsuffix() function, [M-5] **pathjoin( )**
paths, [U] **11.6 File-naming conventions**
paths and directories, [P] **creturn**
pathsearchlist() function, [M-5] **pathjoin( )**
pathsplit() function, [M-5] **pathjoin( )**
pathstatasuffix() function, [M-5] **pathjoin( )**
pathsubsysdir() function, [M-5] **pathjoin( )**
pathsuffix() function, [M-5] **pathjoin( )**
pattern
  matching, [M-5] **strmatch( )**
  of missing values, [R] **misstable**
  of missingness, [MI] **Glossary**, [MI] **intro substantive**, [MI] **mi impute**, [MI] **mi misstable**
patterns, misstable subcommand, [R] **misstable**

patterns of data, [D] **egen**
pause command, [P] **pause**
pausing until key is pressed, [P] **more**, [R] **more**
pc(), egen function, [D] **egen**
PCA, [MV] **Glossary**
pca command, [MV] **pca**
pcamat command, [MV] **pca**
pcarrow, graph twoway subcommand, [G] **graph twoway pcarrow**
pcarrowi, graph twoway subcommand, [G] **graph twoway pcarrowi**
pcbarrow, graph twoway subcommand, [G] **graph twoway pcarrow**
pccapsym, graph twoway subcommand, [G] **graph twoway pccapsym**
pchart command, [R] **qc**
pchi command, [R] **diagnostic plots**
pci, graph twoway subcommand, [G] **graph twoway pci**
p-conformability, [M-6] **Glossary**
pcorr command, [R] **pcorr**
pcscatter, graph twoway subcommand, [G] **graph twoway pcscatter**
PCSE (panel-corrected standard error), [XT] **xtpcse postestimation**, [XT] **xtpcse postestimation**
pcspike, graph twoway subcommand, [G] **graph twoway pcspike**
pctile(), egen function, [D] **egen**
_pctile command, [D] **pctile**
pctile command, [D] **pctile**
PDF files (Mac only), [R] **translate**
Pearson
  coefficient similarity measure, [MV] *measure_option*
  goodness-of-fit test, [R] **logistic postestimation**, [R] **poisson postestimation**
  product-moment correlation coefficient, [R] **correlate**
  residual, [R] **binreg postestimation**, [R] **glm postestimation**, [R] **logistic postestimation**, [R] **logit postestimation**
penalized log-likelihood function, [ST] **Glossary**, [ST] **stcox**
percentiles,
  create
    dataset of, [D] **collapse**
    variable containing, [D] **codebook**, [D] **egen**, [D] **pctile**
  displaying, [R] **centile**, [R] **lv**, [R] **summarize**, [R] **table**, [R] **tabstat**
pergram command, [TS] **pergram**
_perhapsequilc() function, [M-5] **_equilrc( )**
_perhapsequilr() function, [M-5] **_equilrc( )**
_perhapsequilrc() function, [M-5] **_equilrc( )**
periodogram, [G] **graph other**, [TS] **Glossary**, [TS] **pergram**
permname macro extended function, [P] **macro**

permutation
matrix and vector, [M-1] **permutation**,
[M-5] **invorder( )**, [M-6] **Glossary**
tests, [R] **permute**
permutations, [M-5] **cvpermute( )**
permute prefix command, [R] **permute**
persistfv, set subcommand, [R] **set**
persistvtopic, set subcommand, [R] **set**
person time, [ST] **stptime**
personal command, [P] **sysdir**
PERSONAL directory, [P] **sysdir**, [U] **17.5 Where does Stata look for ado-files?**
pharmaceutical statistics, [R] **pk**, [R] **pksumm**
pharmacokinetic data, [U] **26.22 Pharmacokinetic data**,
see pk (pharmacokinetic data)
pharmacokinetic plots, [G] **graph other**
Phillips–Perron test, [TS] **pperron**
phtest, estat subcommand, [ST] **stcox PH-assumption tests**
pi( ) function, [M-5] **sin( )**
pi, value of, [U] **11.3 Naming conventions**,
[U] **13.4 System variables (_variables)**
_pi built-in variable, [U] **11.3 Naming conventions**
pie, graph subcommand, [G] **graph pie**
pie charts, [G] **graph pie**
piece macro extended function, [P] **macro**
piecewise
cubic functions, [R] **mkspline**
linear functions, [R] **mkspline**
Pillai's trace statistic, [MV] **canon**, [MV] **Glossary**,
[MV] **manova**, [MV] **mvtest means**
pinnable, set subcommand, [R] **set**
_pinv( ) function, [M-5] **pinv( )**
pinv( ) function, [M-5] **pinv( )**
pk (pharmacokinetic data), [R] **pk**, [R] **pkcollapse**,
[R] **pkcross**, [R] **pkequiv**, [R] **pkexamine**,
[R] **pkshape**, [R] **pksumm**
pkcollapse command, [R] **pkcollapse**
pkcross command, [R] **pkcross**
pkequiv command, [R] **pkequiv**
pkexamine command, [R] **pkexamine**
.pkg filename suffix, [R] **net**
pkshape command, [R] **pkshape**
pksumm command, [R] **pksumm**
Plackett–Luce model, [R] **rologit**
platforms for which Stata is available,
[U] **5.1 Platforms**
play, graph subcommand, [G] **graph play**
play( ) option, [G] *play_option*
playsnd, set subcommand, [R] **set**
plot, definition, [G] *pstyle*
plot, ml subcommand, [R] **ml**
plot region, [G] *region_options*
suppressing border around, [G] *region_options*
plotregion( ) option, [G] *region_options*
plotregionstyle, [G] *plotregionstyle*

plottypes
base, [G] *advanced_options*
derived, [G] *advanced_options*
plugin, loading, [P] **plugin**
plugin option, [P] **plugin**, [P] **program**
plural( ) string function, [D] **functions**
PLUS directory, [P] **sysdir**, [U] **17.5 Where does Stata look for ado-files?**
PMM imputation, see imputation, predictive mean matching
PNG, [G] *png_options*
pnorm command, [R] **diagnostic plots**
point estimate, [SVY] **Glossary**
pointers, [M-2] **ftof**, [M-2] **pointers**,
[M-5] **findexternal( )**, [M-6] **Glossary**
points, connecting, [G] *connect_options*,
[G] *connectstyle*
poisson( ) function, [D] **functions**
poisson( ) function, [M-5] **normal( )**
Poisson
distribution,
cdf, [D] **functions**
confidence intervals, [R] **ci**
cumulative, [D] **functions**
regression, see Poisson regression
inverse cumulative, [D] **functions**
inverse reverse cumulative, [D] **functions**
reverse cumulative, [D] **functions**
probability mass function, [D] **functions**
poisson command, [R] **poisson**, [R] **poisson postestimation**; [R] **nbreg**, also see postestimation command
Poisson regression, [R] **glm**, [R] **nbreg**, [R] **poisson**,
[SVY] **svy estimation**, [XT] **xtgee**,
[XT] **xtmepoisson**
fixed-effects, [XT] **xtpoisson**
model, [XT] **Glossary**
population-averaged, [XT] **xtpoisson**; [XT] **xtgee**
postestimation, [XT] **xtpoisson postestimation**
random-effects, [XT] **xtpoisson**
zero-inflated, [R] **zip**
zero-truncated, [R] **ztp**
poissonp( ) function, [D] **functions**, [M-5] **normal( )**
poissontail( ) function, [D] **functions**,
[M-5] **normal( )**
polar coordinates, [D] **range**
polyadd( ) function, [M-5] **polyeval( )**
polyderiv( ) function, [M-5] **polyeval( )**
polydiv( ) function, [M-5] **polyeval( )**
polyeval( ) function, [M-5] **polyeval( )**
polyinteg( ) function, [M-5] **polyeval( )**
polymorphism, [P] **class**
polymult( ) function, [M-5] **polyeval( )**
polynomial smoothing, see local polynomial smoothing
polynomials, [M-5] **polyeval( )**
orthogonal, [R] **orthog**
polyroots( ) function, [M-5] **polyeval( )**
polysolve( ) function, [M-5] **polyeval( )**

polytomous
    logistic regression, [SVY] **svy estimation**
    outcome model, [R] **asclogit**, [R] **asmprobit**,
        [R] **asroprobit**, [R] **clogit**, [R] **mlogit**,
        [R] **mprobit**, [R] **nlogit**, [R] **ologit**, [R] **oprobit**,
        [R] **rologit**, [R] **slogit**
polytrim() function, [M-5] **polyeval( )**
pooled
    estimates, [ST] **epitab**
    estimator, [XT] **Glossary**
pooling step, [MI] **intro substantive**, [MI] **mi estimate**,
    [MI] **mi estimate postestimation**, [MI] **mi**
    estimate using
population-averaged model, [XT] **Glossary**,
    [XT] **xtcloglog**, [XT] **xtgee**, [XT] **xtlogit**,
    [XT] **xtnbreg**, [XT] **xtpoisson**, [XT] **xtreg**
population
    attributable risk, [ST] **epitab**
    marginal means, [R] **margins**
    pyramid, [G] **graph twoway bar**
    standard deviation, *see* subpopulation, standard
        deviations
populations,
    diagnostic plots, [R] **diagnostic plots**
    examining, [R] **histogram**, [R] **lv**, [R] **stem**,
        [R] **summarize**, [R] **table**
    standard, [R] **dstdize**
    testing equality of, [R] **ksmirnov**, [R] **kwallis**,
        [R] **signrank**
    testing for normality, [R] **sktest**, [R] **swilk**
portmanteau
    statistic, [TS] **Glossary**
    test, [TS] **corrgram**, [TS] **wntestq**
post, ereturn subcommand, [P] **ereturn**, [P] **makecns**
post command, [P] **postfile**
post hoc tests, [R] **oneway**
postclose command, [P] **postfile**
posterior probabilities, [MV] **Glossary**
postestimation, *see* estimation, tests after
postestimation command, [I] **postestimation**
    commands, [P] **estat programming**, [R] **estat**,
    [R] **estimates**, [R] **hausman**, [R] **lincom**,
    [R] **linktest**, [R] **lrtest**, [R] **margins**, [R] **nlcom**,
    [R] **predict**, [R] **predictnl**, [R] **suest**, [R] **test**,
    [R] **testnl**, [ST] **stcurve**, [TS] **fcast compute**,
    [TS] **fcast graph**, [TS] **irf**, [TS] **vargranger**,
    [TS] **varlmar**, [TS] **varnorm**, [TS] **varsoc**,
    [TS] **varstable**, [TS] **varwle**, [TS] **veclmar**,
    [TS] **vecnorm**, [TS] **vecstable**
postfile command, [P] **postfile**
PostScript, [G] **graph export**, [G] *text*;
    [G] *eps_options*, [G] *ps_options*
poststratification, [SVY] **Glossary**,
    [SVY] **poststratification**
postutil
    clear command, [P] **postfile**
    dir command, [P] **postfile**
poverty indices, [R] **inequality**

power, [M-2] **op_arith**, [M-2] **op_colon**,
    [M-5] **matpowersym( )**, [ST] **Glossary**
    Cox proportional hazards regression, [ST] **stpower**
    cox, [ST] **stpower**
    exponential survival, [ST] **stpower exponential**,
    [ST] **stpower**
    exponential test, [ST] **stpower exponential**,
    [ST] **stpower**
    log-rank, [ST] **stpower logrank**, [ST] **stpower**
power of a test, [R] **sampsi**
power, raise to, function, *see* arithmetic operators
power transformations, [R] **boxcox**, [R] **lnskew0**
P–P plot, [R] **diagnostic plots**
pperron command, [TS] **pperron**
pragma, [M-2] **pragma**, [M-6] **Glossary**
prais command, [TS] **Glossary**, [TS] **prais**, [TS] **prais**
    postestimation, [XT] **xtpcse**
Prais–Winsten regression, *see* prais command
precision, [U] **13.11 Precision and problems therein**
predetermined variable, [XT] **Glossary**
predict, estat subcommand, [R] **exlogistic**
    postestimation
_predict command, [P] **_predict**
predict command, [I] **estimation commands**,
    [I] **postestimation commands**, [MV] **factor**
    postestimation, [MV] **pca postestimation**,
    [R] **predict**, [R] **regress postestimation**,
    [SVY] **svy postestimation**, [U] **20.9 Obtaining**
    predicted values; [P] **ereturn**, [P] **_estimates**,
    [TS] **dfactor postestimation**, [TS] **dvech**
    postestimation, [TS] **sspace postestimation**
predicted
    marginals, [R] **margins**
    population margins, [R] **margins**
prediction, standard error of, [R] **glm**, [R] **predict**,
    [R] **regress postestimation**
predictions, [R] **margins**, [R] **predict**, [SVY] **svy**
    postestimation
    nonlinear, [R] **predictnl**
    obtaining after estimation, [P] **_predict**
predictive
    marginal means, [R] **margins**
    margins, [R] **margins**, [U] **20.14 Obtaining**
    marginal means, adjusted predictions, and
    predictive margins
    mean matching imputation, *see* imputation,
        predictive mean matching
predictnl command, [R] **predictnl**, [SVY] **svy**
    postestimation
prefix command, [R] **bootstrap**, [R] **fracpoly**,
    [R] **jackknife**, [R] **mfp**, [R] **nestreg**,
    [R] **permute**, [R] **simulate**, [R] **stepwise**, [R] **xi**,
    [U] **11.1.10 Prefix commands**
Pregibon delta beta influence statistic, [R] **logistic**
preprocessor commands, [R] **#review**
preserve command, [P] **preserve**
preserving user's data, [P] **preserve**
prevalence studies, *see* case–control data, *see* cross-
    sectional studies

prevented fraction, [ST] **epitab**, [ST] **Glossary**
prewhiten, [XT] **Glossary**
primary sampling unit, [SVY] **Glossary**,
    [SVY] **svydescribe**, [SVY] **svyset**
priming values, [TS] **Glossary**
principal
    component analysis, [MV] **Glossary**, [MV] **pca**
    factors analysis, [MV] **factor**
print,
    graph set subcommand, [G] **graph set**
    graph subcommand, [G] **graph print**
print command, [R] **translate**
printcolor, set subcommand, [R] **set**, [G] **set**
    printcolor
printf( ) function, [M-5] **printf( )**
printing, logs (output), [R] **translate**, [U] **15 Saving**
    and printing output—log files
printing graphs, [G] **graph print**, [G] *pr_options*
    exporting options, [G] **graph set**
    settings, [G] **graph set**
private, [M-2] **class**
probability weight, *see* sampling weight, *also see* survey
    data
probit command, [R] **probit**, [R] **probit**
    postestimation, *also see* postestimation command
probit model with
    endogenous regressors, [SVY] **svy estimation**
    sample selection, [SVY] **svy estimation**
probit regression, [R] **probit**, [SVY] **svy estimation**
    alternative-specific multinomial probit,
        [R] **asmprobit**
    alternative-specific rank-ordered, [R] **asroprobit**
    bivariate, [R] **biprobit**
    generalized estimating equations, [XT] **xtgee**
    generalized linear model, [R] **glm**
    heteroskedastic, [R] **hetprob**
    multinomial, [R] **mprobit**
    ordered, [R] **oprobit**
    population-averaged, [XT] **xtprobit**; [XT] **xtgee**
    postestimation, [XT] **xtprobit postestimation**
    random-effects, [XT] **xtprobit**
    two-equation, [R] **biprobit**
    with endogenous regressors, [R] **ivprobit**
    with grouped data, [R] **glogit**
    with panel data, [XT] **xtprobit**; [XT] **xtgee**
    with sample selection, [R] **heckprob**
procedure codes, [D] **icd9**
processors, set subcommand, [R] **set**
procoverlay command, [MV] **procrustes**
    postestimation
procrustes command, [MV] **procrustes**
Procrustes
    rotation, [MV] **Glossary**, [MV] **procrustes**
    transformation, [MV] **Glossary**, [MV] **procrustes**
product, [M-2] **op_arith**, [M-2] **op_colon**,
    [M-2] **op_kronecker**, [M-5] **cross( )**,
    [M-5] **crossdev( )**, [M-5] **quadcross( )**

product-moment correlation, [R] **correlate**
    between ranks, [R] **spearman**
production frontier model, [R] **frontier**, [XT] **xtfrontier**
production function, [XT] **Glossary**
profiles, estat subcommand, [MV] **ca**
    postestimation
program
    define command, [P] **plugin**, [P] **program**,
        [P] **program properties**
    dir command, [P] **program**
    drop command, [P] **program**
    list command, [P] **program**
program properties, [P] **program properties**
programmer's commands and utilities, [MI] **mi select**,
    [MI] **styles**, [MI] **technical**
programming, [P] **syntax**
    cluster analysis, [MV] **cluster programming**
        utilities
    cluster subcommands, [MV] **cluster programming**
        subroutines
    cluster utilities, [MV] **cluster programming**
        subroutines
    dialog, [P] **dialog programming**
    estat, [P] **estat programming**
    functions, [M-4] **programming**
    Mac, [P] **window programming**
    menus, [P] **window programming**
    rotations, [MV] **rotate**
    use, [M-1] **ado**
    Windows, [P] **window programming**
programs,
    adding comments to, [P] **comments**
    debugging, [P] **trace**
    dropping, [P] **discard**
    looping, [P] **continue**
    user-written, [R] **sj**, [R] **ssc**
programs, clear subcommand, [D] **clear**
projection matrix, diagonal elements of, [R] **logistic**
    postestimation, [R] **logit postestimation**,
    [R] **probit postestimation**, [R] **regress**
    postestimation, [R] **rreg**
promax
    power rotation, [MV] **Glossary**
    rotation, [MV] **rotate**, [MV] **rotatemat**
proper( ) string function, [D] **functions**
proper imputation method, [MI] **intro substantive**
proper values, [M-5] **eigensystem( )**
properties, [P] **program properties**
properties macro extended function, [P] **macro**
proportion command, [R] **proportion**,
    [R] **proportion postestimation**, *also see*
    postestimation command
proportional
    hazards models, *see* survival analysis
    odds model, [R] **ologit**, [R] **slogit**
    sampling, [D] **sample**, [R] **bootstrap**

proportions, [R] **ci**, [R] **sampsi**
  estimating, [R] **proportion**
  survey data, [SVY] **svy estimation**,
    [SVY] **svy: tabulate oneway**,
    [SVY] **svy: tabulate twoway**
  testing equality of, [R] **bitest**, [R] **prtest**
prospective study, [ST] **epitab**, [ST] **Glossary**
protected, [M-2] **class**
proximity, [MV] **Glossary**
proximity matrix, [MV] **Glossary**
prtest command, [R] **prtest**
prtesti command, [R] **prtest**
pseudo *R*-squared, [R] **maximize**
pseudofunctions, [D] **dates and times**, [D] **functions**
pseudoinverse, [M-5] **pinv( )**
pseudolikelihood, [SVY] **Glossary**
pseudosigmas, [R] **lv**
psi function, [D] **functions**
pstyle, [G] *pstyle*
pstyle( ) option, [G] **graph twoway scatter**,
  [G] *pstyle*, [G] *rspike_options*
PSU, *see* primary sampling unit
ptrace, mi subcommand, [MI] **mi ptrace**
.ptrace files, [MI] **mi impute mvn**, [MI] **mi ptrace**
public, [M-2] **class**
push, window subcommand, [P] **window programming**
pwcorr command, [R] **correlate**
pwd( ) function, [M-5] **chdir( )**
pwd command, [D] **cd**
pweight, *see* sampling weight
[pweight=*exp*] modifier, [U] **11.1.6 weight**,
  [U] **20.18.3 Sampling weights**
pyramid, population, [G] **graph twoway bar**

Q

Q statistic, [TS] **wntestq**
qc charts, *see* quality control charts
qchi command, [R] **diagnostic plots**
QDA, [MV] **discrim qda**, [MV] **Glossary**
qda, discrim subcommand, [MV] **discrim qda**
qfit, graph twoway subcommand, [G] **graph twoway qfit**
qfitci, graph twoway subcommand, [G] **graph twoway qfitci**
qladder command, [R] **ladder**
qnorm command, [R] **diagnostic plots**
qofd( ) function, [D] **dates and times**, [D] **functions**,
  [M-5] **date( )**
Q–Q plot, [R] **diagnostic plots**
qqplot command, [R] **diagnostic plots**
QR decomposition, [M-5] **qrd( )**
qrd( ) function, [M-5] **qrd( )**
qrdp( ) function, [M-5] **qrd( )**
_qreg command, [R] **qreg**
qreg command, [R] **qreg**, [R] **qreg postestimation**,
  also see postestimation command
_qrinv( ) function, [M-5] **qrinv( )**

qrinv( ) function, [M-5] **qrinv( )**
_qrsolve( ) function, [M-5] **qrsolve( )**
qrsolve( ) function, [M-5] **qrsolve( )**
quad precision, [M-5] **runningsum( )**, [M-5] **sum( )**,
  [M-5] **mean( )**, [M-5] **quadcross( )**
quadchk command, [XT] **quadchk**
quadcolsum( ) function, [M-5] **sum( )**
quadcorrelation( ) function, [M-5] **mean( )**
quadcross( ) function, [M-5] **quadcross( )**
quadcrossdev( ) function, [M-5] **quadcross( )**
quadmeanvariance( ) function, [M-5] **mean( )**
quadrant( ) function, [M-5] **sign( )**
quadratic
  discriminant analysis, [MV] **discrim qda**,
    [MV] **Glossary**
  terms, [SVY] **svy postestimation**
quadrature, [XT] **Glossary**
quadrowsum( ) function, [M-5] **sum( )**
_quadrunningsum( ) function, [M-5] **runningsum( )**
quadrunningsum( ) function, [M-5] **runningsum( )**
quadsum( ) function, [M-5] **sum( )**
quadvariance( ) function, [M-5] **mean( )**
qualitative dependent variables, [R] **asclogit**,
  [R] **asmprobit**, [R] **asroprobit**, [R] **binreg**,
  [R] **biprobit**, [R] **brier**, [R] **clogit**, [R] **cloglog**,
  [R] **cusum**, [R] **exlogistic**, [R] **glm**, [R] **glogit**,
  [R] **heckprob**, [R] **hetprob**, [R] **ivprobit**,
  [R] **logistic**, [R] **logit**, [R] **mlogit**, [R] **mprobit**,
  [R] **nlogit**, [R] **ologit**, [R] **oprobit**, [R] **probit**,
  [R] **rocfit**, [R] **rologit**, [R] **scobit**, [R] **slogit**,
  [SVY] **svy estimation**, [XT] **xtcloglog**,
  [XT] **xtgee**, [XT] **xtlogit**, [XT] **xtnbreg**,
  [XT] **xtpoisson**, [XT] **xtprobit**
quality control
  charts, [R] **qc**, [R] **serrbar**
  plots, [G] **graph other**
quantile command, [R] **diagnostic plots**
quantile–normal plots, [R] **diagnostic plots**
quantile plots, [G] **graph other**, [R] **diagnostic plots**
quantile–quantile plots, [G] **graph other**,
  [R] **diagnostic plots**
quantile regression, [R] **qreg**
quantiles, [D] **pctile**, *also see* percentiles, *see*
  percentiles, displaying
quantiles, estat subcommand, [MV] **mds postestimation**
quarter( ) function, [D] **dates and times**,
  [D] **functions**, [M-5] **date( )**
quarterly( ) function, [D] **dates and times**,
  [D] **functions**, [M-5] **date( )**
quartimax rotation, [MV] **Glossary**, [MV] **rotate**,
  [MV] **rotatemat**
quartimin rotation, [MV] **Glossary**, [MV] **rotate**,
  [MV] **rotatemat**
Quattro Pro, reading data from, [U] **21.4 Transfer programs**
query,
  cluster subcommand, [MV] **cluster programming utilities**

query, *continued*
  estimates subcommand, [R] **estimates store**
  file subcommand, [P] **file**
  graph subcommand, [G] **graph query**
  icd9 subcommand, [D] **icd9**
  icd9p subcommand, [D] **icd9**
  mi subcommand, [MI] **mi describe**
  ml subcommand, [R] **ml**
  net subcommand, [R] **net**
  odbc subcommand, [D] **odbc**
  translator subcommand, [R] **translate**
  transmap subcommand, [R] **translate**
  update subcommand, [R] **update**
  webuse subcommand, [D] **webuse**
query, [M-3] **mata set**
query command, [R] **query**
query graphics, [G] **set graphics**, [G] **set printcolor**,
    [G] **set scheme**
query memory command, [D] **memory**
querybreakintr( ) function, [M-5] **setbreakintr( )**
quick reference, [D] **missing values**, [I] **data types**,
    [I] **estimation commands**, [I] **file extensions**,
    [I] **format**, [I] **immediate commands**,
    [I] **missing values**, [I] **postestimation**
    commands, [I] **prefix commands**, [I] **reading**
    data
quietly prefix, [P] **quietly**
quitting Stata, *see* exit command
quotes
  to delimit strings, [U] **18.3.5 Double quotes**
  to expand macros, [U] **18.3.1 Local macros**
quotes to expand macros, [P] **macro**

R

r( )
  function, [D] **functions**
  saved results, [R] **saved results**, [U] **18.8 Accessing**
    results calculated by other programs,
    [U] **18.10.1 Saving results in r()**
  scalars, macros, matrices, functions, [P] **discard**,
    [P] **return**
r(functions) macro extended function, [P] **macro**
r(macros) macro extended function, [P] **macro**
r(matrices) macro extended function, [P] **macro**
r(scalars) macro extended function, [P] **macro**
r1title( ) option, [G] *title_options*
r2title( ) option, [G] *title_options*
radians, [D] **functions**
raise to a power function, [U] **13.2.1 Arithmetic**
    operators
Ramsey test, [R] **regress postestimation**
random
  number function, [D] **functions**, [D] **generate**
  numbers, [M-5] **runiform( )**
    normally distributed, [D] **functions**, [D] **generate**
  order, test for, [R] **runtest**

random, *continued*
  sample, [D] **sample**, [R] **bootstrap**, [U] **21.3 If you**
    run out of memory
  variates, [M-5] **runiform( )**
  walk, [TS] **Glossary**
random-coefficients
  linear regression, [XT] **xtrc**
  model, [XT] **Glossary**
  postestimation, [XT] **xtrc postestimation**
random-effects model, [R] **anova**, [R] **loneway**,
    [XT] **Glossary**, [XT] **xtabond**, [XT] **xtcloglog**,
    [XT] **xtdpd**, [XT] **xtdpdsys**, [XT] **xtgee**,
    [XT] **xthtaylor**, [XT] **xtintreg**, [XT] **xtivreg**,
    [XT] **xtlogit**, [XT] **xtnbreg**, [XT] **xtpoisson**,
    [XT] **xtprobit**, [XT] **xtreg**, [XT] **xtregar**,
    [XT] **xttobit**
random-number seed, [MI] **mi impute**
range
  operators, [M-2] **op_range**
  subscripts, *see* subscripts
  vector, [M-5] **range( )**
range( ) function, [M-5] **range( )**
range chart, [R] **qc**
range command, [D] **range**
range of data, [D] **codebook**, [D] **inspect**, [R] **lv**,
    [R] **summarize**, [R] **table**, [XT] **xtsum**
range plots, [G] *rcap_options*
range spikes, [G] *rspike_options*
rangen( ) function, [M-5] **range( )**
rank, [M-5] **rank( )**, [M-6] **Glossary**
rank( ), egen function, [D] **egen**
rank( ) function, [M-5] **rank( )**
rank correlation, [R] **spearman**
rank-order statistics, [D] **egen**, [R] **signrank**,
    [R] **spearman**
rank-ordered logistic regression, [R] **rologit**
ranking data, [R] **rologit**
ranks of observations, [D] **egen**
ranksum command, [R] **ranksum**
Rao's canonical-factor method, [MV] **factor**
rarea, graph twoway subcommand, [G] **graph**
    twoway rarea
rate ratio, [ST] **epitab**, [ST] **stci**, [ST] **stir**,
    [ST] **stptime**, [ST] **stsum**, *see* incidence-rate ratio
ratio command, [R] **ratio postestimation**, [R] **ratio**,
    also see postestimation command
ratios,
  estimating, [R] **ratio**
  survey data, [SVY] **svy estimation**,
    [SVY] **svy: tabulate twoway**
raw data, [U] **12 Data**
.raw filename suffix, [U] **11.6 File-naming conventions**
rbar, graph twoway subcommand, [G] **graph twoway**
    rbar
rbeta( ) function, [D] **functions**, [M-5] **runiform( )**
rbinomial( ) function, [D] **functions**,
    [M-5] **runiform( )**
rc (return codes), *see* error messages and return codes

_rc built-in variable, [P] **capture**, [U] **13.4 System variables (_variables)**

rcap, graph twoway subcommand, [G] **graph twoway rcap**

rcapsym, graph twoway subcommand, [G] **graph twoway rcapsym**

R charts, [G] **graph other**

rchart command, [R] **qc**

rchi2() function, [D] **functions**, [M-5] **runiform( )**

r-class command, [P] **program**, [P] **return**, [U] **18.8 Accessing results calculated by other programs**

r-conformability, [M-5] **normal( )**, [M-6] **Glossary**

rconnected, graph twoway subcommand, [G] **graph twoway rconnected**

rdiscrete() function, [M-5] **runiform( )**

Re() function, [M-5] **Re( )**

read, file subcommand, [P] **file**

reading console input in programs, *see* console, obtaining input from

reading data, [I] **reading data**

reading data from disk, [U] **21 Inputting data**, [U] **21.4 Transfer programs**; [D] **infile**, [D] **infile (fixed format)**, [D] **infile (free format)**, [D] **infix (fixed format)**, [D] **insheet**, *also see* inputting data interactively; combining datasets

real, [M-2] **declarations**, [M-6] **Glossary**

real number to string conversion, [D] **functions**

real part, [M-5] **Re( )**

real() string function, [D] **functions**

recase() string function, [D] **functions**

recast() option, [G] *advanced_options*, [G] *rcap_options*, [G] *rspike_options*

recast command, [D] **recast**

receiver operating characteristic (ROC) analysis, [R] **logistic postestimation**, [R] **roc**, [R] **rocfit**, [R] **rocfit postestimation**

reciprocal averaging, [MV] **ca**

recode() function, [D] **functions**, [U] **25.1.2 Converting continuous variables to categorical variables**

recode command, [D] **recode**

recoding data, [D] **recode**

recoding data autocode() function, [D] **functions**

reconstructed correlations, [MV] **factor postestimation**

record I/O versus stream I/O, [U] **21 Inputting data**

recording sessions, [U] **15 Saving and printing output—log files**

recovariance, estat subcommand, [XT] **xtmelogit postestimation**, [XT] **xtmepoisson postestimation**, [XT] **xtmixed postestimation**

recruitment period, *see* accrual period

rectangularize dataset, [D] **fillin**

recursive
    estimation, [TS] **rolling**
    regression analysis, [TS] **Glossary**

recycled predictions, [R] **margins**

redisplay graph, [G] **graph display**

reexpression, [R] **boxcox**, [R] **ladder**, [R] **lnskew0**

.ref built-in class function, [P] **class**

.ref_n built-in class function, [P] **class**

references, class, [P] **class**

reflection, [MV] **Glossary**

reg3 command, [R] **reg3**, [R] **reg3 postestimation**, *also see* postestimation command

regexm() string function, [D] **functions**

regexr() string function, [D] **functions**

regexs() string function, [D] **functions**

regions
    look of, [G] *areastyle*
    outlining, [G] *region_options*
    shading, [G] *region_options*

register, mi subcommand, [MI] **mi set**

registered variables, *see* variables, registered

regress command, [R] **regress**, [R] **regress postestimation**, [R] **regress postestimation time series**, *also see* postestimation command

regression
    diagnostic plots, [G] **graph other**
    diagnostics, [R] **predict**; [R] **ladder**, [R] **logistic**, [R] **regress postestimation**, [R] **regress postestimation time series**
    function, estimating, [R] **lpoly**
    instrumental variables, [R] **gmm**
    lines, *see* fits
    scoring, [MV] **factor postestimation**

regression (in generic sense), *also see* estimation commands
    accessing coefficients and standard errors, [P] **matrix get**, [U] **13.5 Accessing coefficients and standard errors**
    dummy variables, with, [XT] **xtreg**
    fixed-effects, [XT] **xtreg**
    instrumental variables, [XT] **xtabond**, [XT] **xtdpd**, [XT] **xtdpdsys**, [XT] **xthtaylor**, [XT] **xtivreg**
    random-effects, [XT] **xtgee**, [XT] **xtreg**

regression,
    competing risks, [ST] **stcrreg**
    constrained, [R] **cnsreg**
    creating orthogonal polynomials for, [R] **orthog**
    dummy variables, with, [R] **anova**, [R] **areg**, [R] **xi**
    fixed-effects, [R] **areg**
    fractional polynomial, [R] **fracpoly**, [R] **mfp**
    graphing, [R] **logistic**, [R] **regress postestimation**
    grouped data, [R] **intreg**
    increasing number of variables allowed, [R] **matsize**
    instrumental variables, [R] **ivprobit**, [R] **ivregress**, [R] **ivtobit**, [R] **nlsur**
    linear, *see* linear regression
    system, [R] **mvreg**, [R] **reg3**, [R] **sureg**
    truncated, [R] **truncreg**

regular variables, *see* variables, regular

relational operator, [U] **13.2.3 Relational operators**

relative
    difference function, [D] **functions**
    efficiency, [MI] **Glossary**, [MI] **mi estimate**

relative, *continued*
    risk, [ST] **epitab**
    variance increase, [MI] **Glossary**, [MI] **mi estimate**
reldif() function, [D] **functions**, [M-5] **reldif( )**
release marker, [P] **version**
releases, compatibility of Stata programs across,
    [P] **version**
reliability, [MV] **factor**, [R] **alpha**, [R] **eivreg**,
    [R] **loneway**
reliability theory, *see* survival analysis
remainder function, [D] **functions**
REML (restricted maximum likelihood), [XT] **Glossary**
removing
    directories, [D] **rmdir**
        creating, [D] **mkdir**
    files, [D] **erase**
rename,
    char subcommand, [P] **char**
    cluster subcommand, [MV] **cluster utility**
    graph subcommand, [G] **graph rename**
    irf subcommand, [TS] **irf rename**
    matrix subcommand, [P] **matrix utility**
    mi subcommand, [MI] **mi rename**
rename, [M-3] **mata rename**
rename command, [D] **rename**
rename for mi data, [MI] **mi rename**
rename graph, [G] **graph rename**
renamevar, cluster subcommand, [MV] **cluster
    utility**
renaming variables, [MI] **mi rename**
renpfix command, [D] **rename**
renumber, notes subcommand, [D] **notes**
reordering data, [D] **sort**; [D] **order**, [D] **gsort**
reorganizing data, [D] **reshape**, [D] **xpose**
repeated measures, [MV] **Glossary**
    ANOVA, [R] **anova**
    MANOVA, [MV] **manova**
repeated options, [G] **concept: repeated options**
repeating and editing commands, [R] **#review**,
    [U] **10 Keyboard use**
repeating commands, [D] **by**, [P] **continue**, [P] **foreach**,
    [P] **forvalues**
replace, notes subcommand, [D] **notes**
replace command, [D] **generate**, [MI] **mi passive**,
    [MI] **mi xeq**
replace option, [U] **11.2 Abbreviation rules**
replace0, mi subcommand, [MI] **mi replace0**
replay, estimates subcommand, [R] **estimates
    replay**
replay() function, [D] **functions**; [P] **ereturn**,
    [P] **_estimates**
replicate-weight variable, [SVY] **Glossary**,
    [SVY] **survey**, [SVY] **svy brr**, [SVY] **svy
    jackknife**, [SVY] **svyset**
replicating
    clustered observations, [D] **expandcl**
    observations, [D] **expand**

replication method, [SVY] **svy brr**, [SVY] **svy
    jackknife**, [SVY] **svyset**, [SVY] **variance
    estimation**
report,
    datasignature subcommand, [D] **datasignature**
    duplicates subcommand, [D] **duplicates**
    ml subcommand, [R] **ml**
repost, ereturn subcommand, [P] **return**
_request(*macname*), display directive, [P] **display**
resampling, [SVY] **Glossary**
reserved
    names, [U] **11.3 Naming conventions**
    words, [M-2] **reswords**
reset, mi subcommand, [MI] **mi reset**
reset, translator subcommand, [R] **translate**
RESET test, [R] **regress postestimation**
reset_id, serset subcommand, [P] **serset**
reshape, mi subcommand, [MI] **mi reshape**
reshape command, [D] **reshape**
reshape for mi data, [MI] **mi reshape**
residual-versus-fitted and residual-versus-predicted plots,
    [G] **graph other**
residual-versus-fitted plot, [R] **regress postestimation**
residual-versus-predictor plot, [R] **regress
    postestimation**
residuals, [R] **logistic**, [R] **predict**, [R] **regress
    postestimation**, [R] **rreg**
residuals, estat subcommand, [MV] **factor
    postestimation**, [MV] **pca postestimation**
resistant smoothers, [R] **smooth**
restore,
    estimates subcommand, [R] **estimates store**
    _return subcommand, [P] **_return**
    snapshot subcommand, [D] **snapshot**
restore command, [P] **preserve**
restoring data, [D] **snapshot**
restricted cubic splines, [R] **mkspline**
results, clear subcommand, [D] **clear**
results,
    clearing, [P] **ereturn**, [P] **_estimates**, [P] **_return**
    listing, [P] **ereturn**, [P] **_estimates**, [P] **_return**
    returning, [P] **_return**, [P] **return**
    saved, [R] **saved results**
    saving, [R] **estimates save**; [P] **ereturn**,
        [P] **_estimates**, [P] **postfile**, [P] **_return**,
        [P] **return**
_return
    dir command, [P] **_return**
    drop command, [P] **_return**
    hold command, [P] **_return**
    restore command, [P] **_return**
return
    add command, [P] **return**
    clear command, [P] **return**
    local command, [P] **return**
    matrix command, [P] **return**
    scalar command, [P] **return**

return, [M-2] **return**

return() function, [D] **functions**

return codes, [P] **rmsg**, *see* error messages and return codes

return list command, [R] **saved results**

returning results, [P] **return**

class programs, [P] **class**

reventries, set subcommand, [R] **set**

reverse() string function, [D] **functions**

reversed scales, [G] *axis_scale_options*

#review command, [R] **#review**, [U] **10 Keyboard use**, [U] **15 Saving and printing output—log files**

revkeyboard, set subcommand, [R] **set**

revorder() function, [M-5] **invorder( )**

rgamma() function, [D] **functions**, [M-5] **runiform( )**

rhypergeometric() function, [D] **functions**, [M-5] **runiform( )**

ridge prior, [MI] **mi impute mvn**

right eigenvectors, [M-5] **eigensystem( )**

right suboption, [G] *justificationstyle*

rightmost options, [G] **concept: repeated options**

risk

difference, [ST] **epitab**

factor, [ST] **epitab**, [ST] **Glossary**

pool, [ST] **Glossary**, [ST] **stcox**, [ST] **stcrreg**, [ST] **stset**

ratio, [R] **binreg**, [ST] **epitab**, [ST] **Glossary**

rline, graph twoway subcommand, [G] **graph twoway rline**

rm command, [D] **erase**

_rmcoll command, [P] **_rmcoll**

_rmdcoll command, [P] **_rmcoll**

_rmdir() function, [M-5] **chdir( )**

rmdir() function, [M-5] **chdir( )**

rmdir command, [D] **rmdir**

rmexternal() function, [M-5] **findexternal( )**

rmsg, [P] **creturn**, [P] **error**, [U] **8 Error messages and return codes**

set subcommand, [P] **rmsg**

rmsg, set subcommand, [R] **set**

rnbinomial() function, [D] **functions**, [M-5] **runiform( )**

rnormal() function, [D] **functions**, [M-5] **runiform( )**

robust, Huber/White/sandwich estimator of variance, [P] **_robust**, [R] *vce_option*, [SVY] **variance estimation**, [XT] *vce_options*

alternative-specific conditional logit model, [R] **asclogit**

alternative-specific multinomial probit regression, [R] **asmprobit**

alternative-specific rank-ordered probit regression, [R] **asroprobit**

ARCH, [TS] **arch**

ARIMA and ARMAX, [TS] **arima**

bivariate probit regression, [R] **biprobit**

complementary log-log regression, [R] **cloglog**

conditional logistic regression, [R] **clogit**

robust, Huber/White/sandwich estimator of variance, *continued*

constrained linear regression, [R] **cnsreg**

Cox proportional hazards model, [ST] **stcox**, [ST] **stcrreg**

fit population-averaged panel-data models by using GEE, [XT] **xtgee**

fixed- and random-effects linear models, [XT] **xtreg**

generalized linear models, [R] **glm**

generalized linear models for binomial family, [R] **binreg**

generalized method of moments, [R] **gmm**

heckman selection model, [R] **heckman**

heteroskedastic probit model, [R] **hetprob**

instrumental-variables regression, [R] **ivregress**

interval regression, [R] **intreg**

linear dynamic panel-data estimation, [XT] **xtabond**, [XT] **xtdpd**, [XT] **xtdpdsys**

linear regression, [R] **regress**

linear regression with dummy-variable set, [R] **areg**

logistic regression, [R] **logistic**, [R] **logit**

logit and probit estimation for grouped data, [R] **glogit**

multinomial logistic regression, [R] **mlogit**

multinomial probit regression, [R] **mprobit**

negative binomial regression, [R] **nbreg**

nested logit regression, [R] **nlogit**

Newey–West regression, [TS] **newey**

nonlinear least-squares estimation, [R] **nl**

nonlinear systems of equations, [R] **nlsur**

ordered logistic regression, [R] **ologit**

ordered probit regression, [R] **oprobit**

parametric survival models, [ST] **streg**

Poisson regression, [R] **poisson**

population-averaged cloglog models, [XT] **xtcloglog**

population-averaged logit models, [XT] **xtlogit**

population-averaged negative binomial models, [XT] **xtnbreg**

population-averaged Poisson models, [XT] **xtpoisson**

population-averaged probit models, [XT] **xtprobit**

Prais–Winsten and Cochrane–Orcutt regression, [TS] **prais**

probit model with endogenous regressors, [R] **ivprobit**

probit model with sample selection, [R] **heckprob**

probit regression, [R] **probit**

rank-ordered logistic regression, [R] **rologit**

skewed logistic regression, [R] **scobit**

stereotype logistic regression, [R] **slogit**

tobit model, [R] **tobit**

tobit model with endogenous regressors, [R] **ivtobit**

treatment-effects model, [R] **treatreg**

truncated regression, [R] **truncreg**

zero-inflated negative binomial regression, [R] **zinb**

zero-inflated Poisson regression, [R] **zip**

zero-truncated negative binomial regression, [R] **ztnb**

zero-truncated Poisson regression, [R] **ztp**

robust, other methods of, [R] **qreg**, [R] **rreg**,
    [R] **smooth**
_robust command, [P] **_robust**
robust regression, [R] **regress**, [R] **rreg**, *also see* robust,
    Huber/White/sandwich estimator of variance
robust standard errors, [XT] **Glossary**
robust test for equality of variance, [R] **sdtest**
robvar command, [R] **sdtest**
ROC analysis, [G] **graph other**, [R] **logistic**
    postestimation, [R] **roc**, [R] **rocfit**, [R] **rocfit**
    postestimation
roccomp command, [R] **roc**
rocfit command, [R] **rocfit**, [R] **rocfit postestimation**,
    also see postestimation command
rocgold command, [R] **roc**
rocplot command, [R] **rocfit postestimation**
roctab command, [R] **roc**
Rogers and Tanimoto similarity measure,
    [MV] *measure_option*
roh, [R] **loneway**
rolling command, [TS] **rolling**
rolling regression, [TS] **Glossary**, [TS] **rolling**
rologit command, [R] **rologit**, [R] **rologit**
    postestimation, *also see* postestimation command
rootograms, [G] **graph other**, [R] **spikeplot**
roots of polynomials, [M-5] **polyeval( )**
rotate, estat subcommand, [MV] **canon**
    postestimation
rotate command, [MV] **factor postestimation**,
    [MV] **pca postestimation**, [MV] **rotate**,
    [MV] **rotatemat**
rotatecompare, estat subcommand, [MV] **canon**
    postestimation, [MV] **factor postestimation**,
    [MV] **pca postestimation**
rotated
    factor loadings, [MV] **factor postestimation**
    principal components, [MV] **pca postestimation**
rotation, [MV] **factor postestimation**, [MV] **Glossary**,
    [MV] **pca postestimation**, [MV] **rotate**,
    [MV] **rotatemat**
    procrustes, [MV] **procrustes**
    toward a target, [MV] **procrustes**, [MV] **rotate**,
    [MV] **rotatemat**
round( ) rounding function, [D] **functions**,
    [M-5] **trunc( )**
roundoff error, [U] **13.11 Precision and problems**
    therein; [M-5] **epsilon( )**, [M-5] **edittozero( )**,
    [M-5] **edittoint( )**
row-join operator, [M-2] **op_join**
row of matrix, selecting, [M-5] **select( )**
row operators for data, [D] **egen**
roweq, matrix subcommand, [P] **matrix rownames**
roweq macro extended function, [P] **macro**
rowfirst( ), egen function, [D] **egen**
rowfullnames macro extended function, [P] **macro**
rowlast( ), egen function, [D] **egen**
row-major order, [M-6] **Glossary**
rowmax( ), egen function, [D] **egen**

rowmax( ) function, [M-5] **minmax( )**
rowmaxabs( ) function, [M-5] **minmax( )**
rowmean( ), egen function, [D] **egen**
rowmedian( ), egen function, [D] **egen**
rowmin( ), egen function, [D] **egen**
rowmin( ) function, [M-5] **minmax( )**
rowminmax( ) function, [M-5] **minmax( )**
rowmiss( ), egen function, [D] **egen**
rowmissing( ) function, [M-5] **missing( )**
rownames, matrix subcommand, [P] **matrix**
    rownames
rownames macro extended function, [P] **macro**
rownonmiss( ), egen function, [D] **egen**
rownonmissing( ) function, [M-5] **missing( )**
rownumb( ) matrix function, [D] **functions**, [P] **matrix**
    define
rowpctile( ), egen function, [D] **egen**
rows( ) function, [M-5] **rows( )**
rows of matrix
    appending to, [P] **matrix define**
    names, [P] **ereturn**, [P] **matrix define**, [P] **matrix**
    rownames
    operators, [P] **matrix define**
rowscalefactors( ) function, [M-5] **_equilrc( )**
rowsd( ), egen function, [D] **egen**
rowshape( ) function, [M-5] **rowshape( )**
rowsof( ) matrix function, [D] **functions**, [P] **matrix**
    define
rowsum( ) function, [M-5] **sum( )**
rowtotal( ), egen function, [D] **egen**
rowvector, [M-2] **declarations**, [M-6] **Glossary**
Roy's
    largest root test, [MV] **canon**, [MV] **Glossary**,
    [MV] **manova**, [MV] **mvtest means**
    union-intersection test, [MV] **canon**, [MV] **manova**,
    [MV] **mvtest means**
rpoisson( ) function, [D] **functions**, [M-5] **runiform( )**
rreg command, [R] **rreg postestimation**, [R] **rreg**, *also*
    see postestimation command
rscatter, graph twoway subcommand, [G] **graph**
    twoway rscatter
rseed( ) function, [D] **functions**, [M-5] **runiform( )**
rspike, graph twoway subcommand, [G] **graph**
    twoway rspike
rt( ) function, [D] **functions**, [M-5] **runiform( )**
rtrim( ) string function, [D] **functions**
Rubin's combination rules, [MI] **mi estimate**, [MI] **mi**
    estimate using
run command, [R] **do**, [U] **16 Do-files**
runiform( ) function, [D] **functions**,
    [M-5] **runiform( )**, [R] **set seed**
_runningsum( ) function, [M-5] **runningsum( )**
runningsum( ) function, [M-5] **runningsum( )**
runtest command, [R] **runtest**
Russell and Rao coefficient similarity measure,
    [MV] *measure_option*
rvalue, class, [P] **class**
rvfplot command, [R] **regress postestimation**

RVI, *see* relative variance increase

rvpplot command, [R] **regress postestimation**

S

s()

  function, [D] **functions**

  saved results, [D] **functions**, [P] **return**, [R] **saved results**, [U] **18.8 Accessing results calculated by other programs**, [U] **18.10.3 Saving results in s()**

s(macros) macro extended function, [P] **macro**

s1color scheme, [G] **scheme s1**

s1manual scheme, [G] **scheme s1**

s1mono scheme, [G] **scheme s1**

s1rcolor scheme, [G] **scheme s1**

S_ macros, [R] **saved results**

s2color scheme, [G] **scheme s2**

s2gmanual scheme, [G] **scheme s2**

s2manual scheme, [G] **scheme s2**

s2mono scheme, [G] **scheme s2**

SAARCH, [TS] **arch**

Sammon mapping criterion, [MV] **Glossary**

sample, random, *see* random sample

sample command, [D] **sample**

sample size, [R] **sampsi**

  Cox proportional hazards regression, [ST] **stpower**, [ST] **stpower cox**

  exponential survival, [ST] **stpower**, [ST] **stpower exponential**

  exponential test, [ST] **stpower**, [ST] **stpower exponential**

  log-rank, [ST] **stpower**, [ST] **stpower logrank**

sampling, [D] **sample**, [R] **bootstrap**, [R] **bsample**, [SVY] **Glossary**, [SVY] **survey**, [SVY] **svydescribe**, [SVY] **svyset**, *also see* cluster sampling

  stage, [SVY] **Glossary**

  unit, [SVY] **Glossary**

  weight, [SVY] **Glossary**, [SVY] **survey**, [U] **11.1.6 weight**, [U] **20.18.3 Sampling weights**, with and without replacement, [SVY] **Glossary**

sampsi command, [R] **sampsi**

sandwich/Huber/White estimator of variance, *see* robust, Huber/White/sandwich estimator of variance,

sargan, estat subcommand, [XT] **xtabond postestimation**, [XT] **xtdpd postestimation**, [XT] **xtdpdsys postestimation**

Sargan test, [XT] **xtabond postestimation**, [XT] **xtdpd postestimation**, [XT] **xtdpdsys postestimation**

SAS, reading data from, [U] **21.4 Transfer programs**

SAS XPORT, [D] **fdasave**

save,

  estimates subcommand, [R] **estimates save**

  graph subcommand, [G] **graph save**

  label subcommand, [D] **label**

  snapshot subcommand, [D] **snapshot**

save estimation results, [P] **ereturn**, [P] **_estimates**

save command, [D] **save**

saved results, [P] **_return**, [P] **return**, [R] **saved results**, [U] **18.8 Accessing results calculated by other programs**, [U] **18.9 Accessing results calculated by estimation commands**, [U] **18.10 Saving results**

saveold command, [D] **save**

saving() option, [G] *saving_option*

saving data, [D] **outfile**, [D] **outsheet**, [D] **save**, [D] **snapshot**

saving results, [R] **estimates save**; [P] **ereturn**, [P] **_estimates**, [P] **postfile**, [P] **_return**, [P] **return**

scalar,

  confirm subcommand, [P] **confirm**

  ereturn subcommand, [P] **ereturn**

  return subcommand, [P] **return**

scalar, [M-2] **declarations**, [M-6] **Glossary**

scalar() function, [D] **functions**

scalars, [P] **scalar**

  namespace and conflicts, [P] **matrix**, [P] **matrix define**

scale() option, [G] *scale_option*

scalar command and scalar() pseudofunction, [P] **scalar**

scalar functions, [M-4] **scalar**

scale,

  log, [G] *axis_scale_options*

  range of, [G] *axis_scale_options*

  reversed, [G] *axis_scale_options*

scaling, [MV] **mds**, [MV] **mds postestimation**, [MV] **mdslong**, [MV] **mdsmat**

scatter, graph twoway subcommand, [G] **graph twoway scatter**

scatteri, graph twoway subcommand, [G] **graph twoway scatteri**

scatterplot matrices, [G] **graph matrix**

Scheffé multiple comparison test, [R] **oneway**

scheme() option, [G] *scheme_option*

scheme, set subcommand, [G] **set scheme**, [R] **set**

schemes, [G] *play_option*, [G] **schemes intro**; [G] **scheme economist**, [G] **scheme s1**, [G] **scheme s2**, [G] **scheme sj**, [G] *scheme_option*, [G] **set scheme**

  changing, [G] **graph display**

  creating your own, [G] **schemes intro**

  default, [G] **set scheme**

Schoenfeld residual, [ST] **stcox postestimation**, [ST] **streg postestimation**

Schur

  decomposition, [M-5] **schurd( )**, [M-6] **Glossary**

  form, [M-6] **Glossary**

_schurd() function, [M-5] **schurd( )**

schurd() function, [M-5] **schurd( )**

_schurdgroupby() function, [M-5] **schurd( )**

schurdgroupby() function, [M-5] **schurd( )**

_schurdgroupby_la() function, [M-5] **schurd( )**

_schurd_la() function, [M-5] **schurd( )**
Schwarz information criterion, *see* BIC
scientific notation, [U] **12.2 Numbers**
s-class command, [P] **program**, [P] **return**, [R] **saved results**, [U] **18.8 Accessing results calculated by other programs**
scobit command, [R] **scobit**, [R] **scobit postestimation**, *also see* postestimation command
scope, class, [P] **class**
score, [MV] **Glossary**
score, matrix subcommand, [P] **matrix score**
score, ml subcommand, [R] **ml**
score plot, [MV] **Glossary**, [MV] **scoreplot**
scoreplot command, [MV] **discrim lda postestimation**, [MV] **factor postestimation**, [MV] **pca postestimation**, [MV] **scoreplot**
scores, [R] **predict**
scores, obtaining, [U] **20.17 Obtaining scores**
scoring, [MV] **factor postestimation**, [MV] **pca postestimation**, [P] **matrix score**
scree plot, [MV] **Glossary**, [MV] **screeplot**
screeplot command, [MV] **discrim lda postestimation**, [MV] **factor postestimation**, [MV] **mca postestimation**, [MV] **pca postestimation**, [MV] **screeplot**
scrollbufsize, set subcommand, [R] **set**
scrolling of output, controlling, [P] **more**, [R] **more**
sd(), egen function, [D] **egen**
sd, estat subcommand, [SVY] **estat**
sdtest command, [R] **sdtest**
sdtesti command, [R] **sdtest**
se, estat subcommand, [R] **exlogistic postestimation**, [R] **expoisson postestimation**
_se[], [U] **13.5 Accessing coefficients and standard errors**
search
    help, [R] **hsearch**
    Internet, [R] **net search**
search,
    icd9 subcommand, [D] **icd9**
    icd9p subcommand, [D] **icd9**
    ml subcommand, [R] **ml**
    net subcommand, [R] **net**
    notes subcommand, [D] **notes**
    view subcommand, [R] **view**
search command, [R] **search**, [U] **4 Stata's help and search facilities**
search_d, view subcommand, [R] **view**
searchdefault, set subcommand, [R] **search**, [R] **set**
seasonal
    ARIMA, [TS] **tssmooth**, [TS] **tssmooth shwinters**
    difference operator, [TS] **Glossary**
    lag operator, [U] **11.4.4 Time-series varlists**
    smoothing, [TS] **tssmooth**, [TS] **tssmooth shwinters**
seconds() function, [D] **dates and times**, [D] **functions**, [M-5] **date( )**
seed, set subcommand, [R] **set**, [R] **set seed**
seek, file subcommand, [P] **file**

seemingly unrelated
    estimation, [R] **suest**
    regression, [R] **nlsur**, [R] **reg3**, [R] **sureg**
select() function, [M-5] **select( )**
select, mi subcommand, [MI] **mi select**
selection models, [R] **heckman**, [R] **heckprob**, [SVY] **svy estimation**
selection-order statistics, [TS] **varsoc**
semicolons, [M-2] **semicolons**
semiparametric
    imputation method, *see* imputation, predictive mean matching
    model, [ST] **Glossary**, [ST] **stcox**, [ST] **stcrreg**, [ST] **stset**
semirobust standard errors, [XT] **Glossary**
sensitivity, [R] **logistic**
    model, [R] **regress postestimation**, [R] **rreg**
separate command, [D] **separate**
separating string variables into parts, [D] **split**
seq(), egen function, [D] **egen**
sequential
    imputation, [MI] **mi impute**
    limit theory, [XT] **Glossary**
serial correlation, *see* autocorrelation
serial independence, test for, [R] **runtest**
serrbar command, [R] **serrbar**
serset
    clear command, [P] **serset**
    create command, [P] **serset**
    create_cspline command, [P] **serset**
    create_xmedians command, [P] **serset**
    dir command, [P] **serset**
    drop command, [P] **serset**
    reset_id command, [P] **serset**
    set command, [P] **serset**
    sort command, [P] **serset**
    summarize command, [P] **serset**
    use command, [P] **serset**
sersetread, file subcommand, [P] **serset**
sersetwrite, file subcommand, [P] **serset**
session, recording, [R] **log**, [U] **15 Saving and printing output—log files**
set
    adosize command, [P] **sysdir**, [R] **set**
    autotabgraphs command, [R] **set**
    checksum command, [D] **checksum**, [R] **set**
    charset command, [P] **smcl**
    command, [R] **query**
    conren command, [R] **set**
    copycolor command, [R] **set**
    dockable command, [R] **set**
    dockingguides command, [R] **set**
    doublebuffer command, [R] **set**
    dp command, [D] **format**
    dp command, [R] **set**
    emptycells command, [R] **set**, [R] **set emptycells**
    eolchar command, [R] **set**

set, *continued*

  fastscroll command, [R] **set**

  floatresults command, [R] **set**

  floatwindows command, [R] **set**

  graphics command, [R] **set**

  httpproxy command, [R] **netio**, [R] **set**

  httpproxyauth command, [R] **netio**, [R] **set**

  httpproxyhost command, [R] **netio**, [R] **set**

  httpproxyport command, [R] **netio**, [R] **set**

  httpproxypw command, [R] **netio**, [R] **set**

  httpproxyuser command, [R] **netio**, [R] **set**

  level command, [R] **level**, [R] **set**

  linegap command, [R] **set**

  linesize command, [R] **log**, [R] **set**

  locksplitters command, [R] **set**

  logtype command, [R] **log**, [R] **set**

  matacache command, [R] **set**

  matafavor command, [R] **set**

  matalibs command, [R] **set**

  matalnum command, [R] **set**

  matamofirst command, [R] **set**

  mataoptimize command, [R] **set**

  matastrict command, [R] **set**

  matsize command, [R] **matsize**, [R] **set**

  maxdb command, [R] **db**, [R] **set**

  maxiter command, [R] **maximize**, [R] **set**

  maxvar command, [D] **memory**

  maxvar command, [R] **set**

  memory command, [D] **memory**

  memory command, [R] **set**

  more command, [P] **more**, [R] **more**, [R] **set**

  notifyuser command, [R] **set**

  obs command, [D] **obs**

  obs command, [R] **set**

  odbcmgr command, [R] **set**

  output command, [P] **quietly** [R] **set**

  pagesize command, [R] **more**, [R] **set**

  persistfv command, [R] **set**

  persistvtopic command, [R] **set**

  pinnable command, [R] **set**

  playsnd command, [R] **set**

  printcolor command, [R] **set**

  processors command, [R] **set**

  reventries command, [R] **set**

  revkeyboard command, [R] **set**

  rmsg command, [P] **rmsg**, [R] **set**

  scheme command, [R] **set**

  scrollbufsize command, [R] **set**

  searchdefault command, [R] **search**, [R] **set**

  seed command, [R] **set seed**, [R] **set**

  smoothfonts command, [R] **set**

  timeout1 command, [R] **netio**, [R] **set**

  timeout2 command, [R] **netio**, [R] **set**

  trace command, [P] **trace**, [R] **set**

  tracedepth command, [P] **trace**, [R] **set**

  traceexpand command, [P] **trace**, [R] **set**

  tracehilite command, [P] **trace**, [R] **set**

set, *continued*

  traceindent command, [P] **trace**, [R] **set**

  tracenumber command, [P] **trace**, [R] **set**

  tracesep command, [P] **trace**, [R] **set**

  type command, [D] **generate**, [R] **set**

  update_interval command, [R] **set**, [R] **update**

  update_prompt command, [R] **set**, [R] **update**

  update_query command, [R] **set**, [R] **update**

  varabbrev command, [R] **set**

  varkeyboard command, [R] **set**

  varlabelpos command, [R] **set**

  virtual command, [D] **memory**

  virtual command, [R] **set**

set,

  cluster subcommand, [MV] **cluster programming utilities**

  datasignature subcommand, [D] **datasignature**

  file subcommand, [P] **file**

  irf subcommand, [TS] **irf set**

  mi subcommand, [MI] **mi set**

  serset subcommand, [P] **serset**

  sysdir subcommand, [P] **sysdir**

  translator subcommand, [R] **translate**

  webuse subcommand, [D] **webuse**

set, [M-3] **mata set**

set ado, net subcommand, [R] **net**

set graphics, [G] **set graphics**

set other, net subcommand, [R] **net**

set printcolor, [G] **set printcolor**

set scheme, [G] **schemes intro**, [G] **set scheme**

setbreakintr() function, [M-5] **setbreakintr( )**

set_defaults command, [R] **set_defaults**

setmore() function, [M-5] **more( )**

setmoreonexit() function, [M-5] **more( )**

setting M, [MI] **mi add**, [MI] **mi set**

setting mi data, [MI] **mi set**

settings,

  efficiency, [P] **creturn**

  graphics, [P] **creturn**

  memory, [P] **creturn**

  network, [P] **creturn**

  output, [P] **creturn**

  program debugging, [P] **creturn**

  trace, [P] **creturn**

sfrancia command, [R] **swilk**

shadestyle, [G] ***shadestyle***

shading region, [G] ***region_options***

Shapiro–Francia test for normality, [R] **swilk**

Shapiro–Wilk test for normality, [R] **swilk**

shared object, [P] **class**, [P] **plugin**

shell command, [D] **shell**

Shepard

  diagram, [MV] **Glossary**, [MV] **mds postestimation**

  plot, [MV] **mds postestimation**

shewhart command, [R] **qc**

shift, macro subcommand, [P] **macro**

shownrtolerance option, [R] **maximize**

showstep option, [R] **maximize**

showtolerance option, [R] **maximize**

shwinters, tssmooth subcommand, [TS] **tssmooth shwinters**

Šidák multiple comparison test, [R] **oneway**

sign() function, [D] **functions**, [M-5] **dsign( )**, [M-5] **sign( )**

signature of data, [D] **datasignature**, [P] **_datasignature**, [P] **signestimationsample**

signestimationsample command, [P] **signestimationsample**

significance levels, [R] **level**, [R] **query**, [U] **20.7 Specifying the width of confidence intervals**

signrank command, [R] **signrank**

signtest command, [R] **signrank**

signum function, [D] **functions**

similarity, [MV] **Glossary**

    matrices, [MV] **matrix dissimilarity**, [P] **matrix dissimilarity**

    measures, [MV] **cluster programming utilities**, [MV] **cluster**, [MV] **matrix dissimilarity**, [MV] *measure_option*, [P] **matrix dissimilarity**

        Anderberg coefficient, [MV] *measure_option*

        angular, [MV] *measure_option*

        correlation, [MV] *measure_option*

        Dice coefficient, [MV] *measure_option*

        Gower coefficient, [MV] *measure_option*

        Hamann coefficient, [MV] *measure_option*

        Jaccard coefficient, [MV] *measure_option*

        Kulczynski coefficient, [MV] *measure_option*

        matching coefficient, [MV] *measure_option*

    matrix, [MV] **Glossary**

    measure, [MV] **Glossary**

        Ochiai coefficient, [MV] *measure_option*

        Pearson coefficient, [MV] *measure_option*

        Rogers and Tanimoto coefficient, [MV] *measure_option*

        Russell and Rao coefficient, [MV] *measure_option*

        Sneath and Sokal coefficient, [MV] *measure_option*

        Yule coefficient, [MV] *measure_option*

simulate prefix command, [R] **simulate**

simulations, Monte Carlo, [P] **postfile**, [R] **simulate**; [R] **permute**

simultaneous

    quantile regression, [R] **qreg**

    systems, [R] **reg3**

sin() function, [D] **functions**, [M-5] **sin( )**

sine function, [D] **functions**

single-precision floating point number, [U] **12.2.2 Numeric storage types**

single-imputation methods, [MI] **intro substantive**

singlelinkage, cluster subcommand, [MV] **cluster linkage**

single-linkage clustering, [MV] **cluster**, [MV] **cluster linkage**, [MV] **clustermat**, [MV] **Glossary**

singleton-group data, [ST] **Glossary**, [ST] **stcox**

singleton strata, [SVY] **estat**, [SVY] **variance estimation**

singular value decomposition, [MV] **Glossary**, [P] **matrix svd**; [M-5] **svd( )**, [M-5] **fullsvd( )**

sinh() function, [D] **functions**, [M-5] **sin( )**

SITE directory, [P] **sysdir**, [U] **17.5 Where does Stata look for ado-files?**

size, estat subcommand, [SVY] **estat**

size of

    all text and markers, [G] *scale_option*

    graph, [G] *region_options*

        changing, [G] **graph display**

    markers, [G] *marker_options*

    objects, [G] *relativesize*

    text, [G] *textbox_options*

sizeof() function, [M-5] **sizeof( )**

SJ, *see* Stata Journal and Stata Technical Bulletin

sj, net subcommand, [R] **net**

sj scheme, [G] **scheme sj**

skew(), egen function, [D] **egen**

skewed logistic regression, [R] **scobit**, [SVY] **svy estimation**

skewness, [MV] **mvtest normality**, [R] **summarize**, [TS] **varnorm**; [R] **lnskew0**, [R] **lv**, [R] **sktest**, [R] **tabstat**

_skip(#), display directive, [P] **display**

sktest command, [R] **sktest**

sleep command, [P] **sleep**

slogit command, [R] **slogit**, [R] **slogit postestimation**, *also see* postestimation command

S_ macros, [P] **creturn**, [P] **macro**

Small Stata, [U] **5 Flavors of Stata**

smallestdouble() function, [D] **functions**, [M-5] **mindouble( )**

smc, estat subcommand, [MV] **factor postestimation**, [MV] **pca postestimation**

SMCL, *see* Stata Markup and Control Language

    SMCL tags, [G] *text*

.smcl filename suffix, [U] **11.6 File-naming conventions**

smooth command, [R] **smooth**

smoothers, [TS] **Glossary**, [TS] **tssmooth**, [TS] **tssmooth dexponential**, [TS] **tssmooth exponential**, [TS] **tssmooth hwinters**, [TS] **tssmooth ma**, [TS] **tssmooth nl**, [TS] **tssmooth shwinters**

smoothfonts, set subcommand, [R] **set**

smoothing, [G] **graph twoway lpoly**, [R] **lpoly**, [R] **smooth**

smoothing graphs, [G] **graph other**, [R] **kdensity**, [R] **lowess**

SMR, [ST] **epitab**, [ST] **Glossary**, [ST] **stptime**

snapshot, [D] **snapshot**

    data, [ST] **Glossary**, [ST] **snapspan**, [ST] **stset**

snapshot command, [D] **snapshot**

snapspan command, [ST] **snapspan**

Sneath and Sokel coefficient similarity measure, [MV] *measure_option*

soft missing value, [MI] **Glossary**

solve AX=B, [M-4] **solvers**, [M-5] **cholsolve( )**, [M-5] **lusolve( )**, [M-5] **qrsolve( )**, [M-5] **svsolve( )**, [M-5] **solvelower( )**, [M-5] **solve_tol( )**

_solvelower() function, [M-5] **solvelower( )**

solvelower() function, [M-5] **solvelower( )**

solve_tol() function, [M-5] **solve_tol( )**

_solvetolerance, [M-5] **solve_tol( )**

_solveupper() function, [M-5] **solvelower( )**

solveupper() function, [M-5] **solvelower( )**

_sort() function, [M-5] **sort( )**

sort() function, [M-5] **sort( )**

sort, serset subcommand, [P] **serset**

sort command, [D] **sort**

sort option, [G] *connect_options*

sort order, [D] **describe**, [P] **byable**, [P] **macro**, [P] **sortpreserve**

    for strings, [U] **13.2.3 Relational operators**

sortedby macro extended function, [P] **macro**

sortpreserve option, [P] **sortpreserve**

soundex() function, [D] **functions**, [M-5] **soundex( )**

soundex_nara() function, [D] **functions**, [M-5] **soundex( )**

source code, [M-1] **how**, [M-1] **source**, [M-6] **Glossary**

    view, [P] **viewsource**

Spearman–Brown prophecy formula, [R] **alpha**

spearman command, [R] **spearman**

Spearman's rho, [R] **spearman**

specialized graphs, [G] **graph other**

specification test, [R] **boxcox**, [R] **hausman**, [R] **linktest**, [R] **regress postestimation**, [XT] **xtreg postestimation**

specificity, [MV] **factor**, [R] **logistic**

spectral distribution, [TS] **cumsp**, [TS] **Glossary**, [TS] **pergram**

spectral distribution plots, cumulative, [G] **graph other**

spectrum, [TS] **Glossary**

spell data, [ST] **Glossary**

spherical covariance, [MV] **mvtest covariances**

sphericity, [MV] **Glossary**

Spiegelhalter's Z statistic, [R] **brier**

spike, graph twoway subcommand, [G] **graph twoway spike**

spike plot, [R] **spikeplot**

spikeplot command, [R] **spikeplot**

spline3() function, [M-5] **spline3( )**

spline3eval() function, [M-5] **spline3( )**

splines

    linear, [R] **mkspline**

    restricted cubic, [R] **mkspline**

split command, [D] **split**

split-plot designs, [MV] **manova**, [R] **anova**

splitting time-span records, [ST] **stsplit**

S-Plus, reading data from, [U] **21.4 Transfer programs**

spread, [R] **lv**

spreadsheets, transferring

    from Stata, [D] **outfile**, [D] **outsheet**, [D] **xmlsave**, [U] **21.4 Transfer programs**

    into Stata, [D] **infile**, [D] **infile (fixed format)**, [D] **infile (free format)**, [D] **insheet**, [D] **odbc**, [D] **xmlsave**, [U] **21 Inputting data**, [U] **21.4 Transfer programs**

sprintf() function, [M-5] **printf( )**

SPSS, reading data from, [U] **21.4 Transfer programs**

SQL, [D] **odbc**

sqlfile(), odbc subcommand, [D] **odbc**

sqreg command, [R] **qreg**, [R] **qreg postestimation**, *also see* postestimation command

sqrt() function, [D] **functions**, [M-5] **sqrt( )**

square

    matrix, [M-6] **Glossary**

    root, [M-5] **sqrt( )**, [M-5] **cholesky( )**

    root function, [D] **functions**

squared multiple correlations, [MV] **factor postestimation**

sreturn

    clear command, [P] **return**

    list command, [R] **saved results**

    local command, [P] **return**

SRS, [SVY] **Glossary**

ss() function, [D] **dates and times**, [D] **functions**, [M-5] **date( )**

ssC() function, [D] **dates and times**, [D] **functions**, [M-5] **date( )**

ssc

    copy command, [R] **ssc**

    describe command, [R] **ssc**

    hot command, [R] **ssc**

    install command, [R] **ssc**

    new command, [R] **ssc**

    type command, [R] **ssc**

    uninstall command, [R] **ssc**

SSC archive, [R] **ssc**

SSCP matrix, [MV] **Glossary**

sspace command, [TS] **sspace**

SSU, [SVY] **Glossary**

st, mi subcommand, [MI] **mi XXXset**

_st_addobs() function, [M-5] **st_addobs( )**

st_addobs() function, [M-5] **st_addobs( )**

_st_addvar() function, [M-5] **st_addvar( )**

st_addvar() function, [M-5] **st_addvar( )**

st commands for mi data, [MI] **mi stsplit**, [MI] **mi XXXset**

st data, [ST] **Glossary**

_st_data() function, [M-5] **st_data( )**

st_data() function, [M-5] **st_data( )**

st_dir() function, [M-5] **st_dir( )**

st_dropobsif() function, [M-5] **st_dropvar( )**

st_dropobsin() function, [M-5] **st_dropvar( )**

st_dropvar() function, [M-5] **st_dropvar( )**

st_eclear() function, [M-5] **st_rclear( )**

st_global() function, [M-5] **st_global( )**

st_isfmt() function, [M-5] **st_isfmt( )**

st_islmname() function, [M-5] **st_isname( )**
st_isname() function, [M-5] **st_isname( )**
st_isnumfmt() function, [M-5] **st_isfmt( )**
st_isnumvar() function, [M-5] **st_vartype( )**
st_isstrfmt() function, [M-5] **st_isfmt( )**
st_isstrvar() function, [M-5] **st_vartype( )**
st_keepobsif() function, [M-5] **st_dropvar( )**
st_keepobsin() function, [M-5] **st_dropvar( )**
st_keepvar() function, [M-5] **st_dropvar( )**
st_local() function, [M-5] **st_local( )**
_st_macroexpand() function,
      [M-5] **st_macroexpand( )**
st_macroexpand() function,
      [M-5] **st_macroexpand( )**
st_matrix() function, [M-5] **st_matrix( )**
st_matrixcolstripe() function, [M-5] **st_matrix( )**
st_matrixrowstripe() function, [M-5] **st_matrix( )**
st_nobs() function, [M-5] **st_nvar( )**
st_numscalar() function, [M-5] **st_numscalar( )**
st_nvar() function, [M-5] **st_nvar( )**
st_rclear() function, [M-5] **st_rclear( )**
st_replacematrix() function, [M-5] **st_matrix( )**
st_sclear() function, [M-5] **st_rclear( )**
_st_sdata() function, [M-5] **st_data( )**
st_sdata() function, [M-5] **st_data( )**
st_select() function, [M-5] **select( )**
_st_sstore() function, [M-5] **st_store( )**
st_sstore() function, [M-5] **st_store( )**
_st_store() function, [M-5] **st_store( )**
st_store() function, [M-5] **st_store( )**
st_strscalar() function, [M-5] **st_numscalar( )**
st_subview() function, [M-5] **st_subview( )**
st_tempfilename() function, [M-5] **st_tempname( )**
st_tempname() function, [M-5] **st_tempname( )**
_st_tsrevar() function, [M-5] **st_tsrevar( )**
st_tsrevar() function, [M-5] **st_tsrevar( )**
st_updata() function, [M-5] **st_updata( )**
st_varformat() function, [M-5] **st_varformat( )**
_st_varindex() function, [M-5] **st_varindex( )**
st_varindex() function, [M-5] **st_varindex( )**
st_varlabel() function, [M-5] **st_varformat( )**
st_varname() function, [M-5] **st_varname( )**
st_varrename() function, [M-5] **st_varrename( )**
st_vartype() function, [M-5] **st_vartype( )**
st_varvaluelabel() function, [M-5] **st_varformat( )**
st_view() function, [M-5] **st_view( )**
st_viewobs() function, [M-5] **st_viewvars( )**
st_viewvars() function, [M-5] **st_viewvars( )**
st_vldrop() function, [M-5] **st_vlexists( )**
st_vlexists() function, [M-5] **st_vlexists( )**
st_vlload() function, [M-5] **st_vlexists( )**
st_vlmap() function, [M-5] **st_vlexists( )**
st_vlmodify() function, [M-5] **st_vlexists( )**
st_vlsearch() function, [M-5] **st_vlexists( )**
stability, [TS] **var**, [TS] **var intro**, [TS] **var svar**,
      [TS] **vecstable**
  after VAR or SVAR, [TS] **varstable**

stability, *continued*
  after VEC, [TS] **vec** [TS] **vec intro**,
stack command, [D] **stack**
stacked variables, [MV] **Glossary**
stacking
  data, [D] **stack**
  variables, [MV] **Glossary**
stairstep, connecting points with, [G] *connectstyle*
standard deviations, *also see* subpopulation, standard
      deviations
  creating
    dataset of, [D] **collapse**
    variable containing, [D] **egen**
  displaying, [R] **lv**, [R] **summarize**, [R] **table**,
      [R] **tabulate, summarize()**, [XT] **xtsum**
  testing equality of, [R] **sdtest**
standard error bar charts, [G] **graph other**
standard errors, [SVY] **variance estimation**
standard errors,
  accessing, [P] **matrix get**, [U] **13.5 Accessing
      coefficients and standard errors**
  for general predictions, [R] **predictnl**
  forecast, [R] **predict**, [R] **regress postestimation**
  mean, [R] **ci**, [R] **mean**
  panel-corrected, [XT] **xtpcse**
  prediction, [R] **glm**, [R] **predict**, [R] **regress
      postestimation**
  residuals, [R] **predict**, [R] **regress postestimation**
  robust, *see* robust, Huber/White/sandwich estimator
      of variance, *see* robust
standard strata, *see* direct standardization
standard weights, *see* direct standardization
standardized
  data, [MV] **Glossary**
  means, [R] **mean**
  mortality ratio, *see* SMR
  proportions, [R] **proportion**
  rates, [R] **dstdize**, [ST] **epitab**
  ratios, [R] **ratio**
  residuals, [R] **binreg postestimation**, [R] **glm
      postestimation**, [R] **logistic postestimation**,
      [R] **logit postestimation**, [R] **predict**, [R] **regress
      postestimation**
standardized, variables, [D] **egen**
start() option, [G] **graph twoway histogram**
Stat/Transfer, [U] **21.4 Transfer programs**
Stata
  c()-class results, [M-5] **st_global( )**
  characteristics, [M-5] **st_global( )**, [M-5] **st_dir( )**
  description, [U] **2 A brief description of Stata**
  documentation, [U] **1 Read this—it will help**
  e()-class results, [M-5] **st_global( )**, [M-5] **st_dir( )**,
      [M-5] **st_rclear( )**
  error message, [M-5] **error( )**
  example datasets, [U] **1.2.2 Example datasets**
  execute command, [M-3] **mata stata**, [M-5] **stata( )**
  exiting, *see* exit command
  for Mac, *see* Mac

Stata, *continued*
> for Unix, *see* Unix
> for Windows, *see* Windows
> limits, [U] **5 Flavors of Stata**
> listserver, [U] **3.4 The Stata listserver**
> macros, [M-5] **st_global( )**, [M-5] **st_local( )**,
>   [M-5] **st_dir( )**
> matrices, [M-5] **st_matrix( )**, [M-5] **st_dir( )**
> NetCourseNow, [U] **3.7.1 NetCourses**
> NetCourses, [U] **3.7.1 NetCourses**
> op.varname, *see* Stata, time series–operated variable
> platforms, [U] **5.1 Platforms**
> r( )-class results, [M-5] **st_global( )**, [M-5] **st_dir( )**,
>   [M-5] **st_rclear( )**
> s( )-class results, [M-5] **st_global( )**, [M-5] **st_dir( )**,
>   [M-5] **st_rclear( )**
> scalars, [M-5] **st_numscalar( )**, [M-5] **st_dir( )**
> Small, *see* Small Stata
> Stata/IC, *see* Stata/IC
> Stata/MP, *see* Stata/MP
> Stata/SE, *see* Stata/SE
> supplementary material, [U] **3 Resources for**
>   **learning and using Stata**
> support, [U] **3 Resources for learning and using**
>   **Stata**
> temporary
>   filenames, [M-5] **st_tempname( )**
>   names, [M-5] **st_tempname( )**
> time-series–operated variable, [M-5] **st_tsrevar( )**,
>   [M-6] **Glossary**
> value labels, [M-5] **st_varformat( )**,
>   [M-5] **st_vlexists( )**
> variable
>   formats, [M-5] **st_varformat( )**
>   labels, [M-5] **st_varformat( )**
> web site, [U] **3.2 The Stata web site**
>   **(www.stata.com)**
Stata,
> data file format, technical description, [P] **file**
>   **formats .dta**
> exiting, *see* exit command
> pause, [P] **sleep**
stata, [M-3] **mata stata**
_stata( ) function, [M-5] **stata( )**
stata( ) function, [M-5] **stata( )**
STATA directory, [P] **sysdir**
Stata Journal, [G] **scheme sj**
Stata Journal and *Stata Technical Bulletin*, [U] **3.5 The**
>   **Stata Journal**
> installation of, [R] **net**, [R] **sj**, [U] **17.6 How do I**
>   **install an addition?**
> keyword search of, [R] **search**, [U] **4 Stata's help**
>   **and search facilities**
Stata logo, [G] **graph print**, [G] **pr_options**
Stata Markup and Control Language, [P] **smcl**;
>   [M-5] **display( )**, [M-5] **printf( )**,
>   [M-5] **errprintf( )**

Stata News, [U] **3 Resources for learning and using**
>   **Stata**
Stata Technical Bulletin Reprints, [U] **3.5 The Stata**
>   **Journal**
Stata/IC, [U] **5 Flavors of Stata**
Stata/MP, [U] **5 Flavors of Stata**
Stata/SE, [U] **5 Flavors of Stata**
stata.key file, [R] **search**
Statalist, [U] **3.4 The Stata listserver**
statasetversion( ) function, [M-5] **stataversion( )**
stataversion( ) function, [M-5] **stataversion( )**
state-space model, [TS] **arima**, [TS] **dfactor**,
>   [TS] **dfactor postestimation**, [TS] **Glossary**,
>   [TS] **sspace**, [TS] **sspace postestimation**
static, [M-2] **class**
stationary time series, [TS] **dfgls**, [TS] **dfuller**,
>   [TS] **pperron**, [TS] **var**, [TS] **var intro**,
>   [TS] **vec**, [TS] **vec intro**
statistical
> density functions, [M-5] **normal( )**
> distribution functions, [M-5] **normal( )**
Statistical Software Components (SSC) archive, [R] **ssc**
stats, estimates subcommand, [R] **estimates stats**
statsby prefix command, [D] **statsby**
STB, *see Stata Journal and Stata Technical Bulletin*,
stb, net subcommand, [R] **net**
stbase command, [ST] **stbase**
stci command, [ST] **stci**
stcox, fractional polynomials, [R] **fracpoly**, [R] **mfp**
stcox command, [ST] **stcox**
stcoxkm command, [ST] **stcox PH-assumption tests**
stcrreg command, [ST] **stcrreg**
st_ct, [ST] **st_is**
stcurve command, [ST] **stcurve**
std( ), egen function, [D] **egen**
stdescribe command, [ST] **stdescribe**
steady-state equilibrium, [TS] **Glossary**
steepest descent (ascent), [M-5] **moptimize( )**,
>   [M-5] **optimize( )**
stem command, [R] **stem**
stem-and-leaf displays, [R] **stem**
stepwise estimation, [R] **stepwise**
stepwise prefix command, [R] **stepwise**
.ster filename suffix, [U] **11.6 File-naming**
>   **conventions**
.ster files, [MI] **mi estimate**, [MI] **mi estimate using**
stereotype logistic regression, [R] **slogit**, [SVY] **svy**
>   **estimation**
stfill command, [ST] **stfill**
stgen command, [ST] **stgen**
.sthlp filename suffix, [U] **11.6 File-naming**
>   **conventions**
.sthlp files, [U] **4 Stata's help and search facilities**,
>   [U] **18.11.6 Writing online help**
stir command, [ST] **stir**
st_is 2, [ST] **st_is**
stjoin, mi subcommand, [MI] **mi stsplit**
stjoin command, [ST] **stsplit**

stjoin for mi data, [MI] **mi stsplit**
stmc command, [ST] **strate**
stmh command, [ST] **strate**
stochastic frontier
    model, [R] **frontier**, [XT] **xtfrontier**
    postestimation, [XT] **xtfrontier postestimation**
stopbox, window subcommand, [P] **window programming**
stopping command execution, [U] **10 Keyboard use**
stopping rules, [MV] **Glossary**
    adding, [MV] **cluster programming subroutines**
    Caliński and Harabasz index, [MV] **cluster**, [MV] **cluster stop**
    Duda and Hart index, [MV] **cluster**, [MV] **cluster stop**
    stepsize, [MV] **cluster programming subroutines**
storage types, [D] **codebook**, [D] **compress**, [D] **describe**, [D] **encode**, [D] **format**, [D] **generate**, [D] **recast**, [U] **12.2.2 Numeric storage types**, [U] **12.4.4 String storage types**; [U] **11.4 varlists**
store, estimates subcommand, [R] **estimates store**
storing and restoring estimation results, [R] **estimates store**
stphplot command, [ST] **stcox PH-assumption tests**
stpower
    cox command, [ST] **stpower cox**
    exponential command, [ST] **stpower exponential**
    logrank command, [ST] **stpower logrank**
stptime command, [ST] **stptime**
.stptrace filename suffix, [U] **11.6 File-naming conventions**
str#, [D] **data types**, [U] **12.4.4 String storage types**
str#, [I] **data types**
strata, estat subcommand, [SVY] **estat**
strata with one sampling unit, [SVY] **variance estimation**
strate command, [ST] **strate**
stratification, [R] **asclogit**, [R] **clogit**, [ST] **epitab**, [ST] **stcox**, [ST] **streg**
stratified sampling, [SVY] **Glossary**, [SVY] **survey**, [SVY] **svydescribe**, [SVY] **svyset**
stratified tables, [ST] **epitab**
stratum collapse, [SVY] **svydescribe**
stream I/O versus record I/O, [U] **21 Inputting data**
streg command, [ST] **streg**
streset, mi subcommand, [MI] **mi XXXset**
streset command for mi data, [MI] **mi XXXset**
stress, [MV] **Glossary**, [MV] **mds postestimation**
stress, estat subcommand, [MV] **mds postestimation**
strict stationarity, [TS] **Glossary**
string
    concatenation, [M-5] **invtokens( )**
    display formats, [I] **format**
    duplication, [M-5] **strdup( )**
    functions, [D] **functions**, [M-4] **string**

string, *continued*
    functions, expressions, and operators, [U] **12.4 Strings**, [U] **23 Working with strings**
    pattern matching, [M-5] **strmatch( )**
    to real, convert, [M-5] **strtoreal( )**
string, [M-2] **declarations**, [M-6] **Glossary**
string( ) string function, [D] **functions**
string variables, [D] **data types**, [D] **infile (free format)**, [I] **data types**, [U] **12.4 Strings**, [U] **23 Working with strings**
    converting to numbers, [D] **functions**
    encoding, [D] **encode**
    formatting, [D] **format**
    inputting, [D] **infile**, [U] **21 Inputting data**
    making from value labels, [D] **encode**
    mapping to numbers, [D] **destring**, [D] **encode**, [D] **label**
    parsing, [D] **infile (free format)**, [P] **gettoken**, [P] **tokenize**
    sort order, [U] **13.2.3 Relational operators**
    splitting into parts, [D] **split**
stritrim( ) function, [M-5] **strtrim( )**
strlen( ) function, [D] **functions**, [M-5] **strlen( )**
strlower( ) function, [D] **functions**, [M-5] **strupper( )**
strltrim( ) function, [D] **functions**, [M-5] **strtrim( )**
strmatch( ) function, [D] **functions**, [M-5] **strmatch( )**
strofreal( ) function, [D] **functions**, [M-5] **strofreal( )**
strongly balanced, [XT] **Glossary**
strpos( ) function, [D] **functions**, [M-5] **strpos( )**
strproper( ) function, [D] **functions**, [M-5] **strupper( )**
strreverse( ) function, [D] **functions**, [M-5] **strreverse( )**
strrtrim( ) function, [D] **functions**, [M-5] **strtrim( )**
strtoname( ) function, [D] **functions**, [M-5] **strtoname( )**
_strtoreal( ) function, [M-5] **strtoreal( )**
strtoreal( ) function, [M-5] **strtoreal( )**
strtrim( ) function, [D] **functions**, [M-5] **strtrim( )**
struct, [M-2] **struct**
structural time-series model, [TS] **sspace**
structural vector autoregression, *see* SVAR
structure, [MV] **Glossary**
structure, estat subcommand, [MV] **discrim lda postestimation**, [MV] **factor postestimation**
structures, [M-2] **struct**, [M-5] **liststruct( )**, [M-6] **Glossary**
strupper( ) function, [D] **functions**, [M-5] **strupper( )**
sts
    command, [ST] **sts**, [ST] **sts generate**, [ST] **sts graph**, [ST] **sts list**, [ST] **sts test**, [ST] **stset**
    generate command, [ST] **sts**, [ST] **sts generate**
    graph command, [ST] **sts**, [ST] **sts graph**
    list command, [ST] **sts**, [ST] **sts list**
    test command, [ST] **sts**, [ST] **sts test**
stset, mi subcommand, [MI] **mi XXXset**
stset command, [ST] **stset**

stset command for mi data, [MI] **mi XXXset**

st_show, [ST] **st_is**

stsplit command, [ST] **stsplit**

stsplit for mi data, [MI] **mi stsplit**

stsplit, mi subcommand, [MI] **mi stsplit**

stsum command, [ST] **stsum**

sttocc command, [ST] **sttocc**

studentized residuals, [R] **predict**, [R] **regress postestimation**

Student's *t* distribution

    cdf, [D] **functions**

    confidence interval for mean, [R] **ci**, [R] **mean**

    testing equality of means, [R] **ttest**

stvary command, [ST] **stvary**

style,

    added line, [G] *addedlinestyle*

    area, [G] *areastyle*

    axis, [G] *tickstyle*

    by-graphs, [G] *bystyle*

    clock position, [G] *clockposstyle*

    color, [G] *colorstyle*

    compass direction, [G] *compassdirstyle*

    connect points, [G] *connectstyle*

    flong, *see* flong

    flongsep, *see* flongsep

    grid lines, [G] *gridstyle*

    legends, [G] *legendstyle*

    lines, [G] *linepatternstyle*, [G] *linestyle*, [G] *linewidthstyle*

    lists, [G] *stylelists*

    margins, [G] *marginstyle*

    marker labels, [G] *markerlabelstyle*, [G] *markersizestyle*, [G] *markerstyle*

    markers, [G] *symbolstyle*

    mlong, *see* mlong

    plot, [G] *pstyle*

    text, [G] *textsizestyle*, [G] *textstyle*

    text display angle, [G] *anglestyle*

    text justification, [G] *justificationstyle*

    textboxes, [G] *orientationstyle*, [G] *textboxstyle*

    vertical alignment of text, [G] *alignmentstyle*

    wide, *see* wide

style, [MI] **Glossary**, [MI] **mi convert**, [MI] **styles**

stylelist, [G] *stylelists*

subclass, [M-2] **class**

subdirectories, [U] **11.6 File-naming conventions**

subhazard, [ST] **Glossary**

subhazard ratio, [ST] **Glossary**

subinertia, estat subcommand, [MV] **mca postestimation**

subinstr macro extended function, [P] **macro**

subinstr() function, [D] **functions**, [M-5] **subinstr( )**

subinword() function, [D] **functions**, [M-5] **subinstr( )**

_sublowertriangle() function, [M-5] **sublowertriangle( )**

sublowertriangle() function, [M-5] **sublowertriangle( )**

subpopulation

    differences, [SVY] **survey**, [SVY] **svy postestimation**

    estimation, [SVY] **Glossary**, [SVY] **subpopulation estimation**, [SVY] **svy estimation**

    means, [SVY] **svy estimation**

    proportions, [SVY] **svy estimation**, [SVY] **svy: tabulate oneway**, [SVY] **svy: tabulate twoway**

    ratios, [SVY] **svy estimation**, [SVY] **svy: tabulate oneway**, [SVY] **svy: tabulate twoway**

    standard deviations, [SVY] **estat**

    totals, [SVY] **svy estimation**, [SVY] **svy: tabulate oneway**, [SVY] **svy: tabulate twoway**

subroutines, adding, [MV] **cluster programming utilities**

subscripting matrices, [P] **matrix define**

subscripts, [G] *text*, [M-2] **subscripts**, [M-6] **Glossary**

subscripts in expressions, [U] **13.7 Explicit subscripting**

_substr() function, [M-5] **_substr( )**

substr() function, [D] **functions**, [M-5] **substr( )**

substring function, [D] **functions**

subtitle() option, [G] *title_options*

subtraction, [M-2] **op_arith**, [M-2] **op_colon**

subtraction operator, *see* arithmetic operators

suest command, [R] **suest**, [SVY] **svy postestimation**; [R] **hausman**

sum() function, [D] **functions**, [M-5] **sum( )**

.sum filename suffix, [U] **11.6 File-naming conventions**

sum of vector, [M-5] **runningsum( )**

summarize,

    estat subcommand, [MV] **ca postestimation**, [MV] **discrim estat**, [MV] **discrim knn postestimation**, [MV] **discrim lda postestimation**, [MV] **discrim logistic postestimation**, [MV] **discrim qda postestimation**, [MV] **factor postestimation**, [MV] **mca postestimation**, [MV] **mds postestimation**, [MV] **pca postestimation**, [MV] **procrustes postestimation**, [R] **estat**

    misstable subcommand, [R] **misstable**

    serset subcommand, [P] **serset**

summarize command, [D] **format**, [R] **summarize**; [R] **tabulate, summarize()**

summarizing data, [D] **codebook**, [D] **inspect**, [R] **summarize**, [R] **tabstat**, [SVY] **svy: tabulate twoway**, [XT] **xtsum**; [R] **lv**, [R] **table**, [R] **tabulate oneway**, [R] **tabulate, summarize()**, [R] **tabulate twoway**

summary statistics, *see* descriptive statistics, displaying, *see* descriptive statistics

summary variables, generating, [MV] **cluster generate**

summative (Likert) scales, [R] **alpha**

sums,
  creating dataset containing, [D] **collapse**
  over observations, [D] **egen**, [D] **functions**,
    [R] **summarize**
  over variables, [D] **egen**
sunflower command, [R] **sunflower**
sunflower plots, [R] **sunflower**
Super, class prefix operator, [P] **class**
super-varying variables, [MI] **Glossary**, [MI] **mi**
    varying
.**superclass** built-in class function, [P] **class**
superscripts, [G] *text*
supplementary
  rows or columns, [MV] **Glossary**
  variables, [MV] **Glossary**
support of Stata, [U] **3 Resources for learning and**
    using Stata
suppressing
  graphs, [G] *nodraw_option*
  terminal output, [P] **quietly**
SUR, [TS] **dfactor**
sureg command, [R] **sureg**, [R] **sureg postestimation**,
    also see postestimation command
survey
  data, [MI] **intro substantive**, [SVY] **Glossary**,
    [SVY] **survey**, [SVY] **svydescribe**,
    [SVY] **svyset**, [U] **26.19 Survey data**
  design, [SVY] **Glossary**
  postestimation, [SVY] **svy postestimation**
  prefix command, [SVY] **svy**
  sampling, [ST] **stcox**, [ST] **streg**, [ST] **sts**,
    [SVY] **survey**, [SVY] **svydescribe**,
    [SVY] **svyset**, *see* cluster sampling
survival analysis, [G] **graph other**, [R] **expoisson**,
    [R] **fracpoly**, [R] **gllamm**, [R] **glm**,
    [R] **intreg**, [R] **nbreg**, [R] **oprobit**,
    [R] **poisson**, [R] **zip**, [R] **ztp**, [ST] **Glossary**,
    [ST] **cttost**, [ST] **st**, [ST] **stcox**, [ST] **stcox**
    PH-assumption tests, [ST] **stdescribe**,
    [ST] **streg**, [ST] **stset**, [ST] **sttoct**, [SVY] **svy**
    estimation, [XT] **xtnbreg**, [XT] **xtpoisson**,
    [U] **26.16 Survival-time (failure-time) models**,
    also see Cox proportional hazards model
survival clinical trial, [ST] **stpower**
survival models, [SVY] **svy estimation**
survival-time data, *see* survival analysis
survivor function, [ST] **Glossary**, [ST] **sts**, [ST] **sts**
    generate, [ST] **sts list**, [ST] **sts test**
  graph of, [ST] **stcurve**, [ST] **sts graph**
survivor functions, [G] **graph other**
SVAR, [TS] **Glossary**, [TS] **var intro**, [TS] **var svar**
  postestimation, [R] **regress postestimation time**
    series, [TS] **var svar postestimation**; [TS] **fcast**
    compute, [TS] **fcast graph**, [TS] **irf**, [TS] **irf**
    create, [TS] **vargranger**, [TS] **varlmar**,
    [TS] **varnorm**, [TS] **varsoc**, [TS] **varstable**,
    [TS] **varwle**
svar command, [TS] **var svar**, [TS] **var svar**
    postestimation

SVD, [MV] **Glossary**, *see* singular value decomposition
_**svd**() function, [M-5] **svd( )**
svd() function, [M-5] **svd( )**
svd, matrix subcommand, [P] **matrix svd**
_**svd_la**() function, [M-5] **svd( )**, [M-5] **fullsvd( )**
_**svdsv**() function, [M-5] **svd( )**
svdsv() function, [M-5] **svd( )**
svmat command, [P] **matrix mkmat**
_**svsolve**() function, [M-5] **svsolve( )**
svsolve() function, [M-5] **svsolve( )**
svy: biprobit command, [SVY] **svy estimation**
svy: clogit command, [SVY] **svy estimation**
svy: cloglog command, [SVY] **svy estimation**
svy: cnsreg command, [SVY] **svy estimation**
svy: glm command, [SVY] **svy estimation**
svy: gnbreg command, [SVY] **svy estimation**
svy: heckman command, [SVY] **svy estimation**
svy: heckprob command, [SVY] **svy estimation**
svy: hetprob command, [SVY] **svy estimation**
svy: intreg command, [SVY] **svy estimation**
svy: ivprobit command, [SVY] **svy estimation**
svy: ivregress command, [SVY] **svy estimation**
svy: ivtobit command, [SVY] **svy estimation**
svy: logistic command, [SVY] **svy estimation**,
    [SVY] **svy postestimation**
svy: logit command, [SVY] **svy estimation**
svy: mean command, [SVY] **estat**,
    [SVY] **poststratification**, [SVY] **subpopulation**
    estimation, [SVY] **survey**, [SVY] **svy**,
    [SVY] **svy estimation**, [SVY] **svy**
    postestimation, [SVY] **svydescribe**,
    [SVY] **svyset**
svy: mlogit command, [SVY] **svy estimation**
svy: mprobit command, [SVY] **svy estimation**
svy: nbreg command, [SVY] **svy estimation**
svy: nl command, [SVY] **svy estimation**
svy: ologit command, [SVY] **svy estimation**,
    [SVY] **svy postestimation**
svy: oprobit command, [SVY] **svy estimation**
svy: poisson command, [SVY] **svy estimation**
svy: probit command, [SVY] **svy estimation**
svy: proportion command, [SVY] **svy estimation**
svy: ratio command, [SVY] **direct standardization**,
    [SVY] **svy brr**, [SVY] **svy estimation**,
    [SVY] **svy: tabulate twoway**
svy: regress command, [SVY] **survey**, [SVY] **svy**,
    [SVY] **svy estimation**, [SVY] **svy jackknife**,
    [SVY] **svy postestimation**
svy: scobit command, [SVY] **svy estimation**
svy: slogit command, [SVY] **svy estimation**
svy: stcox command, [SVY] **svy estimation**
svy: streg command, [SVY] **svy estimation**
svy: tabulate command, [SVY] **svy: tabulate**
    oneway, [SVY] **svy: tabulate twoway**
svy: tobit command, [SVY] **svy estimation**
svy: total command, [SVY] **svy brr**, [SVY] **svy**
    estimation
svy: treatreg command, [SVY] **svy estimation**

svy: truncreg command, [SVY] **svy estimation**

svy: zinb command, [SVY] **svy estimation**

svy: zip command, [SVY] **svy estimation**

svy: ztnb command, [SVY] **svy estimation**

svy: ztp command, [SVY] **svy estimation**

svy brr prefix command, [SVY] **svy brr**

svy jackknife prefix command, [SVY] **svy jackknife**

svydescribe command, [SVY] **survey**,
[SVY] **svydescribe**

svymarkout command, [P] **mark**, [SVY] **svymarkout**

svyset, estat subcommand, [SVY] **estat**

svyset, mi subcommand, [MI] **mi XXXset**

svy prefix command, [SVY] **svy**

svyset command, [SVY] **survey**, [SVY] **svyset**

svyset command for mi data, [MI] **mi XXXset**

swap() function, [M-5] **swap( )**

swap, update subcommand, [R] **update**

sweep() matrix function, [D] **functions**, [P] **matrix define**

swilk command, [R] **swilk**

switching styles, [MI] **mi convert**

Sybase, reading data from, [U] **21.4 Transfer programs**

symbolic forms, [R] **anova**

symbolpalette, palette subcommand, [G] **palette**

symbols, [G] **text**, *also see* markers

symbolstyle, [G] ***symbolstyle***

symeigen, matrix subcommand, [P] **matrix symeigen**

_symeigen_la() function, [M-5] **eigensystem( )**

_symeigensystem() function, [M-5] **eigensystem( )**

symeigensystem() function, [M-5] **eigensystem( )**

_symeigensystem_select() functions,
[M-5] **eigensystemselect( )**

symeigensystemselect*() functions,
[M-5] **eigensystemselect( )**

_symeigenvalues() function, [M-5] **eigensystem( )**

symeigenvalues() function, [M-5] **eigensystem( )**

symmetric matrices, [M-5] **issymmetric( )**,
[M-5] **makesymmetric( )**, [M-6] **Glossary**

symmetriconly, [M-6] **Glossary**

symmetry, test of, [R] **symmetry**

symmetry command, [R] **symmetry**

symmetry plots, [G] **graph other**, [R] **diagnostic plots**

symmi command, [R] **symmetry**

symplot command, [R] **diagnostic plots**

syntax, [M-2] **syntax**

syntax command, [P] **syntax**

syntax diagrams explained, [R] **intro**

syntax of Stata's language, [P] **syntax**,
[U] **11 Language syntax**

sysdir
command, [U] **17.5 Where does Stata look for ado-files?**

list command, [P] **sysdir**

macro extended function, [P] **macro**

set command, [P] **sysdir**

sysmiss, *see* missing values

Systat, reading data from, [U] **21.4 Transfer programs**

system
estimators, [R] **gmm**, [R] **ivregress**, [R] **nlsur**,
[R] **reg3**, [R] **sureg**

limits, [P] **creturn**

parameters, [P] **creturn**, [R] **query**,
[R] **set_defaults**, [R] **set**

values, [P] **creturn**

variables, [U] **13.4 System variables (_variables)**

system1 command, [R] **gmm**

sysuse command, [D] **sysuse**

szroeter, estat subcommand, [R] **regress postestimation**

Szroeter's test for heteroskedasticity, [R] **regress postestimation**

T

t distribution
cdf, [D] **functions**

confidence interval for mean, [R] **ci**, [R] **mean**

testing equality of means, [R] **ttest**

%t formats, [D] **format**

%t values and formats, [D] **dates and times**

t1title() option, [G] ***title_options***

t2title() option, [G] ***title_options***

tab
characters, show, [D] **type**

expansion of variable names, [U] **10.6 Tab expansion of variable names**

tab1 command, [R] **tabulate oneway**

tab2 command, [R] **tabulate twoway**

tabdisp command, [P] **tabdisp**

tabi command, [R] **tabulate twoway**

table, estimates subcommand, [R] **estimates table**

table, frequency, *see* frequency table

table command, [MV] **ca postestimation**, [R] **table**

tables, [TS] **irf ctable**, [TS] **irf table**

actuarial, *see* life tables

contingency, [R] **table**, [R] **tabulate twoway**,
[SVY] **svy: tabulate twoway**

epidemiological, *see* epidemiological tables

failure, *see* failure tables

formatting numbers in, [D] **format**

fourfold, *see* fourfold tables

frequency, [R] **tabulate oneway**, [R] **tabulate twoway**; [R] **table**, [R] **tabstat**, [R] **tabulate, summarize()**, [SVY] **svy: tabulate oneway**,
[SVY] **svy: tabulate twoway**

hazard, *see* hazard tables

life, *see* life tables

N-way, [P] **tabdisp**, [R] **table**

of means, [R] **table**, [R] **tabulate, summarize()**

of statistics, [P] **tabdisp**, [R] **table**, [R] **tabstat**

printing, [U] **15 Saving and printing output—log files**

tables of estimation results, [R] **estimates table**

tabodds command, [ST] **epitab**

tabstat command, [R] **tabstat**

tabulate
> one-way, [SVY] **svy: tabulate oneway**
> two-way, [SVY] **svy: tabulate twoway**

tabulate command, [R] **tabulate oneway**,
> [R] **tabulate twoway**

summarize(), [R] **tabulate, summarize()**

tag, duplicates subcommand, [D] **duplicates**

tag(), egen function, [D] **egen**

tan() function, [D] **functions**, [M-5] **sin( )**

tangent function, [D] **functions**

tanh() function, [D] **functions**, [M-5] **sin( )**

TARCH, [TS] **arch**

target rotation, [MV] **Glossary**, [MV] **procrustes**,
> [MV] **rotate**, [MV] **rotatemat**

tau, [R] **spearman**

taxonomy, [MV] **Glossary**

Taylor linearization, *see* linearized variance estimator

tC() pseudofunction, [D] **dates and times**,
> [D] **functions**

tc() pseudofunction, [D] **dates and times**,
> [D] **functions**

td() pseudofunction, [D] **dates and times**,
> [D] **functions**

tden() function, [D] **functions**, [M-5] **normal( )**

TDT test, [R] **symmetry**

technical support, [U] **3.9 Technical support**

technique() option, [R] **maximize**

tempfile
> command, [P] **macro**
> macro extended function, [P] **macro**

tempname, class, [P] **class**

tempname
> command, [P] **macro**, [P] **matrix**, [P] **scalar**
> macro extended function, [P] **macro**

temporary
> files, [P] **macro**, [P] **preserve**, [P] **scalar**
> names, [P] **macro**, [P] **matrix**, [P] **scalar**,
> > [U] **18.7.2 Temporary scalars and matrices**
> variables, [P] **macro**, [U] **18.7.1 Temporary
> > variables**

tempvar
> command, [P] **macro**
> macro extended function, [P] **macro**

termcap(5), [U] **10 Keyboard use**

terminal
> obtaining input from, [P] **display**
> suppressing output, [P] **quietly**

terminfo(4), [U] **10 Keyboard use**

test,
> association, *see* association test
> Breitung, [XT] **xtunitroot**
> Breusch–Pagan Lagrange multiplier, *see* Breusch–
> > Pagan Lagrange multiplier test
> Cox proportional hazards model, assumption,
> > *see* Cox proportional hazards model, test of
> > assumption
> Dickey–Fuller, [TS] **dfgls**, [TS] **dfuller**

test, *continued*
> equality of survivor functions, *see* equality test,
> > survivor functions
> Fisher-type, [XT] **xtunitroot**
> Fisher's exact, *see* Fisher's exact test
> granger causality, [TS] **vargranger**
> Hadri Lagrange multiplier stationarity,
> > [XT] **xtunitroot**
> Harris–Tzavalis, [XT] **xtunitroot**
> Hausman specification, *see* Hausman specification
> > test
> heterogeneity, *see* heterogeneity test
> homogeneity, *see* homogeneity test
> Im–Pesaran–Shin, [XT] **xtunitroot**
> independence, *see* independence test
> Lagrange-multiplier, [TS] **varlmar**, [TS] **veclmar**
> Levin–Lin–Chu, [XT] **xtunitroot**
> log-rank, *see* log-rank test
> Mantel–Haenszel, *see* Mantel–Haenszel test
> McNemar's chi-squared test, *see* McNemar's chi-
> > squared test
> model specification, *see* model specification test
> normality, [TS] **varnorm**, [TS] **vecnorm**
> trend, *see* trend test
> unit root, [XT] **xtunitroot**
> Wald, [TS] **vargranger**, [TS] **varwle**

test, sts subcommand, [ST] **sts test**

test command, [R] **anova postestimation**, [R] **test**,
> [SVY] **svy postestimation**, [U] **20.11 Performing
> hypothesis tests on the coefficients**;
> [SVY] **survey**

test-based confidence intervals, [ST] **epitab**

testnl command, [R] **testnl**, [SVY] **svy
> postestimation**

testparm command, [R] **test**, [SVY] **svy
> postestimation**

tests,
> ARCH effect, [R] **regress postestimation time
> > series**
> association, [R] **tabulate twoway**, [ST] **epitab**,
> > [SVY] **svy: tabulate twoway**
> binomial probability, [R] **bitest**
> Breusch–Pagan, [R] **mvreg**, [R] **sureg**
> differences of two means, [SVY] **svy postestimation**
> equality of
> > coefficients, [R] **test**, [R] **testnl**
> > distributions, [R] **ksmirnov**, [R] **kwallis**,
> > > [R] **ranksum**, [R] **signrank**
> > means, [R] **ttest**
> > medians, [R] **ranksum**
> > proportions, [R] **bitest**, [R] **prtest**
> > variance, [R] **sdtest**
> equality of coefficients, [SVY] **svy postestimation**
> equality of means, [SVY] **svy postestimation**
> equivalence, [R] **pk**, [R] **pkequiv**
> heteroskedasticity, [R] **regress postestimation**
> independence, [R] **tabulate twoway**, [ST] **epitab**,
> > [SVY] **svy: tabulate twoway**

tests, *continued*
  independence of irrelevant alternatives, *see* IIA
  internal consistency, [R] **alpha**
  interrater agreement, [R] **kappa**
  kurtosis, [R] **regress postestimation**, [R] **sktest**
  likelihood-ratio, [R] **lrtest**
  linear hypotheses after estimation, [R] **test**
  marginal homogeneity, [R] **symmetry**
  model coefficients, [R] **lrtest**, [R] **test**, [R] **testnl**,
    [SVY] **svy postestimation**
  model specification, [R] **hausman**, [R] **linktest**
  multivariate, [MV] **mvtest**
  nonlinear, [SVY] **svy postestimation**
  nonlinear hypotheses after estimation, [R] **testnl**
  normality, [MV] **mvtest normality**, [R] **boxcox**,
    [R] **ladder**, [R] **sktest**, [R] **swilk**
  permutation, [R] **permute**
  serial correlation, [R] **regress postestimation time
    series**
  serial independence, [R] **runtest**
  skewness, [R] **regress postestimation**
  symmetry, [R] **symmetry**
  TDT, [R] **symmetry**
  trend, [R] **nptrend**, [R] **symmetry**
  variance-comparison, [MV] **mvtest covariances**,
    [R] **sdtest**
  Wald, [SVY] **svy postestimation**
tests after estimation, *see* estimation, tests after
tetrachoric command, [R] **tetrachoric**
tetrachoric correlation, [MV] **Glossary**
text
  adding, [G] *added_text_options*
  and textboxes, relationship between, [G] *textstyle*
  angle of, [G] *anglestyle*
  captions, [G] *title_options*
  look of, [G] *textboxstyle*, [G] *textstyle*
  note, [G] *title_options*
  resizing, [G] *scale_option*
  running outside of borders, [G] *added_text_options*
  size of, [G] *textbox_options*
  subtitle, [G] *title_options*
  title, [G] *title_options*
  vertical alignment, [G] *alignmentstyle*
text() option, [G] *added_text_options*,
  [G] *aspect_option*
text, in graphs, [G] *text*
textboxes, [G] *textbox_options*
  orientation of, [G] *orientationstyle*
textboxstyle, [G] *textboxstyle*
textsizestyle, [G] *textsizestyle*
textstyle, [G] *textstyle*
th() pseudofunction, [D] **dates and times**,
  [D] **functions**
thickness of lines, [G] *linewidthstyle*
Thomson scoring, [MV] **factor postestimation**
three-stage least squares, [R] **reg3**

tick,
  definition, [G] *tickstyle*
  suppressing, [G] *tickstyle*
ticksetstyle, [G] *ticksetstyle*
tickstyle, [G] *tickstyle*
ties, [MV] **Glossary**
TIFF, [G] *tif_options*
time and date, [M-5] **c( )**
time-domain analysis, [TS] **arch**, [TS] **arima**,
  [TS] **Glossary**
time of day, [P] **creturn**
time-series
  analysis, [D] **egen**, [R] **regress postestimation time
    series**, [TS] **pergram**
  formats, [D] **format**
  functions, [D] **functions**
  operators, [TS] **tsset**
  plots, [G] **graph other**, [G] **graph twoway tsline**
time-series–operated variable, [M-5] **st_data( )**,
  [M-5] **st_tsrevar( )**, [M-6] **Glossary**
time-span data, [ST] **snapspan**
time stamp, [D] **describe**
time variables and values, [D] **dates and times**
time-varying covariates, [ST] **Glossary**
time-varying variance, [TS] **arch**
time-versus-concentration curve, [R] **pk**
timeout1, set subcommand, [R] **netio**, [R] **set**
timeout2, set subcommand, [R] **netio**, [R] **set**
timer
  clear command, [P] **timer**
  list command, [P] **timer**
  off command, [P] **timer**
  on command, [P] **timer**
time-series
  analysis, [P] **matrix accum**
  estimation, [U] **26.14 Models with time-series data**
  operators, [U] **13.9 Time-series operators**
  unabbreviating varlists, [P] **unab**
  varlists, [U] **11.4.4 Time-series varlists**
timing code, [P] **timer**
tin() function, [D] **functions**
title, estimates subcommand, [R] **estimates title**
title() option, [G] *title_options*
titles, [G] *title_options*
  of axis, [G] *axis_title_options*
tlabel() option, [G] *axis_label_options*
tm() pseudofunction, [D] **dates and times**,
  [D] **functions**
tmlabel() option, [G] *axis_label_options*
TMPDIR Unix environment variable, [P] **macro**
tmtick() option, [G] *axis_label_options*
tobit command, [R] **tobit**, [R] **tobit postestimation**,
  also see postestimation command
tobit model with endogenous regressors, [SVY] **svy
  estimation**

tobit regression, [R] **ivtobit**, [R] **tobit**, [SVY] **svy estimation**, *also see* intreg command, truncreg command
  random-effects, [XT] **xttobit**
  postestimation, [XT] **xttobit postestimation**
.toc filename suffix, [R] **net**
Toeplitz() function, [M-5] **Toeplitz( )**
tokenallowhex() function, [M-5] **tokenget( )**
tokenallownum() function, [M-5] **tokenget( )**
tokenget() function, [M-5] **tokenget( )**
tokengetall() function, [M-5] **tokenget( )**
tokeninit() function, [M-5] **tokenget( )**
tokeninitstata() function, [M-5] **tokenget( )**
tokenize command, [P] **tokenize**
tokenoffset() function, [M-5] **tokenget( )**
tokenpchars() function, [M-5] **tokenget( )**
tokenpeek() function, [M-5] **tokenget( )**
tokenqchars() function, [M-5] **tokenget( )**
tokenrest() function, [M-5] **tokenget( )**
tokens() function, [M-5] **tokens( )**
tokenset() function, [M-5] **tokenget( )**
tokenwchars() function, [M-5] **tokenget( )**
tolerance() option, [R] **maximize**
tolerances, [M-1] **tolerance**, [M-5] **solve_tol( )**
top() suboption, [G] *alignmentstyle*
tostring command, [D] **destring**
total command, [R] **total**, [R] **total postestimation**, *also see* postestimation command
total(), egen function, [D] **egen**
total inertia, [MV] **Glossary**
total principal inertia, [MV] **Glossary**
totals, estimation, [R] **total**
totals, survey data, [SVY] **svy estimation**
tq() pseudofunction, [D] **dates and times**, [D] **functions**
trace() function, [D] **functions**, [M-5] **trace( )**, [P] **matrix define**
trace, ml subcommand, [R] **ml**
trace of matrix, [M-5] **trace( )**, [P] **matrix define**
trace option, [R] **maximize**
trace, set subcommand, [P] **creturn**, [P] **trace**, [R] **set**
traceback log, [M-2] **errors**, [M-5] **error( )**, [M-6] **Glossary**
tracedepth, set subcommand, [P] **creturn**, [P] **trace**, [R] **set**
traceexpand, set subcommand, [P] **creturn**, [P] **trace**, [R] **set**
tracehilite, set subcommand, [P] **creturn**, [P] **trace**, [R] **set**
traceindent, set subcommand, [P] **creturn**, [P] **trace**, [R] **set**
tracenumber, set subcommand, [P] **creturn**, [P] **trace**, [R] **set**
tracesep, set subcommand, [P] **creturn**, [P] **trace**, [R] **set**
tracing iterative maximization process, [R] **maximize**
trademark symbol, [G] *text*

transferring data
  copying and pasting, [D] **edit**
  from Stata, [D] **outfile**, [D] **outsheet**, [U] **21.4 Transfer programs**
  into Stata, [U] **21 Inputting data**, [U] **21.4 Transfer programs**; [D] **fdasave**, [D] **infile**, [D] **infile (fixed format)**, [D] **infile (free format)**, [D] **infix (fixed format)**, [D] **insheet**, [D] **odbc**, [D] **xmlsave**
transformation, [MV] **procrustes**
transformations
  to achieve normality, [R] **boxcox**, [R] **ladder**
  to achieve zero skewness, [R] **lnskew0**
transformations,
  log, [R] **lnskew0**
  modulus, [R] **boxcox**
  power, [R] **boxcox**, [R] **lnskew0**
transformed coefficients, [MI] **mi estimate**, [MI] **mi estimate postestimation**, [MI] **mi estimate using**
translate command, [R] **translate**
translate logs, [R] **translate**
translation, file, [D] **filefilter**
translator
  query command, [R] **translate**
  reset command, [R] **translate**
  set command, [R] **translate**
transmap
  define command, [R] **translate**
  query command, [R] **translate**
transmission-disequilibrium test, [R] **symmetry**
transmorphic, [M-2] **declarations**, [M-6] **Glossary**
transpose, [M-2] **op_transpose**, [M-5] **_transpose( )**, [M-5] **transposeonly( )**, [M-6] **Glossary**, *also see* conjugate transpose
_transpose() function, [M-5] **_transpose( )**
_transposeonly() function, [M-5] **transposeonly( )**
transposeonly() function, [M-5] **transposeonly( )**
transposing data, [D] **xpose**
transposing matrices, [P] **matrix define**
transposition, [M-2] **op_transpose**, [M-5] **_transpose( )**, [M-5] **transposeonly( )**
treatment effects, [R] **treatreg**
treatment-effects regression, [SVY] **svy estimation**
treatreg command, [R] **treatreg**, [R] **treatreg postestimation**, *also see* postestimation command
tree, misstable subcommand, [R] **misstable**
trees, [MV] **cluster**, [MV] **cluster dendrogram**
trend test, [ST] **epitab**, [ST] **sts test**
trend, test for, [R] **nptrend**, [R] **symmetry**
triangular matrix, [M-5] **solvelower( )**, [M-6] **Glossary**
trigamma() function, [D] **functions**, [M-5] **factorial( )**
trigonometric functions, [D] **functions**, [M-5] **sin( )**
trim() string function, [D] **functions**
trunc() function, [D] **functions**, [M-5] **trunc( )**
truncated regression, [SVY] **svy estimation**
truncated-normal
  model, stochastic frontier, [R] **frontier**
  regression, [R] **truncreg**

truncating
    real numbers, [D] **functions**
    strings, [D] **functions**
truncation, [ST] **Glossary**
truncreg command, [R] **truncreg**, [R] **truncreg postestimation**, *also see* postestimation command
tsappend command, [TS] **tsappend**
tscale, graph twoway subcommand, [G] **graph twoway tsline**
tscale() option, [G] *axis_scale_options*
tsfill command, [TS] **tsfill**
tsline command, [TS] **tsline**
tsline, graph twoway subcommand, [G] **graph twoway tsline**
tsnorm macro extended function, [P] **macro**
tsreport command, [TS] **tsreport**
tsrevar command, [TS] **tsrevar**
tsrline command, [TS] **tsline**
tsrline, graph twoway subcommand, [G] **graph twoway tsline**
tsset command, [TS] **tsset**
tsset command for mi data, [MI] **mi XXXset**
tsset, mi subcommand, [MI] **mi XXXset**
tssmooth
    commands, introduction, [TS] **tssmooth**
    dexponential command, [TS] **tssmooth dexponential**
    exponential command, [TS] **tssmooth exponential**
    hwinters command, [TS] **tssmooth hwinters**
    ma command, [TS] **tssmooth ma**
    nl command, [TS] **tssmooth nl**
    shwinters command, [TS] **tssmooth shwinters**
tsunab command, [P] **unab**
ttail() function, [D] **functions**, [M-5] **normal( )**
ttest and ttesti commands, [R] **ttest**
ttest command, [MV] **hotelling**
ttick() option, [G] *axis_label_options*
ttitle() option, [G] *axis_title_options*
tuning constant, [R] **rreg**
tutorials, [U] **1.2.2 Example datasets**
tw() pseudofunction, [D] **dates and times**, [D] **functions**
twithin() function, [D] **functions**
two-way multivariate analysis of variance, [MV] **manova**
two-level model, [XT] **Glossary**
two-stage least squares, [R] **gmm**, [R] **ivregress**, [R] **nlsur**, [R] **regress**, [SVY] **svy estimation**, [XT] **xthtaylor**, [XT] **xtivreg**
two-way
    analysis of variance, [R] **anova**
    scatterplots, [R] **lowess**
type, [M-2] **declarations**, [M-6] **Glossary**
type
    command, [D] **type**
    macro extended function, [P] **macro**
    parameter, [D] **generate**
type,
    set subcommand, [D] **generate**, [R] **set**
    ssc subcommand, [R] **ssc**
type, broad, [M-6] **Glossary**
type I error, [ST] **Glossary**
type II error, [ST] **Glossary**

U

U statistic, [R] **ranksum**
unab command, [P] **unab**
unabbreviate
    command names, [P] **unabcmd**
    variable list, [P] **syntax**, [P] **unab**
unabcmd command, [P] **unabcmd**
.uname built-in class function, [P] **class**
unbalanced data, [XT] **Glossary**
uncompress files, [D] **zipfile**
under observation, [ST] **Glossary**
underlining in syntax diagram, [U] **11 Language syntax**
underscore c() function, [D] **functions**
underscore functions, [M-1] **naming**, [M-6] **Glossary**
underscore variables, [U] **13.4 System variables (_variables)**
unhold, _estimates subcommand, [P] **_estimates**
uniform prior, [MI] **mi impute mvn**
uniformly distributed random numbers, [M-5] **runiform( )**
uniformly distributed random variates, [M-5] **runiform( )**
uniformly distributed random-number function, [D] **functions**, [R] **set seed**
uninstall,
    net subcommand, [R] **net**
    ssc subcommand, [R] **ssc**
uniqrows() function, [M-5] **uniqrows( )**
unique options, [G] **concept: repeated options**
unique value labels, [D] **labelbook**
unique values, counting, [R] **table**, [R] **tabulate oneway**
unique values,
    counting, [D] **codebook**
    determining, [D] **inspect**, [D] **labelbook**
uniqueness, [MV] **Glossary**
unit-root
    models, [TS] **vec intro**, [TS] **vec**
    process, [TS] **Glossary**
    test, [TS] **dfgls**, [TS] **dfuller**, [TS] **Glossary**, [TS] **pperron**
unit-root test, [XT] **xtunitroot**
unit vectors, [M-5] **e( )**
unitary matrix, [M-6] **Glossary**
unitcircle() function, [M-5] **unitcircle( )**
univariate
    distributions, displaying, [R] **cumul**, [R] **diagnostic plots**, [R] **histogram**, [R] **ladder**, [R] **lv**, [R] **stem**
    kernel density estimation, [R] **kdensity**

univariate imputation, *see* imputation, univariate
univariate time series, [TS] **arch**, [TS] **arima**,
 [TS] **newey**, [TS] **prais**
Unix,
 keyboard use, [U] **10 Keyboard use**
 specifying filenames, [U] **11.6 File-naming**
 conventions
Unix, pause, [P] **sleep**
_unlink() function, [M-5] **unlink( )**
unlink() function, [M-5] **unlink( )**
unobserved-component model, [TS] **dfactor**,
 [TS] **sspace**
unorder() function, [M-5] **sort( )**
unregister, mi subcommand, [MI] **mi set**
unregistered variables, *see* variables, unregistered
unrestricted FMI test, [MI] **Glossary**, [MI] **mi estimate**,
 [MI] **mi estimate postestimation**
unrestricted transformation, [MV] **Glossary**
unzipfile command, [D] **zipfile**
update
 ado command, [R] **update**
 all command, [R] **update**
 command, [R] **update**
 executable command, [R] **update**
 from command, [R] **update**
 query command, [R] **update**
 swap command, [R] **update**
 utilities command, [R] **update**
update_interval, set subcommand, [R] **set**,
 [R] **update**
update, mi subcommand, [MI] **mi update**,
 [MI] **noupdate option**
update_prompt, set subcommand, [R] **set**,
 [R] **update**
update_query, set subcommand, [R] **set**, [R] **update**
update, view subcommand, [R] **view**
update_d, view subcommand, [R] **view**
UPDATES directory, [P] **sysdir**, [U] **17.5 Where does**
 Stata look for ado-files?
updates to Stata, [R] **adoupdate**, [R] **net**, [R] **sj**,
 [R] **update**, [U] **3.5 The Stata Journal**,
 [U] **3.6 Updating and adding features from the**
 web, [U] **17.6 How do I install an addition?**
upper() string function, [D] **functions**
uppercase, [M-5] **strupper( )**
uppercase-string function, [D] **functions**
_uppertriangle() function, [M-5] **lowertriangle( )**
uppertriangle() function, [M-5] **lowertriangle( )**
upper-triangular matrix, *see* triangular matrix
use,
 cluster subcommand, [MV] **cluster utility**
 estimates subcommand, [R] **estimates save**
 graph subcommand, [G] **graph use**
 serset subcommand, [P] **serset**
use command, [D] **use**
uselabel command, [D] **labelbook**
user interface, [P] **dialog programming**

user-written additions,
 installing, [R] **net**, [R] **ssc**
 searching for, [R] **net search**, [R] **ssc**
using data, [D] **sysuse**, [D] **use**, [D] **webuse**, [P] **syntax**
using graphs, [G] **graph use**
utilities, update subcommand, [R] **update**
utilities, programming, [MV] **cluster utility**
utility routines, [MI] **technical**

V

valofexternal() function, [M-5] **valofexternal( )**
value label macro extended function, [P] **macro**
value labels, [D] **codebook**, [D] **describe**, [D] **encode**,
 [D] **inspect**, [D] **label**, [D] **label language**,
 [D] **labelbook**, [P] **macro**, [U] **12.6.3 Value**
 labels, [U] **13.10 Label values**
 potential problems in, [D] **labelbook**
values, label subcommand, [D] **label**
Vandermonde() function, [M-5] **Vandermonde( )**
VAR, [TS] **dfactor**, [TS] **Glossary**, [TS] **sspace**,
 [TS] **var**, [TS] **var intro**, [TS] **var svar**,
 [TS] **varbasic**
 postestimation, [TS] **fcast compute**, [TS] **fcast**
 graph, [TS] **irf**, [TS] **irf create**, [TS] **var**
 postestimation, [TS] **vargranger**, [TS] **varlmar**,
 [TS] **varnorm**, [TS] **varsoc**, [TS] **varstable**,
 [TS] **varwle**
var command, [TS] **var**, [TS] **var postestimation**
varabbrev command, [P] **varabbrev**
varabbrev, set subcommand, [R] **set**
varbasic command, [TS] **varbasic**, [TS] **varbasic**
 postestimation
vargranger command, [TS] **vargranger**
variable, [M-2] **declarations**, [M-5] **st_data( )**,
 [M-6] **Glossary**
 abbreviation, [P] **varabbrev**
 description, [D] **describe**
 identifying panels, [XT] **xtset**
 labels, [D] **codebook**, [D] **describe**, [D] **label**,
 [D] **label language**, [D] **notes**, [P] **macro**,
 [U] **11.4 varlists**, [U] **12.6.2 Variable labels**
 lists, *see* varlist
 naming convention, [M-1] **naming**
 types, [D] **codebook**, [D] **data types**, [D] **describe**,
 [I] **data types**, [P] **macro**, [U] **11.4 varlists**,
 [U] **12.2.2 Numeric storage types**,
 [U] **12.4.4 String storage types**
 class, [P] **class**
variable, confirm subcommand, [P] **confirm**
variable, label subcommand, [D] **label**
variable label macro extended function, [P] **macro**
variables,
 alphabetizing, [D] **order**
 categorical, *see* categorical data
 changing storage types of, [D] **recast**
 characteristics of, [P] **char**, [P] **macro**,
 [U] **12.8 Characteristics**
 comparing, [D] **compare**

variables, *continued*
    copying, [D] **clonevar**
    creating, [D] **varmanage**
    creating new, [D] **separate**
    describing, [D] **codebook**, [D] **notes**
    determining storage types of, [D] **describe**
    displaying contents of, [D] **edit**, [D] **list**
    documenting, [D] **codebook**, [D] **labelbook**,
       [D] **notes**
    dropping, [D] **drop**
    dummy, *see* indicator variables
    factor, *see* factor variables
    filtering, [D] **varmanage**
    finding, [D] **lookfor**
    generate, summary, or grouping, [MV] **cluster**
       generate
    generating, [ST] **stgen**
    imputed, [MI] **Glossary**, [MI] **mi rename**, [MI] **mi
       reset**, [MI] **mi set**
    in dataset, maximum number of, [D] **describe**,
       [D] **memory**, [U] **6 Setting the size of memory**
    in model, maximum number, [R] **matsize**
    list values of, [P] **levelsof**
    listing, [D] **edit**, [D] **list**; [D] **codebook**,
       [D] **describe**, [D] **labelbook**
    mapping numeric to string, [D] **destring**
    naming, [D] **rename**, [U] **11.2 Abbreviation rules**,
       [U] **11.3 Naming conventions**
    ordering, [D] **sort**
    orthogonalize, [R] **orthog**
    passive, [MI] **Glossary**, [MI] **mi passive**, [MI] **mi
       rename**, [MI] **mi reset**, [MI] **mi set**, [MI] **mi xeq**
    registered, [MI] **Glossary**, [MI] **mi rename**, [MI] **mi
       set**
    regular, [MI] **Glossary**, [MI] **mi rename**, [MI] **mi
       set**
    reordering, [D] **order**
    setting properties of, [D] **varmanage**
    sorting, [D] **varmanage**
    sorting and alphabetizing, [D] **sort**; [D] **gsort**
    standardizing, [D] **egen**
    storage types, *see* storage types
    string, *see* string variables
    system, *see* system variables
    tab expansion of, [U] **10.6 Tab expansion of
       variable names**
    temporary, [P] **macro**
    transposing with observations, [D] **xpose**
    unabbreviating, [P] **syntax**, [P] **unab**
    unique values, [D] **codebook**
    unique values, determining, [D] **inspect**
    unregistered, [MI] **Glossary**, [MI] **mi rename**,
       [MI] **mi set**
    varying and super varying, [MI] **Glossary**, [MI] **mi
       set**, [MI] **mi varying**, [ST] **stvary**
  variables, [U] **11.3 Naming conventions**,
       [U] **13.4 System variables (_variables)**
Variables Manager, [D] **varmanage**

variance
    estimation, [SVY] **Glossary**
    estimators, [R] *vce_option*
    Huber/White/sandwich estimator, [SVY] **variance
       estimation**, *see* robust
    inflation factors, [R] **regress postestimation**
    linearized, [SVY] **variance estimation**
    nonconstant, [SVY] **variance estimation**, *see* robust
    stabilizing transformations, [R] **boxcox**
variance,
    analysis of, [R] **anova**, [R] **loneway**, [R] **oneway**
    displaying, [R] **summarize**, [R] **table**, [R] **tabulate,
       summarize()**, [XT] **xtsum**; [R] **lv**
    estimators, [XT] *vce_options*
    Huber/White/sandwich estimator, *see* robust,
       Huber/White/sandwich estimator of variance
    nonconstant, *see* robust, Huber/White/sandwich
       estimator of variance
    testing equality of, [R] **sdtest**
variance analysis, [MV] **manova**
variance-comparison test, [R] **sdtest**
variance components, [XT] **Glossary**
variance–covariance matrix of estimators, [P] **ereturn**,
    [P] **matrix get**, [R] **correlate**, [R] **estat**
variance–covariance matrix, obtaining,
    [U] **20.8 Obtaining the variance–covariance
    matrix**
variance decompositions, *see* FEVD
variance() function, [M-5] **mean( )**
variance-weighted least squares, [R] **vwls**
variance,
    creating dataset of, [D] **collapse**
    creating variable containing, [D] **egen**
varimax rotation, [MV] **Glossary**, [MV] **rotate**,
    [MV] **rotatemat**
varkeyboard, set subcommand, [R] **set**
varlabelpos, set subcommand, [R] **set**
varlist, [P] **syntax**, [U] **11 Language syntax**,
    [U] **11.4 varlists**
    existing, [U] **11.4.1 Lists of existing variables**
    new, [U] **11.4.2 Lists of new variables**
    time series, [U] **11.4.4 Time-series varlists**
varlmar command, [TS] **varlmar**
varmanage command, [D] **varmanage**
varnorm command, [TS] **varnorm**
varsoc command, [TS] **varsoc**
varstable command, [TS] **varstable**
varwle command, [TS] **varwle**
varying
    estimation sample, [MI] **mi estimate**
    variables, *see* variables, varying and supervarying
varying, mi subcommand, [MI] **mi varying**
vce, estat subcommand, [R] **estat**, [SVY] **estat**
vce() option, [R] *vce_option*, [XT] *vce_options*
vec command, [TS] **vec**, [TS] **vec postestimation**
vec() function, [D] **functions**, [M-5] **vec( )**, [P] **matrix
    define**
vecaccum, matrix subcommand, [P] **matrix accum**

vecdiag() matrix function, [D] **functions**, [P] **matrix define**

vech() function, [M-5] **vec( )**

veclmar command, [TS] **veclmar**

VECM, [TS] **dvech**, [TS] **Glossary**, [TS] **vec**, [TS] **vec intro**
 postestimation, [TS] **fcast compute**, [TS] **fcast graph**, [TS] **irf**, [TS] **irf create**, [TS] **varsoc**, [TS] **vec postestimation**, [TS] **veclmar**, [TS] **vecnorm**, [TS] **vecrank**, [TS] **vecstable**

vecnorm command, [TS] **vecnorm**

vecrank command, [TS] **vecrank**

vecstable command, [TS] **vecstable**

vector, [M-2] **declarations**, [M-6] **Glossary**

vector autoregression, *see* VAR

vector autoregressive forecast, [TS] **fcast compute**, [TS] **fcast graph**

vector autoregressive (VAR) models, [G] **graph other**

vector autoregressive moving average model, [TS] **dfactor**, [TS] **sspace**

vector error-correction model, *see* VECM

vector norm, [M-5] **norm( )**

vectors, *see* matrices

verifying data, [D] **assert**, [D] **count**, [D] **inspect**, *also see* certifying data

verifying mi data are consistent, [MI] **mi update**

version, [M-2] **version**

version command, [P] **version**, [U] **16.1.1 Version**, [U] **18.11.1 Version**
 class programming, [P] **class**

version control, [M-2] **version**, [M-5] **callersversion( )**, *see* version command

version of ado-file, [R] **which**

version of Stata, [M-5] **stataversion( )**, [R] **about**

vertical alignment of text, [G] *alignmentstyle*

view
 ado command, [R] **view**
 ado_d command, [R] **view**
 browse command, [R] **view**
 command, [R] **view**
 help command, [R] **view**
 help_d command, [R] **view**
 net command, [R] **view**
 net_d command, [R] **view**
 news command, [R] **view**
 search command, [R] **view**
 search_d command, [R] **view**
 update command, [R] **view**
 update_d command, [R] **view**
 view_d command, [R] **view**

view matrix, [M-5] **isview( )**, [M-5] **st_subview( )**, [M-5] **st_view( )**, [M-5] **st_viewvars( )**, [M-6] **Glossary**

view source code, [P] **viewsource**

view_d, view subcommand, [R] **view**

viewing previously typed lines, [R] **#review**

viewsource, [M-1] **source**

viewsource command, [P] **viewsource**

vif, estat subcommand, [R] **regress postestimation**

virtual, [M-2] **class**

virtual memory, [D] **memory**, [U] **6.5 Virtual memory and speed considerations**

virtual, set subcommand, [D] **memory**, [R] **set**

void
 function, [M-2] **declarations**, [M-6] **Glossary**
 matrix, [M-2] **void**, [M-6] **Glossary**

vwls command, [R] **vwls**, [R] **vwls postestimation**, *also see* postestimation command

W

Wald tests, [R] **predictnl**, [R] **test**, [R] **testnl**, [SVY] **svy postestimation**, [TS] **vargranger**, [TS] **varwle**, [U] **20.11 Performing hypothesis tests on the coefficients**, [U] **20.11.4 Nonlinear Wald tests**

wardslinkage, cluster subcommand, [MV] **cluster linkage**

Ward's linkage clustering, [MV] **cluster**, [MV] **cluster linkage**, [MV] **clustermat**, [MV] **Glossary**

Ward's method clustering, [MV] **cluster**, [MV] **clustermat**

warning messages, [M-2] **pragma**

waveragelinkage, cluster subcommand, [MV] **cluster linkage**

weakly balanced, [XT] **Glossary**

web site,
 stata.com, [U] **3.2 The Stata web site (www.stata.com)**
 stata-press.com, [U] **3.3 The Stata Press web site (www.stata-press.com)**

webuse
 query command, [D] **webuse**
 set command, [D] **webuse**
 command, [D] **webuse**

week() function, [D] **dates and times**, [D] **functions**, [M-5] **date( )**

weekly() function, [D] **dates and times**, [D] **functions**, [M-5] **date( )**

Weibull distribution, [ST] **streg**

Weibull survival regression, [ST] **streg**

weight, [P] **syntax**

[weight=*exp*] modifier, [U] **11.1.6 weight**, [U] **20.18 Weighted estimation**

weighted data, [U] **11.1.6 weight**, [U] **20.18 Weighted estimation**, *also see* survey data

weighted least squares, [R] **gmm**, [R] **ivregress**, [R] **nlsur**, [R] **regress**, [R] **regress postestimation**, [R] **vwls**

weighted moving average, [TS] **tssmooth**, [TS] **tssmooth ma**

weighted-average linkage clustering, [MV] **cluster linkage**, [MV] **cluster**, [MV] **clustermat**, [MV] **Glossary**

weights, [G] **graph twoway scatter**
 probability, [SVY] **survey**, [SVY] **svydescribe**, [SVY] **svyset**

weights, *continued*
    sampling, *see* probability weights
Welsch distance, [R] **regress postestimation**
which, [M-3] **mata which**
which, class, [P] **classutil**
which command, [U] **17.3 How can I tell if a command is built in or an ado-file?**
which command, [R] **which**
which, classutil subcommand, [P] **classutil**
while command, [M-2] **while**, [M-2] **continue**, [M-2] **break**, [M-2] **semicolons** [P] **while**
White/Huber/sandwich estimator of variance, [SVY] **variance estimation**
white noise, [TS] **Glossary**, [XT] **Glossary**
white-noise test, [TS] **wntestb**, [TS] **wntestq**
White/Huber/sandwich estimator of variance, *see* robust, Huber/White/sandwich estimator of variance, *see* robust
White's test for heteroskedasticity, [R] **regress postestimation**
wide
    data style, [MI] **Glossary**, [MI] **styles**
    technical description, [MI] **technical**
width of % *fmt*, [M-5] **fmtwidth( )**
width( ) option, [G] **graph twoway histogram**
Wilcoxon
    rank-sum test, [R] **ranksum**
    signed-ranks test, [R] **signrank**
Wilcoxon test (Wilcoxon–Breslow, Wilcoxon–Gehan, Wilcoxon–Mann–Whitney), [ST] **sts test**
wildcard, *see* strmatch( ) string function, regexm( ) string function, regexr( ) string function, and regexs( ) string function
Wilks'
    lambda, [MV] **canon**, [MV] **Glossary**, [MV] **manova**, [MV] **mvtest means**
    likelihood-ratio test, [MV] **canon**, [MV] **manova**, [MV] **mvtest means**
window
    fopen command, [P] **window programming**
    fsave command, [P] **window programming**
    manage command, [P] **window programming**
    menu command, [P] **window programming**
    push command, [P] **window programming**
    stopbox command, [P] **window programming**
Windows Metafile, [G] **graph export**
Windows metafiles programming, [P] **automation**
Windows programming, [P] **automation**
Windows,
    filenames, [U] **18.3.11 Constructing Windows filenames by using macros**
    keyboard use, [U] **10 Keyboard use**
    pause, [P] **sleep**
    specifying filenames, [U] **11.6 File-naming conventions**
winexec command, [D] **shell**
Wishart distribution, [MV] **Glossary**
withdrawal, [ST] **Glossary**

within-cell means and variances, [XT] **xtsum**
within estimators, [XT] **Glossary**, [XT] **xthtaylor**, [XT] **xtivreg**, [XT] **xtreg**, [XT] **xtregar**
within matrix, [MV] **Glossary**
within-imputation variability, [MI] **mi estimate**
WLF, *see* worst-linear function
wntestb command, [TS] **wntestb**
wntestq command, [TS] **wntestq**
wofd( ) function, [D] **dates and times**, [D] **functions**, [M-5] **date( )**
Woolf confidence intervals, [ST] **epitab**
word macro extended function, [P] **macro**
word( ) string function, [D] **functions**
wordcount( ) string function, [D] **functions**
workflow, [MI] **workflow**
worst linear function, [MI] **Glossary**, [MI] **mi impute mvn**
write, file subcommand, [P] **file**
writing and reading ASCII text and binary files, [P] **file**
writing data, [D] **outfile**, [D] **outsheet**, [D] **save**
www.stata.com web site, [U] **3.2 The Stata web site (www.stata.com)**
www.stata-press.com web site, [U] **3.3 The Stata Press web site (www.stata-press.com)**

X

xaxis( ) suboption, [G] *axis_choice_options*
X-bar charts, [G] **graph other**
xchart command, [R] **qc**
xcommon option, [G] **graph combine**
xcorr command, [TS] **xcorr**
xeq, mi subcommand, [MI] **mi xeq**
xi prefix command, [R] **xi**
xlabel( ) option, [G] *axis_label_options*
xline( ) option, [G] *added_line_options*
XML, [D] **xmlsave**
xmlabel( ) option, [G] *axis_label_options*
xmlsave command, [D] **xmlsave**
xmluse command, [D] **infile**, [D] **xmlsave**
xmtick( ) option, [G] *axis_label_options*
xpose command, [D] **xpose**
xscale( ) option, [G] *axis_scale_options*
xshell command, [D] **shell**
xsize( ) option, [G] **graph display**, [G] *region_options*
xtabond command, [XT] **xtabond**
xtcloglog command, [XT] **xtcloglog**; [XT] **quadchk**
xtdata command, [XT] **xtdata**
xtdescribe command, [XT] **xtdescribe**
xtdpd command, [XT] **xtdpd**
xtdpdsys command, [XT] **xtdpdsys**
xtfrontier command, [XT] **xtfrontier**
xtgee command, [XT] **xtgee**
xtgls command, [XT] **xtgls**
xthtaylor command, [XT] **xthtaylor**
xtick( ) option, [G] *axis_label_options*
xtile command, [D] **pctile**

xtintreg command, [XT] **xtintreg**; [XT] **quadchk**

xtitle() option, [G] *axis_title_options*

xtivreg command, [XT] **xtivreg**

xtline command, [XT] **xtline**

xtlogit command, [XT] **xtlogit**; [XT] **quadchk**

xtmelogit command, [XT] **xtmelogit**

xtmepoisson command, [XT] **xtmepoisson**

xtmixed command, [XT] **xtmixed**

xtnbreg command, [XT] **xtnbreg**

xtpcse command, [XT] **xtpcse**

xtpoisson command, [XT] **xtpoisson**; [XT] **quadchk**

xtprobit command, [XT] **xtprobit**; [XT] **quadchk**

xtrc command, [XT] **xtrc**

xtreg command, [XT] **xtreg**

xtregar command, [XT] **xtregar**

xtset command, [XT] **xtset**

xtset command for mi data, [MI] **mi XXXset**

xtset, mi subcommand, [MI] **mi XXXset**

xtsum command, [XT] **xtsum**

xttab command, [XT] **xttab**

xttest0 command, [XT] **xtreg postestimation**

xttobit command, [XT] **xttobit**; [XT] **quadchk**

xttrans command, [XT] **xttab**

xtunitroot command, [XT] **xtunitroot**

xvarformat() option, [G] *advanced_options*

xvarlabel() option, [G] *advanced_options*

xxxset, programming, [MI] **technical**

Y

yaxis() suboption, [G] *axis_choice_options*

ycommon option, [G] **graph combine**

year() function, [D] **dates and times**, [D] **functions**, [M-5] **date( )**, [U] **24.5 Extracting components of dates and times**

yearly() function, [D] **dates and times**, [D] **functions**, [M-5] **date( )**

yh() function, [D] **dates and times**, [D] **functions**, [M-5] **date( )**

ylabel() option, [G] *axis_label_options*

yline() option, [G] *added_line_options*

ym() function, [D] **dates and times**, [D] **functions**, [M-5] **date( )**

ymlabel() option, [G] *axis_label_options*

ymtick() option, [G] *axis_label_options*

yofd() function, [D] **dates and times**, [D] **functions**, [M-5] **date( )**

yq() function, [D] **dates and times**, [D] **functions**, [M-5] **date( )**

yscale() option, [G] *axis_scale_options*

ysize() option, [G] **graph display**, [G] *region_options*

ytick() option, [G] *axis_label_options*

ytitle() option, [G] *axis_title_options*

Yule coefficient similarity measure, [MV] *measure_option*

Yule–Walker equations, [TS] **corrgram**, [TS] **Glossary**

yvarformat() option, [G] *advanced_options*

yvarlabel() option, [G] *advanced_options*

yw() function, [D] **dates and times**, [D] **functions**, [M-5] **date( )**

Z

Zellner's seemingly unrelated regression, [R] **sureg**; [R] **reg3**, [R] **suest**

zero matrix, [P] **matrix define**

zero-altered

negative binomial regression, [R] **zinb**

Poisson regression, [R] **zip**

zero-inflated

negative binomial regression, [R] **zinb**, [SVY] **svy estimation**

Poisson regression, [R] **zip**, [SVY] **svy estimation**

zero-skewness transform, [R] **lnskew0**

zero-truncated

negative binomial regression, [R] **ztnb**, [SVY] **svy estimation**

Poisson regression, [R] **ztp**, [SVY] **svy estimation**

zinb command, [R] **zinb**, [R] **zinb postestimation**, *also see* postestimation command

zip command, [R] **zip**, [R] **zip postestimation**, *also see* postestimation command

zipfile command, [D] **zipfile**

ztnb command, [R] **ztnb**, [R] **ztnb postestimation**, *also see* postestimation command

ztp command, [R] **ztp**, [R] **ztp postestimation**, *also see* postestimation command